Communications
in Computer and Information Science 38

Elaine Chew Adrian Childs
Ching-Hua Chuan (Eds.)

Mathematics and Computation in Music

Second International Conference, MCM 2009
John Clough Memorial Conference
New Haven, CT, USA, June 19-22, 2009
Proceedings

Volume Editors

Elaine Chew
Viterbi School of Engineering and Thornton School of Music
University of Southern California, Los Angeles, CA, USA
and
School of Engineering and Applied Sciences and Department of Music
Harvard University, Cambridge, MA, USA
E-mail: echew@usc.edu

Adrian Childs
Hugh Hodgson School of Music
University of Georgia, Athens, GA, USA
E-mail: apchilds@uga.edu

Ching-Hua Chuan
Department of Mathematics and Computer Science
Barry University, Miami Shores, FL, USA
E-mail: chchuan@mail.barry.edu

Library of Congress Control Number: Applied for

CR Subject Classification (1998): H.5.5, J.5, I.1, I.6, G.2

ISSN 1865-0929
ISBN-10 3-642-02393-2 Springer Berlin Heidelberg New York
ISBN-13 978-3-642-02393-4 Springer Berlin Heidelberg New York

springer.com

Printed in Germany

Typesetting: Camera-ready by author, data conversion by Scientific Publishing Services, Chennai, India
Printed on acid-free paper SPIN: 12699223 06/3180 5 4 3 2 1 0

Preface

These proceedings comprise 26 refereed research papers that were presented at the Second International Conference on Mathematics and Computation in Music (MCM 2009), which met in conjunction with the John Clough Memorial Conference during June 19-22, 2009, at Yale University in New Haven, Connecticut, USA.

The International Conference on Mathematics and Computation in Music (MCM) is the flagship conference of the Society for Mathematics and Computation in Music. The inaugural conference of the society took place in 2007 in Berlin. The study of mathematics and music dates back to the time of the ancient Greeks. The rise of computing and the digital age has added computation to this august tradition. MCM aims to provide a dedicated platform for the communication and exchange of ideas amongst researchers in mathematics, informatics, music theory, composition, musicology, and related disciplines.

The John Clough Memorial Conference honors a mathematical music theorist whose research modeled the virtues of collaborative work across the disciplines, and who generously fostered a cooperative attitude with and among younger researchers. The quadrennial conferences that Clough first organized at Buffalo in 1993 positioned neo-Riemannian theory on the map of musical scholarship. The John Clough Memorial Conference carries the spirit of those sessions beyond his passing in 2003, while embracing the entire domain of mathematical music theory.

High-quality contributions — including research papers, invited sessions or panels, tutorials, and exhibits — were solicited in all areas related to the mission of the society.

To promote objectivity and fairness in judging research paper contributions, the peer review process was double-blind, and consisted of two stages. Following the submission of the paper reviews by Program Committee members, authors were given the opportunity to respond to these reviews in order to correct possible misconceptions, so as to produce more accurate assessments of their work. After the author response period, the Program Committee Members could then re-visit their reviews in light of the authors' comments, discuss amongst themselves the merit of the papers, and offer their final recommendations for acceptance.

Of 38 submissions received, 26 were accepted for presentation at the conference and publication in these proceedings. Six more were accepted for presentation at the conference in the form of posters.

One panel and three tutorials were selected for inclusion in the conference program:

Panel: "Embodiment of Mathematical Formulas in Musical Gestures?"
Moderator: Guerino Mazzola
(University of Minnesota, USA)
Speakers: Emmanuel Amiot
(Lycée Arago, France)
Moreno Andreatta
(Inst. de Recherche et Coordination Acoustique/ Musique / Centre National de la Recherche Scientifique, France)
Rachel Hall
(Saint Joseph's University, USA)
Thomas Noll
(Escola Superior de Música de Catalunya, Spain)
Tutorial: "A Tutorial on Mathematical Models in Computer-Aided Music Theory, Analysis, and Composition via OpenMusic"
Leaders: Carlos Agon
(Inst. de Recherche et Coordination Acoustique/ Musique, France)
Moreno Andreatta
(Inst. de Recherche et Coordination Acoustique/ / Musique Centre National de la Recherche Scientifique, France)
Tutorial: "Hands-on Workshop in Geometrical Music Theory"
Leaders: Rachel Hall (Saint Joseph's University, USA)
Tutorial: "Measuring the Complexity of Musical Rhythm: Mathematical and Psychological Models"
Leaders: Godfried T. Toussaint (McGill University, Canada)

As part of the conference, the Beinecke Rare Book and Manuscript Library at Yale University mounted a special display of music and mathematics material from their collection. A related keynote address, "The End of Pythagoreanism: *Musica theorica*, Natural Science, and Aristotle's Philosophy of Mathematics, c.1300-c.1600," was given by David Cohen (Columbia University, USA).

We wish to acknowledge the generous support of Susan Adler and Roberta Hudson of Yale Conference Services; Scott Petersen, Yale Music Department, for technical support; Eric Bianchi, PhD student in Music History at Yale, for curating the exhibit of historical materials; Kathryn James of Yale's Beinecke Library, for arranging the exhibit and reception; Edward Gollin of Williams College for program guidance; and David Cohen of Columbia University for providing a keynote lecture.

Richard Cohn
Ian Quinn
Elaine Chew
Adrian Childs

Organization

MCM 2009

General Chairs

Richard Cohn	Yale University, USA
Ian Quinn	Yale University / Stanford University, USA

Program Chairs

Elaine Chew	University of Southern California / Harvard University, USA
Adrian Childs	University of Georgia, USA

Publications Chair

Ching-Hua Chuan	Barry University, USA

Panels Chair

Anja Volk	University of Utrecht, The Netherlands

Tutorials Chair

Aline Honingh	City University London, UK

Exhibits Chair

Neta Spiro	University of Cambridge, UK

Program Committee

Emmanuel Amiot	Lycée Arago, France
Christina Anagnostopoulou	University of Athens, Greece
Moreno Andreatta	Inst. de Recherche et Coordination Acoustique/Musique Centre National de la Recherche Scientifique, France
Gérard Assayag	Inst. de Recherche et Coordination Acoustique/Musique, France
Chantal Buteau	Brock University, Canada
Ching-Hua Chuan	Barry University, USA
David Clampitt	Ohio State University, USA
Shlomo Dubnov	University of California, San Diego, USA
Morwaread Farbood	New York University, USA
Alexandre François	Tufts University, USA

Emilia Gómez	Escola Superior de Música de Catalunya, Spain
Rachel Hall	Saint Joseph's University, USA
Keiji Hirata	NTT Communication Science Laboratories, Japan
Henkjan Honing	University of Amsterdam, The Netherlands
Aline Honingh	City University, London, UK
Julian Hook	Indiana University, Bloomington, USA
Özgür İzmirli	Connecticut College, USA
Guerino Mazzola	University of Minnesota, USA
David Meredith	Aalborg University, Denmark
Katarina Miljkovic	New England Conservatory, USA
Meinard Müller	Max Planck Institut für Informatik, Germany
Teresa Nakra	The College of New Jersey, USA
Thomas Noll	Escola Superior de Música de Catalunya, Spain
Bryan Pardo	Northwestern University, USA
Richard Parncutt	University of Graz, Austria
Robert Peck	Louisiana State University, USA
John Rahn	University of Washington, USA
Christopher Raphael	Indiana University, Bloomington, USA
William Sethares	University of Wisconsin, USA
Neta Spiro	University of Cambridge, UK
Petri Toiviainen	University of Jyväskylä, Finland
Godfried Toussaint	McGill University, Canada
Charlotte Truchet	Université de Nantes, France
Dmitri Tymoczko	Princeton University, USA
Gerhard Widmer	Johannes Kepler University of Linz, Austria
Geraint Wiggins	Goldsmiths' College, University of London, UK

Society for Mathematics and Computation in Music

President

Guerino Mazzola	University of Minnesota, USA

Vice-President

Moreno Andreatta	Inst. de Recherche et Coordination Acoustique/Musique Centre National de la Recherche Scientifique, France

Secretary

Elaine Chew	University of Southern California / Harvard University, USA

Treasurer

Ian Quinn	Yale University / Stanford University, USA

Co-Editors: Journal of Mathematics and Music

Foreword: In Celebration of Clough's Collaborative Cerebration

For centuries, music theorists have engaged in a dance of attraction and aversion with the physical and mathematical sciences. The alternate sporting and doffing of science is visible in music theory's values, methods, and voices; but, curiously, not in one aspect of its sociology. Perhaps because *Musikwissenschaft* has been slow to detach from its roots in the heroic ethos of high Romanticism, its practitioners have largely toiled on their own. Their reluctance to adopt the collaborative habits characteristic of other theoretical and empirical disciplines has created obstacles to progress on some fundamental issues, at least in the view of some colleagues in the scientific and mathematical communities.[1]

[1] For two expressions of exasperation, see John Backus, "*Die Reihe* — A Scientific Evaluation," *Perspectives of New Music* 1.1 (1962): 160-171, and Eric Regener, "Allen Forte's Theory of Chords," *Perspectives of New Music* 13.1 (1974): 199.

John Clough (1930–2003) had no such reluctance. Trained as a music theorist in the 1950s, Clough did significant systematic work in the 1960s and -70s, notably his important response to Allen Forte's initial proposal of a classification for atonal sets, and two early papers on diatonic set theory (developing some ideas of mathematician Eric Regener).[2]

These latter papers initiated a line of research that leaped forward when Clough began to co-author articles with researchers whose primary training was in mathematics: initially Gerald Myerson, who was for a time a colleague at SUNY Buffalo; eventually the Albuquerque-based Jack Douthett, with whom Clough collaborated continually during his last 15 years.[3]

Clough imported the collaborative habit into his work with other music researchers, co-authoring papers with Lewis Rowell and N. Ramanathan, and with then-PhD students Stefan Ehrenkreutz, John Cuciurean, Nora Engebretsen, and Jonathan Kochavi. (He also co-authored textbooks with Joyce Conley and Claire Boge.) The appearance of his name on a paper always denoted not only sponsorship and guidance (as one finds in scientific disciplines), but also a full engagement and commitment at every stage. Clough welcomed students as equal intellectual partners if they earned that status, and relished what he learned from them.

John Clough's eagerness to encourage the work of younger scholars was by no means limited to his own students. In the summer of 1993, after an extended set of communications exploring the modeling of triadic progressions in chromatic music, Clough invited 15 music scholars to SUNY-Buffalo for a three-day working conference. A subsequent Buffalo gathering in 1997 led to the publication of a specially dedicated topical issue of the *Journal of Music Theory* (volume 42, number 2, 1998), and was succeeded by a third such event in 2001. Upon receiving an invitation to the latter affair, David Lewin wryly remarked, "I sense a certain *Viertaktigkeit.*" After John died in 2003, his co-organizers (David Clampitt, Jack Douthett, and I) determined to sustain the periodicity of these events. The University of Chicago provided the funds for the first John Clough Memorial Conference in June 2005, and its generous sponsorship was matched by Yale University when I assumed a position there in the fall of that year.

---

[2] John Clough, "Pitch-Set Equivalence and Inclusion: A Comment on Forte's Theory of Set-Complexes." *Journal of Music Theory* 9 (1965): 163-171; "Aspects of Diatonic Sets." *Journal of Music Theory* 23 (1979): 45-61; "Diatonic Interval Sets and Transformational Structures," *Perspectives of New Music* 18 (1979-80): 461-482. A complete list of Clough's writings is given at http://www.music.buffalo.edu/theory/cloughpub.shtml.

[3] A recent summary of this line of research can be found in introductions to two recent volumes dedicated to extending it: David Clampitt's "The Legacy of John Clough in Mathematical Music Theory," *Journal of Mathematics and Music* 1.2 (July 2007): 73-78; and Norman Carey, Jack Douthett, and Martha M. Hyde's introduction to *Music Theory and Mathematics: Chords, Collections, and Transformations*, ed. Jack Douthett et. al. (Rochester: University of Rochester Press, 2008): 1-8.

Four months before John Clough passed away, Robert Peck attracted an international group of mathematical music theorists to Baton Rouge, Louisiana, for a special session of the American Mathematical Society. Although John was too ill to attend, the spirit of his personality, as much as the content of his thought, infused the proceedings. The synchronization of Clough's *Viertaktigkeit* with the projected binary periodicity of the meetings of the recently formed Society for Mathematics and Computation in Music suggested the one-time merger of the two events, even though it would produce an event with a different scale and tone than its predecessors. That suggestion was solidified into a commitment as soon as I began to imagine John's excitement, had he survived to witness the founding of this scholarly society—a commingling of musicians, mathematicians, and systems scientists, with membership and leadership from both sides of the Atlantic, producing a thrice-annual periodical, and a periodic conference with proceedings from a major publisher. It is difficult to imagine that, for a mathematical music theorist as dedicated, equanimous, and magnanimous as John Clough, dreams could get any wilder than that.

Richard Cohn
Battell Professor of the Theory of Music
Yale University

Table of Contents

Hamiltonian Cycles in the Topological Dual of the Tonnetz

Giovanni Albini and Samuele Antonini

Department of Mathematics, University of Pavia, Italy

Abstract. The Hamiltonian cycles in the topological dual of the Tonnetz (i.e. the successions of triads connected only through PLR-transformations which visit every minor and major triad only once) will be introduced, enumerated on, studied, and classified both from a theoretical and analytical point of view.

Keywords: Neo-Riemannian Theories, Triads, Nineteenth Century Harmony, PLR-transormations, Tonnetz, GIS, Tone-network, Graph theory, Hamiltonian cycles.

1 Introduction

In [3], Richard Cohn explains the practical uses of Neo-Riemannian theories by showing that they are " an efficient technology and descriptive language for making and communicating new discoveries about the properties of triads and related structures, and the relational systems in which they participate." Recently this framework has been almost exclusively studied from a theoretical and analytical point of view. The aim of the present paper is to show some cycles of the topological dual of the Tonnetz (i.e. some successions of triads connected only through PLR-transformations) which could be useful as a compositional device. Properties of minimal cycles of this graph have been widely studied (we can mention [1] and [4]), but no attention has yet been given to the Hamiltonian class of cycles.

In mathematics a Hamiltonian cycle (or circuit)[1] is a closed path through the vertices of a graph which includes every vertex exactly once. So Hamiltonian cycles in the topological dual of the Tonnetz represent complete sequences through all twenty-four major and minor triads using PLR-transformations in which each major and minor triad is used only once. These cycles are exclusively triadic and overall completely chromatic, since every pitch class is used exactly six times. As we shall see, the succession can also be more or less diatonic, depending on the patterns of the transformations that are employed. So these classes of cycles could be a useful compositional device to define harmonic structures that are triadic (and in some cases locally diatonic) but without any real tonal center. In fact some contemporary composers, like Paul Glass in his "Corale I for Margareth", for string orchestra (1995) or young composer Jeremy Vaughan in his "Violin Sonata" (2008), used successions of triads

[1] It's worthwhile to know that the Hamiltonian cycles has been used in Music Theory also in some formalizations of the Art of Change-Ringing.

E. Chew, A. Childs, and C.-H. Chuan (Eds.): MCM 2009, CCIS 38, pp. 1–10, 2009.

connected almost completely by Neo-Riemannian transformations, showing a particular interest on an explicit use of major and minor triads that is different than the one of tonality.

After a formal generalization of the Tonnetz starting from Lewin's Generalized Interval System (*GIS*), and after the proof of an useful theorem about the automorphism group of the topological dual of the well tempered Tonnetz (section 2), all the Hamiltonian cycles in that graph will be identified and classified (section 3). The properties of the most 'symmetric' of them, which was even used by Beethoven in his Ninth Symphony, will then be analyzed even deeper (section 4). Their properties and their utility as a compositional device is therefore underlined (section 5).

2 Tone-Networks, the Tonnetz and Its Topological Dual

In a definition similar to the one given in [9], a Generalized Interval System (*GIS*) can be defined as follows:

2.1 Definition. A *GIS* is a couple *(X, G)* where *X* is a set of pitches, and *G* is an abelian group which acts freely[2] on *X* and whose action is transitive.[3] *X* is the *pitch set*, and *G* is the *group of intervals*.

2.2 Definition. We call *(Pc, I)* the *GIS* with *Pc* being the set of the twelve well tempered pitch classes, and *I* being the group of intervals in the octave that is isomorphic to the cyclic group *Z/12Z*.

2.3 Definition. Given a *GIS (X, G)* and *H* as a subset of *G*, a *tone-network* is a simple vertex labeled graph which has exactly one vertex for each of the elements of *X*, and an edge between two vertices x_1 and x_2 if, and only if, there exists g_1 in *H* which maps x_1 in x_2 , or there exists g_2 in *H* which maps x_2 in x_1 .

Notice that g_1 and g_2 are the inverse of one other in the group *G*, and that it is not necessary for them to both belong to *H*, because the tone-network is a simple graph, so no more than one edge can be between two vertices. To give both of them while defining a to ne-network is important for having a co mplete vision of the model in the spcific instance of a set of pitch classes, as it shall be done for the Tonnetz.

Now, reminded that a graph is vertex-transitive if and only if every pair of vertices is equivalent under some element of its automorphism group (i.e. no vertex can be distinguished from any other) the following Theorem can be given.

2.4 Theorem. A tone-network is always vertex-transitive.

Proof. *The group of intervals (*G*), whose action on the set of vertices is given by its action on their labelling, always map adjacent vertices in adjacent vertices, so* G *is a subgroup of the automorphism group of the tone-network. So for every ordered couple of vertices there is an element of the automorphism group which maps the first in the second.*

---

[2] A group action is called free if any element of the set is fixed only by the identity of the group.

[3] A group action is transitive if it possesses only a single orbit.

The previous result is interesting because it underlines that in each *GIS*, and consequently in every tone-network built on it, there is theoretically no way to distinguish two pitches/vertices and it shows a property of particular interest while providing the abstract case of pitch classes as we shall do.

Now the tonnetz can be easily defined:

2.5 Definition. The *Tonnetz*, or more simply *Ton*, is a tone-network defined on the *GIS (Pc, I)* as it has been given in definition 2.2, where *H* contains only the following intervals: major and minor thirds, the fifth and their inverses.

Reminded that the topological dual of a simple planar graph is a simple graph that has vertices each of which corresponds to a face of the first graph and that are connected if the corresponding faces have an edge in common, and noticed that *Ton* has no edges crossing if embedded in the torus, the topological dual of the Tonnetz can be defined:

2.6 Definition. *D(Ton)*, the topological dual of the Tonnetz, is a simple labeled graph whose vertices are labeled with the triads defined by the pitch classes that bound the corresponding face.

Because two vertices are connected in *D(Ton)* if the corresponding triads share two pitch classes, also edges can be labelled through the basic Neo-Riemannian operators *P*, *L* and *R*. The pictures of *Ton* and *D(Ton)* are given in Figure 1 and Figure 2.[4]

Before giving the last theorem, let's introduce another object necessary to its proof.

2.7 Definition. Given the *GIS* (*Pc*, *I*), we will call *T-I* the group acting on *Pc* and generated by the group of intervals *G* (the twelve translations) and by the inversion element which fixes *C* and *F#* and exchange *C#* and *B*, *D* and *A#*, *D#* and *A*, *E* and *G#*, *F* and *G*.

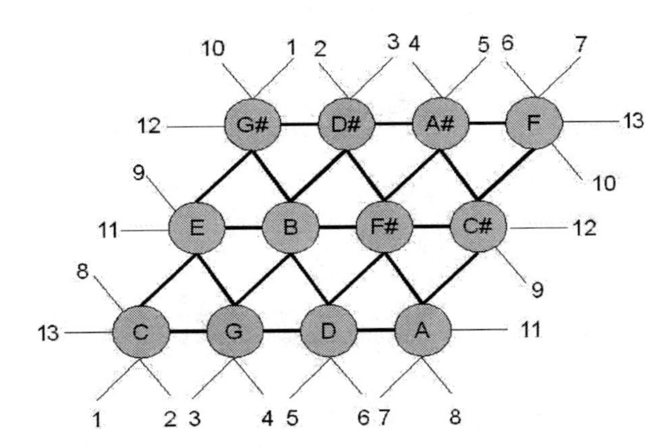

Fig. 1. A planar view of *Ton*. Numbers outside the graph show adjacency.

[4] Triads will be represented giving the fundamental and the sign + for major triads and the sign – for minor ones.

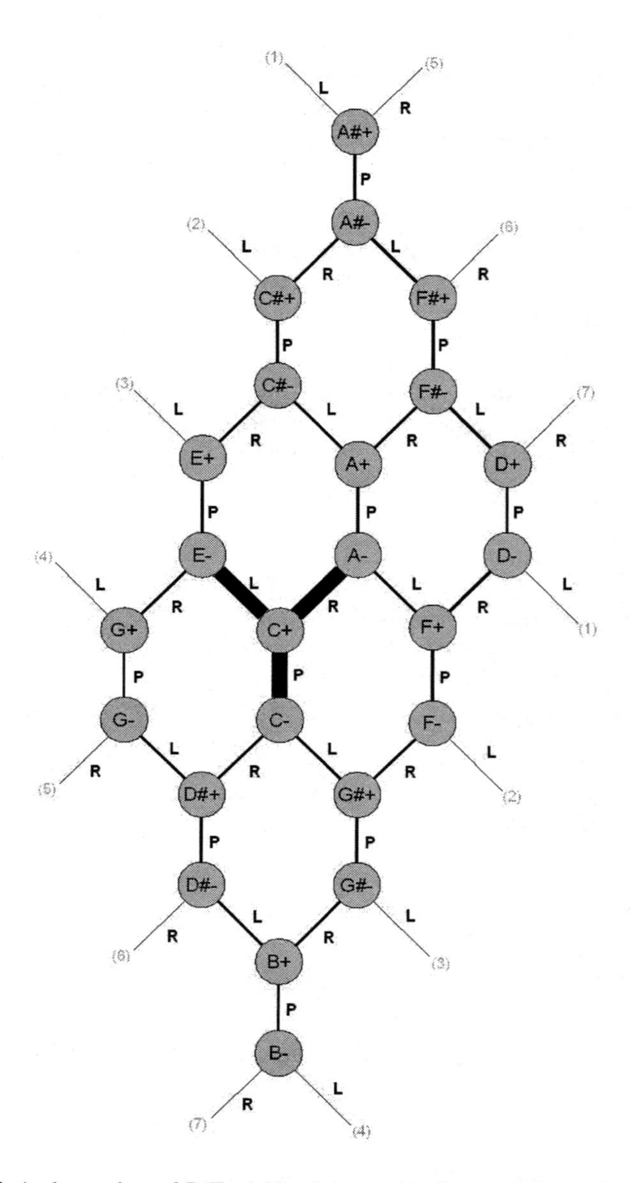

Fig. 2. A planar view of *D(Ton)*. Numbers outside the graph show adjacency.

2.8 Lemma. $T\text{-}I$ is isomorphic to D_{12}.

Proof. *Translations correspond to rotations, and the inversion element is the element of order two.*

The proof that the action through labelling of *T-I* on *Ton* maintain adjacency is left to the reader.[5]

Let's now concentrate on the main result of the section, that's original, and that will be useful for counting Hamiltonian cycles in *D(Ton)*.

2.9 Theorem. *Aut(DTon)*, the automorphism group[6] of *D(Ton)*, is isomorphic to the dihedral group D_{12}.

Proof. *Considering its labelling there is no doubt that* Aut(DTon) *has at least* 24 *elements, because of the action of* T-I *on triads that comes out directly from its actions on pitch classes. It also guarantess that* Aut(DTon) *is vertex transitive. If it is showed that* Aut(DTon) *has no more elements than these, the theorem would be proved, since* T-I *is isomorphic to* D_{12}. *Because of the vertex transitivity of the graph it is sufficient to consider one vertex (we shall take* C+*) and to show that given an element* f *of* Aut(DTon) *which does not fix that vertex, the composition between* f *and the element* g *of* T-I *that put back* f(C+) *in* C+ *can be only the identity element. In fact, in that case,* f *would be the inverse of an element of* T-I, *so it would belong to it itself. So the theorem would be proved if it is showed that only the identity element can fix a vertex. Because* C+ *is only connected to* C-, A- *and* E-, *is sufficient to show that none of the permutations of these three vertices, apart of the identity, maintain the adjacency between all the vertices of* Dton, *and to show that if* C+, C-, A- *and* E- *are fixed so all the vertices would be fixed too.*

Before introducing the Hamiltonian cycles in *D(Ton)* it would be interesting to focus on the omnipresence of the group D_{12} while considering the *GIS* of the twelve well tempered pitch classes as defined in the definition 2.2. It can be shown that the group *T-I*, which both act on the pitch classes and, consequently, on the triads, and the *PLR-group*, which just act on triads, with the automorphism groups of *Ton* and *DTon*, *Aut(Ton)* and *Aut(DTon)*, are all isomorphic to the dihedral group D_{12}. This isomorphism between groups which represents aural and visual transformations can lead to some cognitive considerations on how we represent them mentally in a similiar way. In [10] Hugo Riemann anticipated this concept: " In this fashion, the hearing of changes in pitch level is transformed into a vision of changes in location, and we already have a presentiment of the ultimate identification of the essence of visual and aural imagination."

3 Hamiltonian Cycles in *D(Ton)*

A software[7] has been used to find the Hamiltonian cycles in *D(Ton)*: the result are the *62* cycles shown in Figure 3.

[5] It is also possible, with a proof similar to the one of theorem 2.9, to prove that T-I is isomorphic to the automoprhism group of *Ton*.

[6] The automorphism group of a graph is the group of bijective mappings from the vertices of the graph to the vertices of the same graph which preserve adjacency.

[7] *Groups and Graph*, version 3.2 for MacOSX (2006), by William Kocay and William Palmer.

The number which gives the name to the cycle depends on the order of output of the software.

Now we can begin to study them considering the action of *Aut(DTon)*, classifying them in terms of the succession of transformations (independently from the direction of the path covered) instead of triads. In fact, given a Hamiltonian cycle, if *n* elements of *Aut(DTon)* transform it to itself, then there are exactly *24/n* different Hamiltonian cycles sharing the same model of transformation. Eight models can be recognized, named *H1*, ... , *H8*:

H1: Only the cycle *#41*.
Characterized by the repetition of the model *LR* (or, in the opposite direction, *RL*). This cycle is mapped into itself by all the elements of *Aut(DTon)*. It will be analyzed further in the fourth section.

H2: The cycles *#32* and *#45*.
Characterized by the repetition of the model *PRLR* (or *RLRP*). These cycles are mapped into themselves by *12* elements of *Aut(DTon)*.

H3: The cycles *#4*, *#13* and *#62*.
Characterized by the repetition of the model *LPLPLR* (or *RLPLPL*). These cycles are mapped into themselves by *8* elements of *Aut(DTon)*.

H4: The cycles *#33*, *#38*, *#40* and *#44*.
Characterized by the repetition of the model *PRPRPRLR* (or *RLRPRPRP*). These cycles are mapped into themselves by *6* elements of *Aut(DTon)*.

H5: The cycles *#34*, *#39*, *#42* and *#43*.
Characterized by the repetition of the model *PRPRLRLR* (or *RLRLRPRP*). These cycles are mapped into themselves by *6* elements of *Aut(DTon)*.

H6: The cycles *#6*, *#7*, *#8*, *#9*, *#10*, *#19*, *#21*, *#22*, *#27*, *#30*, *#31* and *#58*.
Characterized by the repetition of the model *LPLPLRPLPLPRPLPLPRPLPLPR* (or *RPLPLPRPLPLPRPLPLPRLPLPL*). These two cycles are mapped into themselves by *2* elements of *Aut(DTon)*.

H7: The cycles *#3*, *#12*, *#15*, *#17*, *#26*, *#28*, *#35*, *#46*, *#51*, *#52*, *#56* and *#61*.
Characterized by the repetition of the model *PLRPLPRLPLRLPRLRPRLRPLRL* (or *LRLPRLRPRLRPLRLPLRPLPRLP*). These two cycles are mapped into themselves by *2* elements of *Aut(DTon)*.

H8: The *24* remaining cycles.
Characterized by the model *LRLPLRLPRLRLPLRPLPRPLRPR* (or *RPRLPRPLPRLPLRLRPLRLPLRL*). Only the idendity map them into themselves.

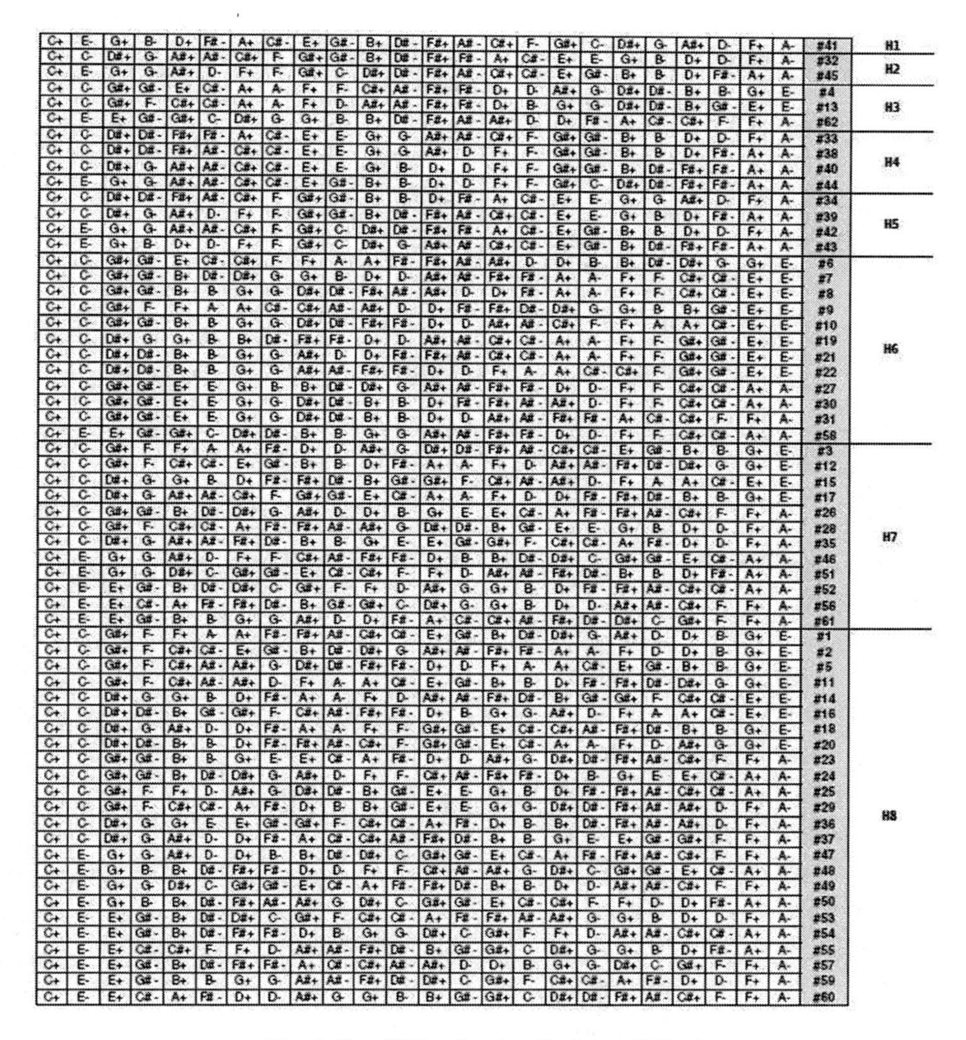

C+	E-	G+	B-	D+	F#-	A+	C#-	E+	G#-	B+	D#-	F#+	A#-	C#+	F-	G#+	C-	D#+	G-	A#+	D-	F+	A-	#41	H1
C+	C-	D#+	G-	A#+	A#-	C#+	F-	G#+	G#-	B+	D#-	F#+	F#-	A+	C#-	E+	E-	G+	B-	D+	D-	F+	A-	#32	H2
C+	E-	G+	G-	A#+	D-	F+	F-	G#+	C-	D#+	D#-	F#+	A#-	C#+	C#-	E+	G#-	B+	B-	D+	F#-	A+	A-	#45	
C+	C-	G#+	G#-	E+	C#-	A+	A-	F+	F-	C#+	A#-	F#+	F#-	D+	D-	A#+	G-	D#+	D#-	B+	B-	G+	E-	#4	H3
C+	C-	G#+	F-	C#+	C#-	A+	A-	F+	D-	A#+	A#-	F#+	F#-	D+	B-	G+	G-	D#+	D#-	B+	G#-	E+	E-	#13	
C+	E-	E+	G#-	G#+	C-	D#+	G-	G+	B-	B+	D#-	F#+	A#-	A#+	D-	D+	F#-	A+	C#-	C#+	F-	F+	A-	#62	
C+	C-	D#+	D#-	F#+	F#-	A+	C#-	E+	E-	G+	G-	A#+	A#-	C#+	F-	G#+	G#-	B+	B-	D+	D-	F+	A-	#33	H4
C+	C-	D#+	D#-	F#+	A#-	C#+	C#-	E+	E-	G+	G-	A#+	D-	F+	F-	G#+	G#-	B+	B-	D+	F#-	A+	A-	#38	
C+	C-	D#+	G-	A#+	A#-	C#+	C#-	E+	E-	G+	B-	D+	D-	F+	F-	G#+	G#-	B+	D#-	F#+	F#-	A+	A-	#40	
C+	E-	G+	G-	A#+	A#-	C#+	C#-	E+	G#-	B+	B-	D+	D-	F+	F-	G#+	C-	D#+	D#-	F#+	F#-	A+	A-	#44	
C+	C-	D#+	D#-	F#+	A#-	C#+	F-	G#+	G#-	B+	B-	D+	F#-	A+	C#-	E+	E-	G+	G-	A#+	D-	F+	A-	#34	H5
C+	C-	D#+	G-	A#+	D-	F+	F-	G#+	G#-	B+	D#-	F#+	A#-	C#+	C#-	E+	E-	G+	B-	D+	F#-	A+	A-	#39	
C+	E-	G+	G-	A#+	A#-	C#+	F-	G#+	C-	D#+	D#-	F#+	F#-	A+	C#-	E+	G#-	B+	B-	D+	D-	F+	A-	#42	
C+	E-	G+	B-	D+	D-	F+	F-	G#+	C-	D#+	G-	A#+	A#-	C#+	C#-	E+	G#-	B+	D#-	F#+	F#-	A+	A-	#43	
C+	C-	G#+	G#-	E+	C#-	C#+	F-	F+	A-	A+	F#-	F#+	A#-	A#+	D-	D+	B-	B+	D#-	D#+	G-	G+	E-	#6	H6
C+	C-	G#+	G#-	B+	D#-	D#+	G-	G+	B-	D+	D-	A#+	A#-	F#+	F#-	A+	A-	F+	F-	C#+	C#-	E+	E-	#7	
C+	C-	G#+	G#-	B+	B-	G+	G-	D#+	D#-	F#+	A#-	A#+	D-	D+	F#-	A+	A-	F+	F-	C#+	C#-	E+	E-	#8	
C+	C-	G#+	F-	F+	A-	A+	C#-	C#+	A#-	A#+	D-	D+	F#-	F#+	D#-	D#+	G-	G+	B-	B+	G#-	E+	E-	#9	
C+	C-	G#+	G#-	B+	B-	G+	G-	D#+	D#-	F#+	F#-	D+	D-	A#+	A#-	C#+	F-	F+	A-	A+	C#-	E+	E-	#10	
C+	C-	D#+	G-	G+	B-	B+	D#-	F#+	F#-	D+	D-	A#+	A#-	C#+	C#-	A+	A-	F+	F-	G#+	G#-	E+	E-	#19	
C+	C-	D#+	D#-	B+	B-	G+	G-	A#+	D-	D+	F#-	F#+	A#-	C#+	C#-	A+	A-	F+	F-	G#+	G#-	E+	E-	#21	
C+	C-	D#+	D#-	B+	B-	G+	G-	A#+	A#-	F#+	F#-	D+	D-	F+	A-	A+	C#-	C#+	F-	G#+	G#-	E+	E-	#22	
C+	C-	G#+	G#-	E+	E-	G+	B-	B+	D#-	D#+	G-	A#+	A#-	F#+	F#-	D+	D-	F+	F-	C#+	C#-	A+	A-	#27	
C+	C-	G#+	G#-	E+	E-	G+	G-	D#+	D#-	B+	B-	D+	F#-	F#+	A#-	A#+	D-	F+	F-	C#+	C#-	A+	A-	#30	
C+	C-	G#+	G#-	E+	E-	G+	G-	D#+	D#-	B+	B-	D+	D-	A#+	A#-	F#+	F#-	A+	C#-	C#+	F-	F+	A-	#31	
C+	E-	E+	G#-	G#+	C-	D#+	D#-	B+	B-	G+	G-	A#+	A#-	F#+	F#-	D+	D-	F+	F-	C#+	C#-	A+	A-	#58	
C+	C-	G#+	F-	F+	A-	A+	F#-	D+	D-	A#+	G-	D#+	D#-	F#+	A#-	C#+	C#-	E+	G#-	B+	B-	G+	E-	#3	H7
C+	C-	G#+	F-	C#+	C#-	E+	G#-	B+	B-	D+	F#-	A+	A-	F+	D-	A#+	A#-	F#+	D#-	D#+	G-	G+	E-	#12	
C+	C-	D#+	G-	G+	B-	D+	F#-	F#+	D#-	B+	G#-	G#+	F-	C#+	A#-	A#+	D-	F+	A-	A+	C#-	E+	E-	#15	
C+	C-	D#+	G-	A#+	A#-	C#+	F-	G#+	G#-	E+	C#-	A+	A-	F+	D-	D+	F#-	F#+	D#-	B+	B-	G+	E-	#17	
C+	C-	G#+	G#-	B+	D#-	D#+	G-	A#+	D-	D+	B-	G+	E-	E+	C#-	A+	F#-	F#+	A#-	C#+	F-	F+	A-	#26	
C+	C-	G#+	F-	C#+	C#-	A+	F#-	F#+	A#-	A#+	G-	D#+	D#-	B+	G#-	E+	E-	G+	B-	D+	D-	F+	A-	#28	
C+	C-	D#+	G-	A#+	A#-	F#+	D#-	B+	B-	G+	E-	E+	G#-	G#+	F-	C#+	C#-	A+	F#-	D+	D-	F+	A-	#35	
C+	E-	G+	G-	A#+	D-	F+	F-	C#+	A#-	F#+	F#-	D+	B-	B+	D#-	D#+	C-	G#+	G#-	E+	C#-	A+	A-	#46	
C+	E-	G+	G-	D#+	C-	G#+	G#-	E+	C#-	C#+	F-	F+	D-	A#+	A#-	F#+	D#-	B+	B-	D+	F#-	A+	A-	#51	
C+	E-	E+	G#-	B+	D#-	D#+	C-	G#+	F-	F+	D-	A#+	G-	G+	B-	D+	F#-	F#+	A#-	C#+	C#-	A+	A-	#52	
C+	E-	E+	C#-	A+	F#-	F#+	D#-	B+	G#-	G#+	C-	D#+	G-	G+	B-	D+	D-	A#+	A#-	C#+	F-	F+	A-	#56	
C+	E-	E+	G#-	B+	B-	G+	G-	A#+	D-	D+	F#-	A+	C#-	C#+	A#-	F#+	D#-	D#+	C-	G#+	F-	F+	A-	#61	
C+	C-	G#+	F-	F+	A-	A+	F#-	F#+	A#-	C#+	C#-	E+	G#-	B+	D#-	D#+	G-	A#+	D-	D+	B-	G+	E-	#1	H8
C+	C-	G#+	F-	C#+	C#-	E+	G#-	B+	D#-	D#+	G-	A#+	A#-	F#+	F#-	A+	A-	F+	D-	D+	B-	G+	E-	#2	
C+	C-	G#+	F-	C#+	A#-	A#+	G-	D#+	D#-	F#+	F#-	D+	D-	F+	A-	A+	C#-	E+	G#-	B+	B-	G+	E-	#5	
C+	C-	G#+	F-	C#+	A#-	A#+	D-	F+	A-	A+	C#-	E+	G#-	B+	B-	D+	F#-	F#+	D#-	D#+	G-	G+	E-	#11	
C+	C-	D#+	G-	G+	B-	D+	F#-	A+	A-	F+	D-	A#+	A#-	F#+	D#-	B+	G#-	G#+	F-	C#+	C#-	E+	E-	#14	
C+	C-	D#+	D#-	B+	G#-	G#+	F-	C#+	A#-	F#+	F#-	D+	B-	G+	G-	A#+	D-	F+	A-	A+	C#-	E+	E-	#16	
C+	C-	D#+	G-	A#+	D-	D+	F#-	A+	A-	F+	F-	G#+	G#-	E+	C#-	C#+	A#-	F#+	D#-	B+	B-	G+	E-	#18	
C+	C-	D#+	D#-	B+	B-	D+	F#-	F#+	A#-	C#+	F-	G#+	G#-	E+	C#-	A+	A-	F+	D-	A#+	G-	G+	E-	#20	
C+	C-	G#+	G#-	B+	B-	G+	E-	E+	C#-	A+	F#-	D+	D-	A#+	G-	D#+	D#-	F#+	A#-	C#+	F-	F+	A-	#23	
C+	C-	G#+	G#-	B+	D#-	D#+	G-	A#+	D-	F+	F-	C#+	A#-	F#+	F#-	D+	B-	G+	E-	E+	C#-	A+	A-	#24	
C+	C-	G#+	F-	F+	D-	A#+	G-	D#+	D#-	B+	G#-	E+	E-	G+	B-	D+	F#-	F#+	A#-	C#+	C#-	A+	A-	#25	
C+	C-	G#+	F-	C#+	C#-	A+	F#-	D+	B-	B+	G#-	E+	E-	G+	G-	D#+	D#-	F#+	A#-	A#+	D-	F+	A-	#29	
C+	C-	D#+	G-	G+	E-	E+	G#-	G#+	F-	C#+	C#-	A+	F#-	D+	B-	B+	D#-	F#+	A#-	A#+	D-	F+	A-	#36	
C+	C-	D#+	G-	A#+	D-	D+	F#-	A+	C#-	C#+	A#-	F#+	D#-	B+	B-	G+	E-	E+	G#-	G#+	F-	F+	A-	#37	
C+	E-	G+	G-	A#+	D-	D+	B-	B+	D#-	D#+	C-	G#+	G#-	E+	C#-	A+	F#-	F#+	A#-	C#+	F-	F+	A-	#47	
C+	E-	G+	B-	B+	D#-	F#+	F#-	D+	D-	F+	F-	C#+	A#-	A#+	G-	D#+	C-	G#+	G#-	E+	C#-	A+	A-	#48	
C+	E-	G+	G-	D#+	C-	G#+	G#-	E+	C#-	A+	F#-	F#+	D#-	B+	B-	D+	D-	A#+	A#-	C#+	F-	F+	A-	#49	
C+	E-	G+	B-	B+	D#-	F#+	A#-	A#+	G-	D#+	C-	G#+	G#-	E+	C#-	C#+	F-	F+	D-	D+	F#-	A+	A-	#50	
C+	E-	E+	G#-	B+	D#-	D#+	C-	G#+	F-	C#+	C#-	A+	F#-	F#+	A#-	A#+	G-	G+	B-	D+	D-	F+	A-	#53	
C+	E-	E+	G#-	B+	D#-	F#+	F#-	D+	B-	G+	G-	D#+	C-	G#+	F-	F+	D-	A#+	A#-	C#+	C#-	A+	A-	#54	
C+	E-	E+	C#-	C#+	F-	F+	D-	A#+	A#-	F#+	D#-	B+	G#-	G#+	C-	D#+	G-	G+	B-	D+	F#-	A+	A-	#55	
C+	E-	E+	G#-	B+	D#-	F#+	F#-	A+	C#-	C#+	A#-	A#+	D-	D+	B-	G+	G-	D#+	C-	G#+	F-	F+	A-	#57	
C+	E-	E+	G#-	B+	B-	G+	G-	A#+	A#-	F#+	D#-	D#+	C-	G#+	F-	C#+	C#-	A+	F#-	D+	D-	F+	A-	#59	
C+	E-	E+	C#-	A+	F#-	D+	D-	A#+	G-	G+	B-	B+	G#-	G#+	C-	D#+	D#-	F#+	A#-	C#+	F-	F+	A-	#60	

Fig. 3. The *62* Hamiltonian Cycles in *D(Ton)*

4 The Hamiltonian Cycle *#41* and Beethoven's Ninth Symphony

Before studying the trivial cycle *#41*, let us introduce some useful definitions.

4.1 Definition. We call *diatonic pitch class set*, a subset of seven elements of *Pc* that can be ordered through perfect fifths. It will be named both with the minor and major scale that can be built with its pitches.

4.2 Example. The set $\{C, D, E, F, G, A, B\}$ is a diatonic pitch class set, since its pitch classes can be ordered by perfect fifths in the following way: *F*, *C*, *G*, *D*, *A*, *E*, *B*. It will be called the *C major / A minor* diatonic set.

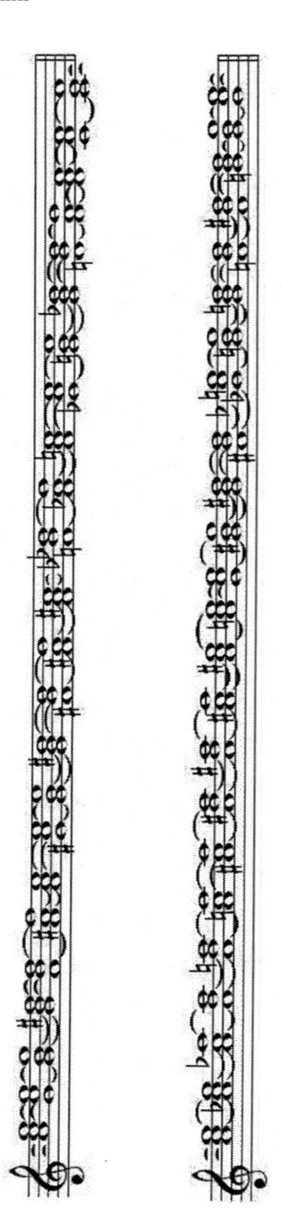

Fig. 4. Cycles *#41* (H1, most symmetrical), and *#1* (H8, least symmetrical)

On every diatonic set only six major and minor triads can be built. In Figure 5 the Neo-Riemannian relations between the six triads of the *C major / A minor* diatonic set are shown.

It is easy to see that two diatonic sets that are at a distance of a perfect fifth (for example *C major / A minor* and *G major / E minor*) share six pitch classes and four

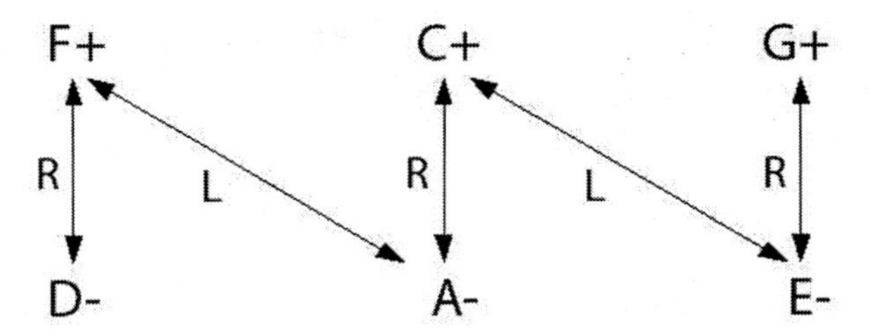

Fig. 5. The six triads of the *C major / A minor* diatonic set

triads. So using alternating *L* and *R* transformations it is possible to move through triads in the most gradual way from a diatonic point of view, as the succession smoothly covers all the common triads of closely related diatonic systems.

Between measures 143 and 176 of the second movement of his "Ninth Smphony", Ludwig Van Beethoven, after a perfect cadence on *C+*, covers 19 triads applying the trivial cycle *#41*, from *C+* to *A+*. It is interesting to notice that, after that E- is reached through a chromatic passage, its tonal center is confirmed. But *E-* is the triad just before *C+* in cycle *#41*, so the chromatic passage could be seen as a jump to complete the cycle. He takes the longest cyclical journey to connect closely related tonalities and closely related triads from a Neo-Riemannian point of view.

5 Hamiltonian Cycles as a Compositional Tool

Hamiltonian cycles of *D(Ton)* are not only characterized by Neo-Riemannian transformations, but they also guarantee two interesting conditions, both contradicting tonality: 1) they are characterized by an overall pitch class completeness, since every pitch class is used exactly six times; 2) as cycles, there are no favourite starting or ending triads. So they allow the 'simplicity' of the parsimonious voice leading[8], and an use of triads that cannot be seen as tonal. Composer Giovanni Albini started to study them looking for successions of triads satisfying conditions 1) and 2), and has used them in his three "Corali", for string quartet, for orchestra, and for violin and guitar. Each of them is based on a different Hamiltonian cycle, treated in different ways. So the present paper is mainly born from a compositional necessity.

The Authors think that Hamiltonian cycles could also be an useful compositional tool from a general point of view, for example considering different *GIS* or different tone-networks. By starting the article with the *GIS*, and basing the definition of a tone-network on it, the purpose is to give a theoretical framework upon which it is possible to build other networks based on the same *GIS* or on other ones (as the Authors would like to do in future papers). The Hamiltonian cycles built on their dual

[8] Simplicity is intended from a Riemannian point of view. In [10] Riemann wrote: «Let attention be drawn here to the definite inclination of the interpreting mind to find its way easily through the confusion of endless possibilities of tonal combinations (in melody and harmony) by means of preferring simple relationship over more complicated ones. This Principle of Greatest Possible Economy for the Musical Imagination moves directly toward the rejection of more complicated structures.»

could be interesting from a compositional point of view, again guaranteeing conditions 1) and 2), and a specific voice leading depending on the tone-network.

References

1. Cohn, R.: Maximally Smooth Cycles, Hexatonic Systems, and the Analysis of Late-Romantic Triadic Progressions. Music Analysis 15(1), 9–40 (1996)
2. Cohn, R.: Neo-Riemannian Operations, Parsimonious Trichords, and Their "Tonnetz" Representations. Journal of Music Theory 41(1), 1–66 (1997)
3. Cohn, R.: Introduction to Neo-Riemannian Theory: A Survey and a Historical Perspective. Journal of Music Theory 42(2), 167–180 (1998)
4. Cohn, R.: Square Dancese with Cubes. Journal of Music Theory 42(2), 167–180 (1998)
5. Crans, A.S., Fiore, T.M., Satyendra, R.: Musical Actions of Dihedral Groups (2008), http://arxiv.org/pdf/0711.1873
6. Harary, F.: Graph Theory. Addison-Wesley, Reading (1969)
7. Harary, F., Palmer, E.M.: Graphical Enumeration. Academic Press, London (1973)
8. Garey, M.R., Johnson, D.S.: Computers and Intractability: A Guide to the Theory of NP-Completeness. W.H. Freeman, New York (1983)
9. Lewin, D.: Generalized musical Intervals and Transformations. Yale University Press (1987)
10. Riemann, H.: Ideas for a Study On the Imagination of Tone. Journal of Music Theory 36(1), 81–117 (1992)

The Continuous Hexachordal Theorem

Brad Ballinger[1], Nadia Benbernou[2], Francisco Gomez[3], Joseph O'Rourke[4], and Godfried Toussaint[5]

[1] Davis School for Independent Study, Davis, USA
[2] Department of Mathematics and Computer Science and Artificial Intelligence Laboratory (CSAIL), Massachusetts Institute of Technology, Cambridge, USA
[3] Department of Applied Mathematics, Universidad Politecnica de Madrid, Madrid, Spain
[4] Department of Computer Science, Smith College, Northampton, USA
[5] Centre for Interdisciplinary Research in Music Media and Technology, The Schulich School of Music, McGill University, Montreal, Canada

Abstract. The Hexachordal Theorem may be interpreted in terms of scales, or rhythms, or as abstract mathematics. In terms of scales it claims that the complement of a chord that uses half the pitches of a scale is homometric to—i.e., has the same interval structure as—the original chord. In terms of onsets it claims that the complement of a rhythm with the same number of beats as rests is homometric to the original rhythm. We generalize the theorem in two directions: from points on a discrete circle (the mathematical model encompassing both scales and rhythms) to a continuous domain, and simultaneously from the discrete presence or absence of a pitch/onset to a continuous strength or weight of that pitch/onset. Athough this is a significant generalization of the Hexachordal Theorem, having all discrete versions as corollaries, our proof is arguably simpler than some that have appeared in the literature.

We also establish the natural analog of what is sometimes known as Patterson's second theorem: if two equal-weight rhythms are homometric, so are their complements.

1 Introduction

1.1 Basic Definitions

We are concerned with cyclic musical rhythms consisting of k *onsets* (pulses, beats) and $n-k$ *rests*, represented by n evenly spaced points on a circle, with arithmetic $\bmod n$, i.e., in the group $\mathbb{Z}_n$. This representation has been used as early as the 13th century, as accounted by Wright [Wri78], but it has been used recently again; see [Vuz85], [Tou05], among others. Alternately, the k onsets (points) may be considered as k pitches making up a musical chord or scale selected from a universe of n pitches [Tym06]. Such sets of points on a circle are called *cyclotomic* sets in the crystallography literature [Pat44], [Bue78]. We will emphasize the rhythms model in this paper, but all results hold equally in the pitch model or the crystallography model.

E. Chew, A. Childs, and C.-H. Chuan (Eds.): MCM 2009, CCIS 38, pp. 11–21, 2009.

Every pair of the points on the circle determines an inter-onset duration interval (the *geodesic* between the pair of points around the circle) [Bue78]. The histogram of this multiset of distances in the context of musical scales and chords is called its *interval content* [Lew59]. Two rhythms which are congruent to each other obviously have the same interval content. Here by *congruence* we mean geometrical congruence, i.e., equivalence under rotation or reflection. However, two rhythms with the same histograms need not be congruent. Two sets of points with the same multiset of distances are said to be *homometric*, a term introduced by Patterson in 1939 [Pat44], who first discovered them. In the music literature, two pitch-class sets (or two rhythms) with the same intervalic content are termed as having the Z-relation or *isomeric* relation [For77].

One of the fundamental theorems in this area is the so-called *Hexachordal Theorem*, which states that complementary sets with $k{=}n/2$ (and n even) are homometric. Two examples are shown in Figs. 1 and 2. In Fig. 1, the $k{=}4$ onsets occur at $(0, 1, 4, 7)$, and the complementary rhythm has onsets precisely where the first rhythm has rests: $(2, 3, 5, 6)$. The histogram of intervals is identical.

Fig. 2 shows two complementary $(n, k){=}(12, 6)$ rhythms, again with identical histograms.

An important convention we follow is that the pair of onsets separated by the diameter $d = n/2$ contributes two counts to the interval d in the histogram. This

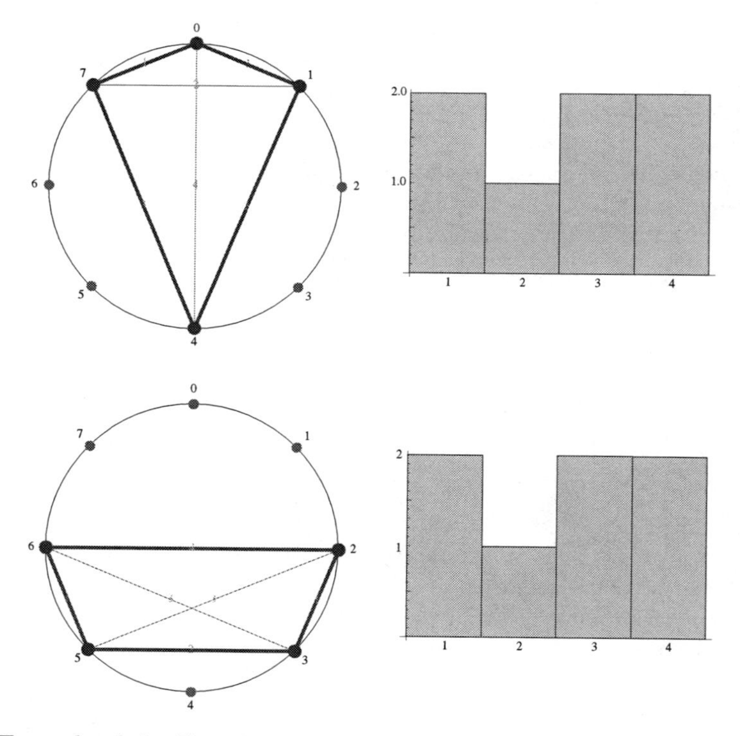

Fig. 1. Example of the Hexachordal Theorem, $(n, k){=}(8, 4)$. Note that the distance $d{=}4$ is counted twice.

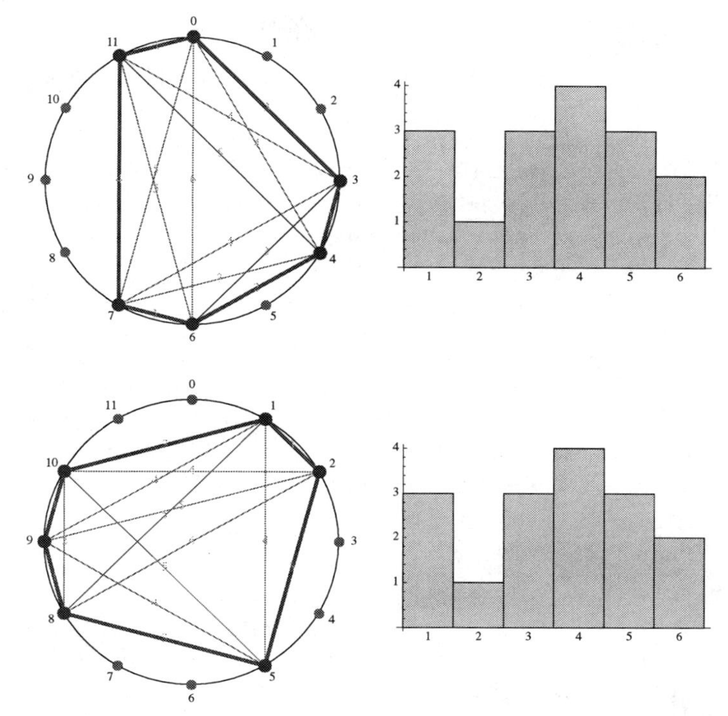

Fig. 2. Another example of the Hexachordal Theorem, $(n, k)=(12, 6)$. Note that the distance $d=6$ is counted twice.

convention simplifies the proofs but changes nothing substantively. This issue is further addressed in Section 2.5.

The term "hexachordal" derives from Schönberg's use of 6-note chords in a 12-tone chromatic scale, and the name "hexachordal" has been retained even though the theorem holds for arbitrary even n.

1.2 History

The earliest proof of the Hexachordal Theorem in the music literature is, to our knowledge, due to Lewin. In 1959 he published a paper [Lew59] on the intervalic relations of two chords that contained an embryonic proof of the Hexachordal Theorem; such a proof was refined in a subsequent paper [Lew60]. In 1974 Regener [Reg74] found an elementary simple proof of this theorem based on the combinatorics of pitch-class sets. Many other proofs have appeared since then, often rivalling in conciseness. Short proofs can be found, for instance, in the work of Mazzola [Maz03] or Jedrzejewski [Jed06]. Amiot [Ami07] gave an elegant, short proof based on the discrete Fourier transform. Perhaps, one of the simplest proofs, in the sense of using no structures such as groups or discrete Fourier transforms, was discovered by Blau [Bla99]. His proof relies on a straightforward analysis of the situation in which two complementary hexachords switch two neighbouring elements.

The music theorists appear to be unaware that this theorem was known to crystallographers about thirty years earlier [Pat44]. It seems to have been proved by Patterson [Pat44] around 1940, but he did not publish a proof. In the crystallography literature the theorem is called Patterson's second theorem [Bue76]. The first published proof in the crystallography literature is due to Buerger [Bue76]; it is based on image algebra, and is non-intuitive. A much simpler and elegant elementary proof was later found by Iglesias [Igl81]. Another simple proof, purely based on geometry, has been recently discovered by Senechal [Sen08].

The Hexachordal Theorem has been generalized in various ways, for example, considering rhythms of different cardinalities; see [Lew76], [Lew87], [Igl81], [Mor90], [Sod95], [AG00] for several directions of generalization. We believe the proof we present in Sec. 2.3 below is not only simple, but also establishes a significant generalization from discrete rhythms to continuous rhythms.

1.3 Outline

We will first introduce weighted rhythms as a generalization of usual rhythms. This generalization will consist of associating certain weights to the onsets and rests of a rhythm. Next we will state and prove the Hexachordal Theorem in terms of such weighted rhythms. We will then generalize the Hexachordal Theorem to a continuous version of it , where rhythms will be considered as continuous functions on the interval $[0, 1]$. From this version we will prove again the discrete Hexachordal Theorem as a straightforward corollary of the continuous version.

2 The Continuous Hexachordal Theorem

2.1 Weighted Rhythms

In order to state our generalization of the Hexachordal Theorem, we introduce a different viewpoint. Each onset i is assigned a *weight* of $w_i = 1$, and each rest is assigned a weight of 0. Thus, the rhythm in Fig. 1 (top) has a weight signature $(1, 1, 0, 0, 1, 0, 0, 1)$. The total weight of a rhythm R is $W(R) = \sum_{i=0}^{n-1} w_i$, the number of onsets k in R. The complementary rhythm $\overline{R}$ is obtained by complementing the weights with respect to 1: $\overline{w_i} = 1 - w_i$. Let H_R be the histogram of intervals determined by rhythm R. This records, for each possible interval distance d, the number of times it occurs in the rhythm. In Fig. 1, we have:

Height:	2	1	2	2
Distance d:	1	2	3	4

This may be viewed as a function of the interval distance d: $H_R(d)$ is the height of the histogram at distance d. With this notation, the Hexachordal Theorem may be stated as follows:

Theorem 1. *If R is a rhythm on n points, n even, and $W(R) = n/2$, then R and $\overline{R}$ are homometric: for all distances d, $H_{\overline{R}}(d) = H_R(d)$.*

Before proceeding to the continuous domain, we need Lemma 1 below, which expresses the histogram function in terms of the weights. This lemma is known in the music literature as the "common-tone theorem" [Joh03]. See [JK06] for a proof in the context of group theory. For the sake of completeness, we include our own proof.

Lemma 1. $H_R(d) = \sum_{i=0}^{n-1} w_i w_{i+d}$.

Proof. Point i is separated by a distance d from the point at $i+d$, where we interpret addition mod n, i.e., in $\mathbb{Z}_n$. If both are onsets, then $w_i = w_{i+d} = 1$, and $w_i w_{i+d} = 1$. If either point is a rest, then $w_i w_{i+d} = 0$. Thus, for each fixed d, summing $w_i w_{i+d}$ over all i counts 1 for each occurrence of d.

We now argue that each pair of points realizing a distance d contributes just once to the sum. A pair $(i, i+d)$ would contribute twice if $i + 2d = i$ so that $(i+d, i)$ would be counted as well. Because d is a shortest path, we have $d \leq n/2$. Thus, $i+2d \leq i+n$, and this equals i (in $\mathbb{Z}_n$) only when $d = n/2$ is the diameter. Our convention is indeed to count a pair realizing the diameter twice.

Consider, for example, the $n = 12$ example in Fig. 2 (top). For $d = n/2 = 6$, both $w_0 w_6$ and $w_6 w_{12} = w_6 w_0$ contribute to $H_R(6) = 2$. Indeed, the reason we follow the convention of double-counting each realization of the diameter is that it naturally fits this weight viewpoint. This point will be revisited in Section 2.5.

2.2 The Continuous Generalizations

We generalize in two directions. First, the circle of n discrete points is generalized to a continuous circle of points. We take its circumference to be 1 without loss of generality. Second, the discrete set of weights w_i is generalized to a real-number weight $f(x) \in [0, 1]$ for $x \in [0, 1]$. Here x specifies a point on the circle, measured by distance clockwise from the zero-position (conventionally at the 12 o'clock position as in Figures 1 and 2), and $f(x)$ the weight of that point. So now the total weight $W(R) = \int_0^1 f(x)\, dx$. Note the maximum possible total weight of any rhythm is achieved by the constant "rhythm" with weight $f(x) = 1$ for all x, in which case $W(R) = 1$.

We define the complement of a rhythm analogously to the discrete case:

Definition 1. *For each point x in rhythm R with weight $f(x)$, the corresponding point x in the complementary rhythm $\overline{R}$ has weight $\overline{f(x)} = 1 - f(x)$.*

The histogram $H_R(d)$ is generalized to a function over the domain $d \in [0, \frac{1}{2}]$. We need the continuous analog of Lemma 1. In fact, we take the analog of that lemma as the definition of the histogram in the continuous domain:

Definition 2. $H_R(d) = \int_0^1 f(x) f(x+d)\, dx$.

For example, if two points x and $x + d$ each have weight $\frac{1}{2}$, they contribute $\frac{1}{4}$ to the height of H_R at distance d.

2.3 Continuous Hexachordal Theorem and Proof

The Continuous Hexachordal Theorem says that for any rhythm on the continuous circle as described above, if the rhythm has weight $\frac{1}{2}$, then it is homometric to its complement. More formally, it may be stated as:

Theorem 2. *If R is a integrable rhythm on the continuous circle, and $W(R) = \frac{1}{2}$, then for all distances d, $H_{\overline{R}}(d) = H_R(d)$.*

Proof. The proof fixes d and establishes that $H_{\overline{R}}(d) = H_R(d)$. From the histogram Definition 2 we have:

$$H_{\overline{R}}(d) = \int_0^1 \overline{f(x)}\,\overline{f(x+d)}\;dx.$$

From the complement Definition 1 this is:

$$= \int_0^1 [1 - f(x)][1 - f(x+d)]\;dx.$$

Multiplying out terms yields:

$$= \int_0^1 (1 - f(x) - f(x+d) + f(x)f(x+d))\;dx.$$

Separating integrals gives:

$$= \int_0^1 1\;dx - \int_0^1 f(x)\;dx - \int_0^1 f(x+d)\;dx + \int_0^1 f(x)f(x+d)\;dx$$

The first integral is just 1, and the second two[1] are each $\frac{1}{2}$ by the assumption of the theorem that $W(R) = \frac{1}{2}$:

$$\begin{aligned} &= 1 - \frac{1}{2} - \frac{1}{2} + \int_0^1 f(x)f(x+d)\;dx \\ &= \int_0^1 f(x)f(x+d)\;dx \\ &= H_R(d) \end{aligned}$$

The last step again follows from the Definition 2, and so we have established that $H_{\overline{R}}(d) = H_R(d)$ for all d, i.e., the histograms are identical and R is homometric to $\overline{R}$.

The weight function $f(x)$ need not be a *continuous function* in the technical mathematical sense.[2] We only need that it be integrable,[3] i.e., a function for which an appropriate "area under the function graph" may be defined.

[1] Shifting x to $x + d$ shifts the graph of $f(\,)$ but does not change the area underneath it.
[2] A function f is *continuous* if, for all c in the domain, $\lim_{x \to c} f(x) = f(c)$.
[3] For example, *Lebesgue integrable* suffices.

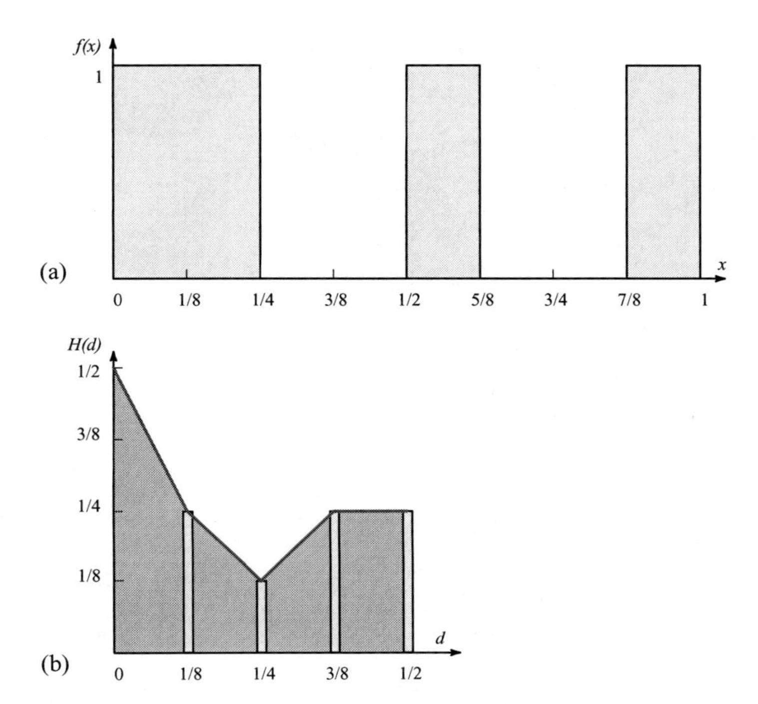

Fig. 3. (a) Weight step function $f(x)$ corresponding to Fig. 1 (top), $(n, k)=(8, 4)$. (b) Corresponding histogram integral $H(d)$.

We should note that the above proof can be directly discretized to yield a parallel proof of the Discrete Hexachordal Theorem. Instead, we show below that the freedom to use any integrable weight function renders the Discrete Hexachordal Theorem 1 an immediate corollary of the Continuous Hexachordal Theorem 2.

2.4 Discrete Theorem as Corollary

Suppose a discrete rhythm R has weights $(w_0, w_1, \ldots, w_{n-1})$, with each weight either 1 or 0. Then define the *step function* $f(x) = w_i$ for $\frac{i}{n} \leq x < \frac{i+1}{n}$. For example, Fig. 3(a) shows the step function corresponding to the top rhythm in Fig. 1, whose discrete weights are $(1, 1, 0, 0, 1, 0, 0, 1)$. Note that the total weight/area is $4 \cdot \frac{1}{8} = \frac{1}{2}$, which accords with the discrete weight of $\frac{1}{2}n = \frac{1}{2}8 = 4$.

We formalize this correspondence between continuous and discrete as follows:

Corollary 1. *The Discrete Hexachordal Theorem 1 follows from the Continuous Hexachordal Theorem 2.*

Proof. We use the notation

$$\chi_A(x) = \begin{cases} 1, & \text{for all } x \in A \\ 0, & \text{otherwise} \end{cases}$$

to represent the 1/0 characteristic function of a set A.

We convert the discrete rhythm $(w_0, w_1, \ldots, w_{n-1})$ into the continuous rhythm

$$f(x) = \sum_{i=0}^{n-1} \left(w_i \cdot \chi_{\left[\frac{i}{n}, \frac{i+1}{n}\right)} \right).$$

This has the feature, mentioned above, that for all $x \in \left[\frac{i}{n}, \frac{i+1}{n}\right.$, we have $f(x) = w_i$.

Because of the horizontal compression involved in this conversion, the discrete histogram contribution $H_R(d) = \sum_{i=0}^{n-1} w_i w_{i+d}$ corresponds to the continuous histogram contribution

$$\begin{aligned}
H_R\left(\frac{d}{n}\right) &= \int_0^1 f(x) f\left(x + \frac{d}{n}\right) \, dx \\
&= \int_0^1 \left. \sum_{i=0}^{n-1} \left(w_i \cdot \chi_{\left[\frac{i}{n}, \frac{i+1}{n}\right)} \right) \right] f\left(x + \frac{d}{n}\right) \, dx \\
&= \sum_{i=0}^{n-1} \left[w_i \int_0^1 \chi_{\left[\frac{i}{n}, \frac{i+1}{n}\right)} \cdot f\left(x + \frac{d}{n}\right) \, dx \right] \\
&= \sum_{i=0}^{n-1} \left. w_i \int_{\frac{i}{n}}^{\frac{i+1}{n}} f\left(x + \frac{d}{n}\right) \, dx \right] \\
&= \sum_{i=0}^{n-1} \left. w_i \int_{\frac{i+d}{n}}^{\frac{i+d+1}{n}} f(x) \, dx \right] \\
&= \sum_{i=0}^{n-1} \left. w_i \int_{\frac{i+d}{n}}^{\frac{i+d+1}{n}} w_{i+d} \, dx \right] \\
&= \frac{1}{n} \sum_{i=0}^{n-1} w_i w_{i+d}
\end{aligned}$$

So, the continuous histogram is proportional to the discrete histogram at integral values of d (see Fig. 3(b)), and the conclusion of the Continuous Hexachordal Theorem 2 that R is homometric to $\overline{R}$ implies the same in the discrete case, which is precisely the claim of the Discrete Hexachordal Theorem 1.

2.5 Double-Counting Diameter Intervals

We return to the the issue of double-counting an interval that equals the diameter ($d = n/2$ in the discrete case or $d = \frac{1}{2}$ in the continuous case) in the histogram $H_R(d)$. In music the diameter in the case of an equal-temperament scale corresponds to a tritone. Recall from Definition 2 that the continuous histogram is defined by the equation $H_R(d) = \int_0^1 f(x)f(x+d)\,dx$. Applying this for $d = \frac{1}{2}$ to the step function $f(x)$ in Figure 3 results in

$$H_R(\frac{1}{2}) = \int_0^1 f(x)f(x+\frac{1}{2})\,dx.$$

When $x \in [0, \frac{1}{8})$, the product $f(x)f(x+\frac{1}{2})$ is 1. And also when $x \in [\frac{1}{2}, \frac{5}{8})$, the product is again 1, because $x + \frac{1}{2}$ wraps around to $[0, \frac{1}{8})$. For all other x, the product is 0. So $H_R(\frac{1}{2}) = 2 \cdot \frac{1}{8} = \frac{1}{4}$, which corresponds to the height 2 for $d = 4$ in the discrete case in Figure 1. Thus, the continuous histogram analog also "double-counts" the diameter $d = \frac{1}{2}$.

Moreover, we can see that this is the natural definition, by considering $d = \frac{1}{2} - \varepsilon$ for some small $\varepsilon > 0$. The same integral leads to $H_R(\frac{1}{2} - \varepsilon) = 2(\frac{1}{8} - \varepsilon)$ which goes to $\frac{1}{4}$ as $\varepsilon \to 0$. Thus, the height $H_R(\frac{1}{2})$ is consistent with the limit for $d < \frac{1}{2}$. Stipulating that $d = \frac{1}{2}$ should be treated specially would destroy this natural correspondence.

2.6 Patterson's First Theorem

Patterson's first Theorem [Pat44] goes beyond the $k = n/2$ precondition of the Discrete Hexachordal Theorem 1. It may be stated as: two homometric (n, k)-rhythms have homometric complements. In our continuous generalizations, two rhythms with the same number k of onsets have the same weight. So the generalization is:

Theorem 3. *If R_1 and R_2 are two integrable rhythms on the continuous circle with equal weights, $W(R_1) = W(R_2)$, and they are homometric, i.e., for all distances d, $H_{R_1}(d) = H_{R_2}(d)$, then their complements are homometric: $H_{\overline{R_1}}(d) = H_{\overline{R_2}}(d)$.*

Proof. Let the weight function of R_1 be $f(x)$ and that of R_2 be $g(x)$. Fix a distance d. We compute $H_{\overline{R_1}}(d)$ and show it is equal to $H_{\overline{R_2}}(d)$. From Definitions 2 and 1, we have

$$\begin{aligned} H_{\overline{R_1}}(d) &= \int_0^1 \overline{f(x)}\,\overline{f(x+d)}\,dx \\ &= \int_0^1 (1 - f(x))(1 - f(x+d))\,dx \end{aligned}$$

Multiplying out terms and separating integrals yields

$$\begin{aligned} &= \int_0^1 1\,dx - 2\int_0^1 f(x)\,dx + \int_0^1 f(x)f(x+d)\,dx \\ &= 1 - 2\int_0^1 f(x)\,dx + \int_0^1 f(x)f(x+d)\,dx \end{aligned}$$

Now, because $W(R_1) = W(R_2)$, we have $\int_0^1 f(x)dx = \int_0^1 g(x)dx$, and because R_1 and R_2 are homometric, we have $\int_0^1 f(x)f(x+d)dx = \int_0^1 g(x)g(x+d)dx$:

$$= 1 - 2\int_0^1 g(x)\,dx + \int_0^1 g(x)g(x+d)\,dx$$

However, we know, by the same reasoning, that this expression is

$$= \int_0^1 \overline{g(x)}\,\overline{g(x+c)}\,dx$$

And we have therefore established that the complementary rhythms are homometric:

$$\int_0^1 \overline{f(x)}\,\overline{f(x+d)}\,dx = \int_0^1 \overline{g(x)}\,\overline{g(x+d)}\,dx$$
$$H_{\overline{R_1}}(d) = H_{\overline{R_2}}(d)$$

3 Open Problems

Our results may be interpreted in terms of *polyphonic rhythms*, in which several instruments are linearly combined [OTT08]. For instance, to model three identical drums playing together, interpret the weight $f(x) = \frac{1}{3}$ to mean that one drum is struck on a particular beat, while the weight $f(x) = 1$ would mean all three are struck. It would be interesting to explore whether homometric polyphonic rhythms have a musical significance.

We know that two sets of points with different cardinalities and different weights may be homometric, but we neither understand the constraints here mathematically nor know if there is any musical interpretation of such sets.

Theorem 2 generalizes to weights in $[0, 1]$ on a sphere, with distances measured by geodesics, and with $W(R) = \frac{1}{2}$ corresponding to the integral over a hemisphere equalling $\frac{1}{2}$. The discrete analog is "distance regular" points on a sphere, e.g., the vertices of a Platonic solid. Is there any musical analog for spheres in any dimension?

Acknowledgements

The authors like to thank the anonymous referees for their useful comments.

References

[AG00] Althuis, T.A., Göbel, F.: Z-related pairs in microtonal systems. Memorandum 1524, University of Twente, The Netherlands (April 2000)

[Ami07] Amiot, E.: David Lewin and maximally even sets. Journal of Mathematics and Music 1(3), 157–172 (2007)

[Bla99] Blau, S.K.: The hexachordal theorem: A mathematical look at interval relations in twelve-tone composition. Mathematics Magazine 72(4), 310–313 (1999)

[Bue76] Buerger, M.J.: Proofs and generalizations of Patterson's theorems on homometric complementary sets. Zeitschrift für Kristallographie 143, 79–98 (1975)
[Bue78] Buerger, M.J.: Interpoint distances in cyclotomic sets. The Canadian Mineralogist 16, 301–314 (1978)
[For77] Forte, A.: The Structure of Atonal Music. The Yale University Press, Madison (1977)
[Igl81] Iglesias, J.E.: On Patterson's cyclotomic sets and how to count them. Zeitschrift für Kristallographie 156, 187–196 (1981)
[Jed06] Jedrzejewski, G.: Mathematical Theory of Music. Editions Delatour France (2006)
[JK06] Jaming, P., Kolountzakis, M.: Reconstruction of functions from their triple correlations. New York Journal of Mathematics 9, 149–164 (2003)
[Joh03] Johnson, T.: Foundations of Diatonic Theory. Key College Publishing (2003)
[Lew59] Lewin, D.: Intervallic relations betwen two collections of notes. Journal of Music Theory 3(2), 298–301 (1959)
[Lew60] Lewin, D.: The intervallic content of a collection of notes, intervallic relations between a collection of notes and its complement: An application to schoenberg's hexachordal pieces. Journal of Music Theory 4(1), 98–101 (1960)
[Lew76] Lewin, D.: On the interval content of invertible hexachords. Journal of Music Theory 20(2), 185–188 (Autumn 1976)
[Lew87] Lewin, D.: Generalized Musical Intervals and Transformations. Yale University Press (1987)
[Maz03] Mazzola, G.: The Topos of Music. Birkhäuser, Basel (2003)
[Mor90] Morris, R.D.: Pitch-class complementation and its generalizations. Journal of Music Theory 34(2), 175–245 (Autumn 1990)
[OTT08] O'Rourke, J., Taslakian, P., Toussaint, G.: A pumping lemma for homometric rhythms. In: Proc. 20th Canad. Conf. Comput. Geom., August 2008, pp. 99–102 (2008)
[Pat44] Lindo Patterson, A.: Ambiguities in the x-ray analysis of crystal structures. Physical Review 64(5-6), 195–201 (1944)
[Reg74] Regener, E.: On Allen Forte's theory of chords. Perspectives of New Music 13(1), 191–212 (Autumn-Winter 1974)
[Sen08] Senechal, M.: A point set puzzle revisited. European Journal of Combinatorics 29(1), 1933–1944 (2008)
[Sod95] Soderberg, S.: Z-related sets as dual inversions. Journal of Music Theory 39(1), 77–100 (Spring 1995)
[Tou05] Toussaint, G.T.: The geometry of musical rhythm. In: Akiyama, J., Kano, M., Tan, X. (eds.) JCDCG 2004. LNCS, vol. 3742, pp. 198–212. Springer, Heidelberg (2005)
[Tym06] Tymoczko, D.: The geometry of musical chords. Science 313(72), 72–74 (2006)
[Vuz85] Vuza, D.: Sur le rythme périodique. Revue Roumaine de Linguistique - Cahiers de Linguistique Théorique et Appliquée 22(1), 103173–103188 (1985)
[Wri78] Wright, O.: The Modal System of Arab and Persian Music AD 1250-1300. Oxford University Press, Oxford (1978)

Speech Rhythms and Metric Frames

Fernando Benadon

American University
fernando@american.edu

Abstract. I present some conceptual and computational tactics related to the metric analysis of speech rhythms. An utterance can be considered metered when it approaches isochrony at the level of the syllable (note) and/or foot (beat). Since the timing patterns of spoken speech resemble those of music, we can apply knowledge of musical meter and expressive timing to the study of speech. However, speech rhythms tend to be more amorphous than musical rhythms, which makes the task of modeling meter in speech far from straightforward. The lack of a score or implicit rhythmic template leads to a meter-finding methodology that juggles the oftentimes incompatible outcomes of different metric frameworks: quantization as opposed to categorical perception, and subdivision isochrony as opposed to beat isochrony.

Keywords: speech, rhythm, meter, categorical perception, duration ratios.

1 Introduction

The goal of this paper is to present some conceptual and computational tactics related to the metric analysis of speech rhythms. The parsing of speech by syllable stress is similar to metric grouping in music. The two domains also inhabit similar temporal regions. The kinship is evident not only in various forms of song, but also in explicitly speech-based works by composers such as Steve Reich, Hermeto Pascoal, Jason Moran, and many others. Despite a widespread interest in the temporal similarities between speech and music, there is almost no published work on the music-temporal structure of speech.

The literature on speech rhythms is vast and complex [1]. With titles such as "Triple Threats to Duple Rhythm" [2], articles on poetry analysis might be a good place to begin our investigation. But they, like most works in metrical phonology, are more interested in patterns of stress than in patterns of duration [3]. On the flipside, we find precise durational measurements in phonetics and phonology. However, it is often difficult to discern how their findings translate to a music-based conception of rhythm. (What are we to make of the observation that the vowel in "bids" is on average 25 ms longer than the vowel in "bits"? [4]) When direct comparisons between speech and musical rhythms are made, they take the form of statistical correlations rather than rhythmic analyses [5].

In the following paragraphs I will show that the timing patterns of speech can be analyzed in musically relevant ways. Before proceeding, I should emphasize that there is no agreed upon methodology for marking syllable onsets. Nick Campbell [6]

E. Chew, A. Childs, and C.-H. Chuan (Eds.): MCM 2009, CCIS 38, pp. 22–31, 2009.

notes: "We can make no claim that clear definable boundaries exist between all phoneme-sized speech sounds, and in many cases the assignment of a label to a portion of speech can be quite arbitrary" (p. 302). A common approach, used here, is to tag the first vowel in the syllable.

2 Assessing Local Meter

The study of speech rhythms amounts to a tough workout in rhythmic analysis. Deviations from metricality are also frequent in music, although there is often a template, such as a score or implicit mental schema, serving as a point of comparison. Such a template is rarely evident in speech.

For speech to be metered, it should abide by the well-formedness prerequisites typically enforced in music. Meter is well-formed when its underlying subdivisions are (nominally) isochronous, and rhythm is well-formed when its onsets align with one of the subdivisions. An onset that is off course can be nudged left or right onto the coordinate we think it belongs. This can be achieved by quantization or categorization.

Let us examine a short spoken phrase in order to illustrate these two processes. We begin with the first beat (foot) in Figure 1.

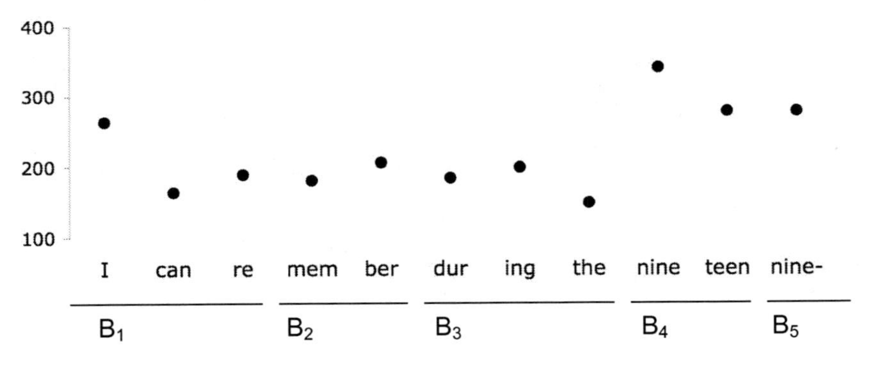

Fig. 1. Syllable IOIs (in milliseconds) for the five-beat phrase "I can remember during the nineteen nineties," recorded from the radio by the author. The beat is equivalent to a foot, which "starts with a [syllable] stress and contains everything that follows that stress up to, but not including, the next stress" [7].

The timing pattern for B_1 falls squarely between two possible grids, one triple and one quadruple (long-short-short). To choose the one that will provide the better fit, we can turn to the literature on categorical rhythm perception. On Desain and Honing's [8] categorization experiments, most subjects transcribed this pattern as a 2-1-1—that is, as a quadruple grid composed of one eighth plus two sixteenths.[1] But this response was by no means unanimous; the pattern sits right on a category boundary between triple and quadruple in the authors' time-clumping map.

[1] The exact IOIs were 421-263-316, which are proportionally similar to B_1's 263-164-189.

Another way to assess the grid of B_1 is to quantize it. While various algorithms exist, the basic strategy involves looking for a goodness of fit between the performance data and one of several competing grids. In the language of preference rules, we would say that we prefer a grid that minimizes error distances between sounded onsets and metrical subdivisions. Figure 2 places B_1's IOIs next to a triple and a quadruple grid. The dashed error tails in the graph show that the two candidate grids produce roughly equal amounts of total error, with the triple grid (74 ms) having a very slight edge over the quadruple grid (80 ms).

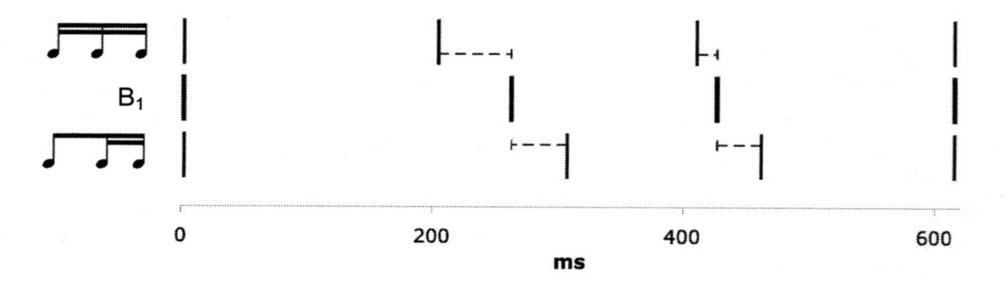

Fig. 2. B_1 IOIs in relation to two metronomic grids: triple and quadruple

Logically, the deviation error can be minimized with larger subdivision cardinalities (provided the subdivision IOIs do not fall below the 100 ms threshold; [9]). This strategy is tested in Figure 3, where the onsets are now compared to quintuplet and sextuplet grids. The total error for the former is 74 ms; for the latter, 61 ms—the lowest error yet.

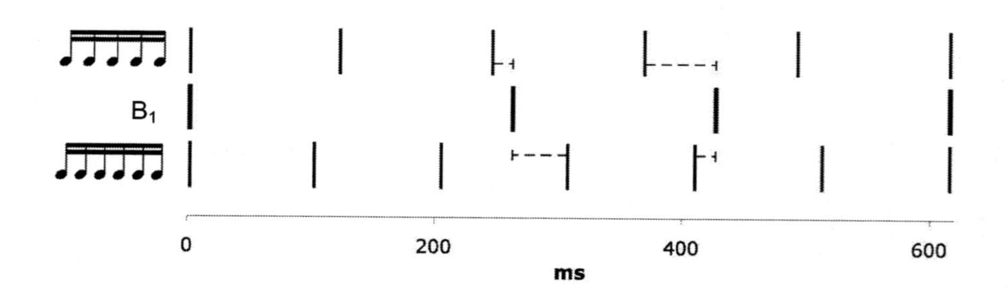

Fig. 3. B_1 IOIs in relation to more fine-grained grids: quintuple and sextuple

The improved goodness of fit that the sextuple grid provides should be rejected on three grounds. One: the first syllable ("I") is 263 ms long and therefore hardly divisible into three parts, as the sextuple grid asks of us. Two: the error difference (12 ms) between the triple and sextuple grid is too small a reward for the computational price being paid. I would rather endure an extra 12 ms in total error and subdivide 1+1+1 (triple) instead of 3+1+2 (sextuple). Three: the error tails now point in two directions, requiring us to hear an anticipation followed by a delay. It seems computationally advantageous to stick to one consistent mode of deviation, as was the case with the triple grid. The moral is that goodness-of-fit scores are helpful but should be evaluated in light of relevant cognitive factors.

In sum, B_1 can either lean toward the quadruple grid (based on categorical perception) or toward the triple grid (based on quantization). Neither tendency is strong enough to rule out the other. This leaves us with a first beat that is either inherently non-metric or only temporarily non-metric until further evidence becomes available in the upcoming beats.

Now we turn our attention to B_2. Our task is twofold: to determine its meter and to reconcile its timing pattern with that of B_1.

The meter of B_2 is most likely duple. The first syllable ("mem") is 25 ms shorter than the second one ("ber"), yielding a .88 ratio. Just as we compared the earlier three-syllable rhythm to Desain and Honing's categorical perception data, we can gauge the meter of this two-syllable rhythm by referring it to experiments that test the perception of two-note rhythms.[2] Specifically, we need to confirm that the mild short-long asymmetry of B_2 can be coded as an even duple rather than a short-long triplet (1+2).[3] In a tapping experiment devised by Povel [10], subjects heard sequences of two-note rhythms that varied in their short-long ratio. As predicted, subjects simplified most ratios in the direction of .5—that is, toward a triple 1-to-2 ratio. However, there was disagreement regarding the .8 ratio. Some subjects exaggerated the duration difference in the direction of triple (.5), while others lessened the difference in the direction of duple (1.0). The ratio of B_2 (.88) is closer to 1.0 than the experiment's highest ratio (.8), which tips the scale in favor of duple meter.

Further experimental support for a duple B_2 comes from time-shrinking studies, which show that listeners tend to underestimate the duration of the second time interval in a short-long ratio [11]. For instance, the sequence (in ms) 160-190 is perceived as being in a 1:1 ratio even though the actual ratio is .89. The 1:1 effect persists even with ratios as uneven as .72. These observations leave us with little doubt concerning the even (duple) partition of B_2.

3 Glued at the Subdivision

How concordant are our conclusions from both beats? B_1 is 620 ms; it is either quadruple (by the time clumping map) or triple (by quantization). B_2 is 390 ms; it is most likely duple (by time-shrinking and quantization). A good metric frame for this two-beat span should support a running isochrony at the subdivision level. Figure 4 compares three scenarios. The quadruple-plus-duple combo shown in (a) is a poor choice because B_1's average sixteenth-note is 155 ms (620÷4) and B_2's is 195 ms (390÷2)—a 26% change. We can reduce the discrepancy with a triple division of B_1, as shown in (b). (Note that this is not a triplet, but a group of three sixteenths.) This metric sequence exhibits only a 6% change in average subdivision speed: from 207 to 195

[2] Strictly speaking, a "two-note" rhythm is really a three-note rhythm consisting of two IOIs. Likewise, a "three-note" rhythm consists of four IOIs: three attacks plus a "downbeat."

[3] There is no need to check for quintuple (2+3) or septuple (3+4), since these subdivisions are inadmissibly small given the size of B_2.

ms. For those reluctant to relinquish a quadruple hearing of B_1, (c) offers an adjusted measurement window that is smaller than the full beat. Leaving out the first syllable, the total duration of B_1's last two syllables is 350 ms, or an average of 175 ms per sixteenth. This constitutes an 11% change from the end of B_1 to B_2. Hence the level of metric concordance depends on the size and location of the measurement windows. I will not pursue this idea further due to space constraints; consider flexible windows duly placed on the pile of soon-to-be explored metric conundrums.

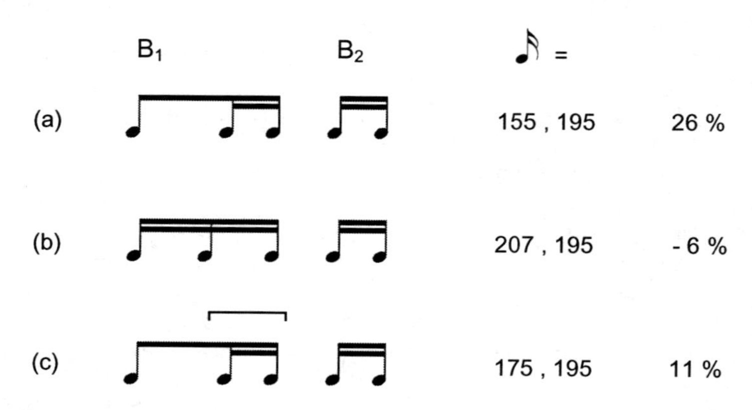

Fig. 4. Evenly subdivided beats and their resulting subdivision durations. The full duration of B_1 is subdivided in (a) and (b), whereas (c) employs a narrower window for improved conformity with B_2. The preferred metric analysis is (b): 3/16 + 2/16.

Let us continue on to B_3, the three-syllable foot "during the." At this point in the game, the reader will be glad to bypass the detailed report in favor of a quick diagnosis: this beat is basically triple (its IOIs are roughly equal). Does it maintain the subdivision tempo established by the previous two beats? Dividing the total size of B_3 by 3 yields metronomic subdivisions of 177 ms, a 9% decrease from the subdivisions in B_2 and a 15% decrease from B_1. Though we need not answer it now, we should ask the question we have been dodging: How much tempo drift are we willing to tolerate before we deem the sequence non-metrical?

B_4, the duple foot "nine-teen," has the same duration as B_1. But B_1 has three short syllables and B_4 has two long ones. On the face of it, the metric solution seems simple: give B_4 eighth-notes (or four sixteenths) and B_1 eighth-note triplets. We will see later that this kind of tuplet approach can be beneficial because it retains isochrony at the beat level when subdivision isochrony begins to wobble. But in this case, it is unclear whether the kinship between B_1 and B_4 also satisfies the timing patterns of the intervening beats. The ideal meter should weigh the global needs of all beats in the phrase. It also should take into account the sequential unfolding of events. From the sixteenths in B_3 to those in B_4, there is a 12% decrease in duration. Our analytical tolerance for this percentage drift is assuaged by recalling that subdivision speeds have been increasing almost linearly since B_1: 207-195-177-155. This trend may be heard as a gradual accelerando of an otherwise fixed subdivision grid. Had the beats been ordered differently (e.g., B_3-B_1-B_4-B_2), we would have walked a different analytical path, leading either to a different metric solution or perhaps to a blind alley.

4 Beyond the Eighth-Note

The average syllable duration in spontaneous American English is roughly 200 ms [1, p. 113]. Not unexpectedly, the syllables in our preceding example have clustered around that value (cf. Fig. 1). This explains why the transcriptions have favored a binary set of subdivision values: the sixteenth-note and the eighth-note. For a syllable to be divisible into three parts, it should be around 300 ms or greater [9]. Therefore, slower speaking rates expand the subdivision palette, as Figure 5 shows.

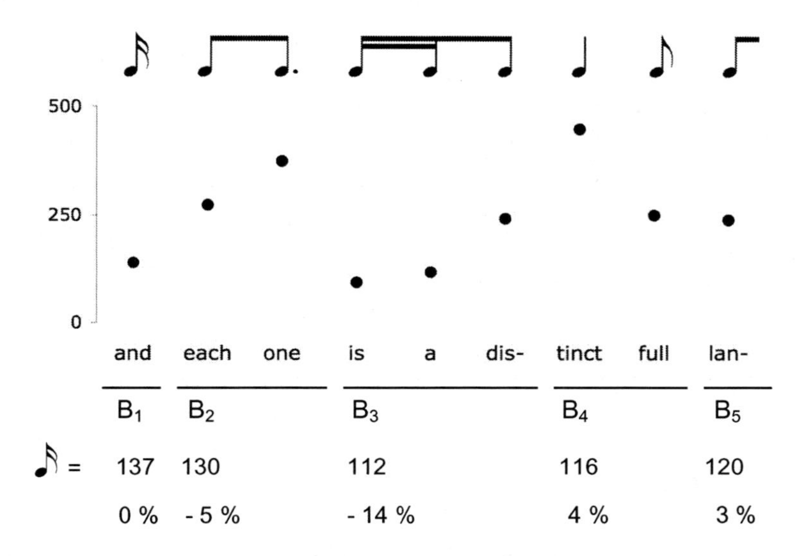

Fig. 5. Lalla Ward reading "and each one is a distinct full language" [12]. Syllables longer than 300 ms open the door for three-way ("one") and four-way ("tinct") subdivision. The chain of mixed meters is held together by a fairly steady—often implicit—average sixteenth-note subdivision from beat to beat.

5 The Tuplets

My main argument thus far has been that speech can be modeled as metered when its subdivisions approach isochrony. The process often produces mixed meters sharing a common subdivision value. When subdivision isochrony between beats is tenuous, we might reconcile them as an overall tempo curve, as we did earlier. There is another way to prop up meter when it falters at the subdivision.

One perceptually feasible alternative is to switch our focus from subdivision isochrony to beat isochrony. The reason I view beat-level isochrony as a second resort rather than as the norm is that most sentences contain feet with unequal syllable counts, resulting in variable beat durations best explained by mixed meters.[4] In some cases,

[4] Some linguists once believed that speakers adjust syllable length to maintain interstress isochrony between feet containing different numbers of syllables. The claim has since been refuted [13].

however, two beats may be the same size even if their subdivision count is different. This translates into non-isochrony at the subdivision, for if two beats are the same size and contain a different number of subdivisions, they cannot share the same subdivision value. The alternative to subdivision isochrony is beat isochrony via the tuplet.

The first two beats in Figure 6 are quadruple and share a common subdivision duration (within 10%, values not shown). The third beat is triple. Seeking subdivision isochrony—a running sixteenth throughout—increases the sixteenth-note of the third beat by almost 30%. Since the size of the third beat is within 5% of the first two, modeling the third beat as three triplets (rather than three sixteenths) helps to preserve a sense of isochrony.

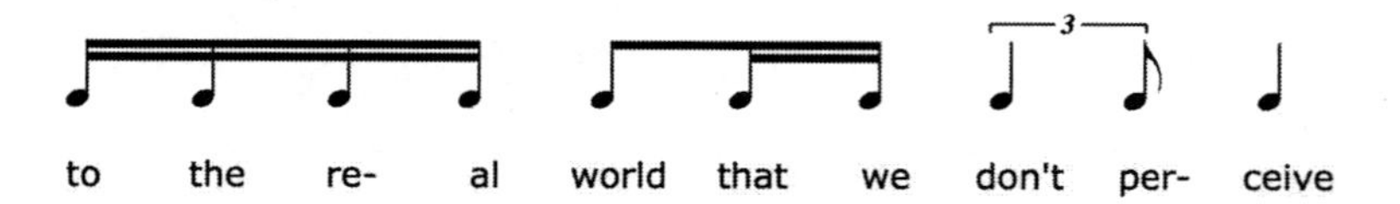

Fig. 6. John Searle reading from [14]. Sixteenths on the third beat would not agree with those in the preceding beats, so we slip into something more comfortable.

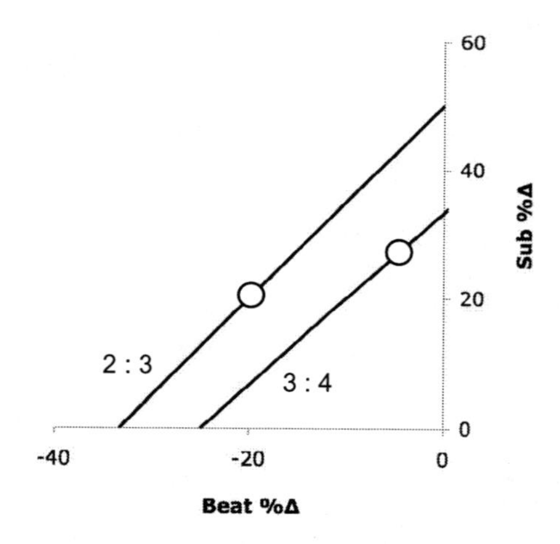

Fig. 7. Percentage change relationship between subdivision size and beat size. The lines correspond to tuplet groupings.

Figure 7 shows the inverse relationship between subdivision and beat isochrony. The y-axis plots percentage change in subdivision duration between two beats (usually adjacent). A change of 0% means that there is perfect subdivision isochrony between them. If the subdivision duration changes by a large enough amount from one beat to the next, the new duration might lock into a tuplet value. The lower line plots a quadruple-to-triple shift, such as the one in Figure 6. The third beat from that phrase is marked here with a circle. Increasing the subdivision difference decreases the beat size difference according to different tuplet configurations. A point near either axis bodes well for meter: we get little change either in subdivision or in beat size. Meter breaks down when a point is roughly half-way along a tuplet line. Returning to our unanswered question concerning allowable metric drift, we might propose a ± 10% threshold for either the subdivision or the beat. A deviation of more than 10% at *both* levels debilitates the sense of meter. For instance, the circle on the upper triple-to-duple line corresponds to the middle beat in Figure 8. All of the phrase's five beats are snug 3/16's, except for the smaller third beat. If we treat it as a beat of 2/16 in order to seek subdivision concordance with the surrounding beats, we encounter a 21% increase in subdivision duration. If we assign a tuplet (2-in-place-of-3) in order to seek concordance at the beat level, we encounter a 20% drop in beat duration. Where to turn? This beat may be lost in the land of no meter.

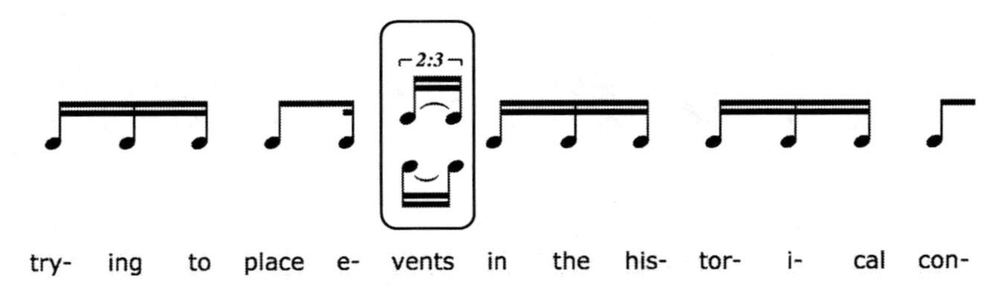

Fig. 8. Noam Chomsky reading from [15]. Of the two boxed-in metrically feasible options, neither one complies with the surrounding isochrony.

Figure 7 included only the quadrant where subdivision size increases, beat size decreases, and the tuplet's numerator is smaller than its denominator. Figure 9 zooms out to reveal all possible combinations. Five tuplet lines (and their reciprocals) are shown: 1:2, 4:7, 2:3, 3:4, and 4:5.[5] A tuplet is most useful as it crosses the y-axis, where it yields little or no change in beat size. For instance, suppose that a beat with 100-ms subdivisions is followed by another containing 150-ms subdivisions, a 50% increase (lower dashed line). We can model this deviation according to different tuplet frameworks, each yielding a different amount of error. With a 4:5 tuplet (circle) we get a beat that is 20% too big. The 2:3 tuplet (triangle) gives the right fit.

For every subdivision change, there is a corresponding tuplet configuration that provides a perfect fit in beat size. The catch, of course, is that only a small handful of tuplet ratios are user friendly, and only one of these will match our desired subdivision count given the syllables in the foot. For instance, a 70% subdivision increase (upper dashed line in Fig. 9) can be counterbalanced by a 10:17 tuplet. Clearly this is not a viable metric solution. What other, more reasonable fractions cross the 70% horizontal coordinate? The 4:7 tuplet (square) offers an appealing beat difference of -4%; also workable is the 2:3 tuplet (diamond), although at 12% its beat difference is significantly larger. Figure 10 illustrates how these two tuplet options might work in different phrases with hypothetical (but feasible) timing patterns.

[5] The reciprocal of the tuplet ratio equals the slope of the line; for instance, the 2:3 tuplet line has a slope of 1.5. Only ratios between 1.0 and 2.0 (and their reciprocals) are given. I include 1:2 (and 2:1) to provide a visual frame, even though this ratio is not generally thought of as a tuplet.

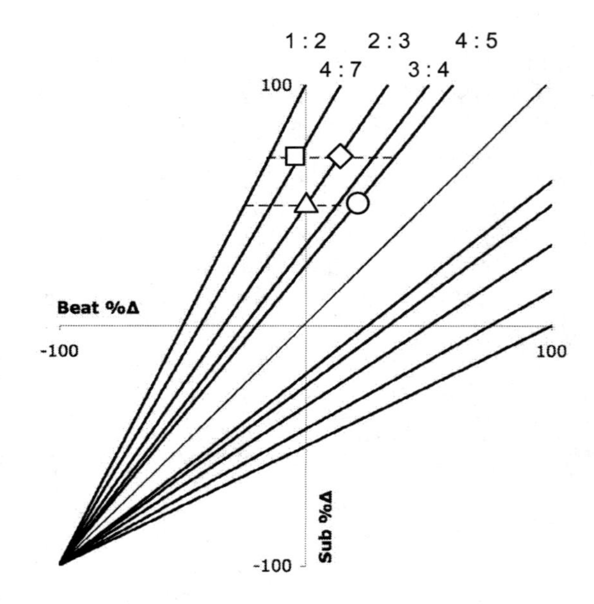

Fig. 9. An expansion of Figure 7

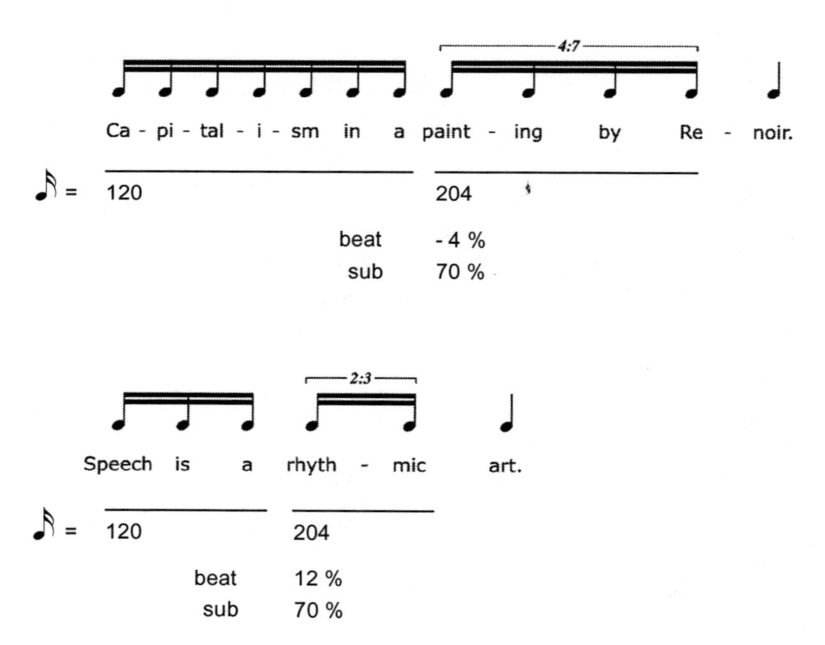

Fig. 10. The number of subdivisions in the beat helps determine which tuplet form is most appropriate. The top and bottom phrases correspond to the square and diamond in Figure 9, respectively. Both phrases undergo the same increase in subdivision duration (70%), resulting in different beat size deviations depending on the tuplet count.

6 Conclusion

Rather than lead us to a yes-or-no decision on the metrical status of a given speech pattern, the various approaches described above suggest a more nuanced view that weighs beat proximity, degree of isochrony within the beat, type of isochrony between beats, and magnitude of deviation. My future work will integrate these approaches into a model that can compute concordance scores for different metric solutions of a given speech sequence. The results could be useful for composers, whose speech-based works are guided by musical intuition; music theorists, who have no duration-based tools for comparing text that has been set to music with its speech state; and popular music scholars who have noted speech-like rhythms in jazz and blues music.

References

1. Patel, A.D.: Music, Language, and the Brain. Oxford University Press, New York (2008)
2. Weismiller, E.R.: Triple Threats to Duple Rhythm. In: Kiparsky, P., Youmans, G. (eds.) Phonetics and Phonology: Rhythm and Meter, pp. 261–290. Academic Press, San Diego (1989)
3. Goldsmith, J.A.: Autosegmental Metrical Phonology. Basil Blackwell, Oxford (1990)
4. Jacewicz, E., Fox, R.A., Salmons, J.: Vowel Duration in Three American Dialects. American Speech 82(4), 367–385 (2007)
5. Huron, D., Ollen, J.: Agogic Contrast in French and English Themes: Further Support for Patel and Daniele. Music Perception 21, 267–271 (2003)
6. Campbell, N.: Timing in Speech: A Multi-Level Process. In: Horne, M. (ed.) Prosody: Theory and Experiment: Studies Presented to Gösta Bruce, pp. 281–334. Kluwer Academic Publishers, Dordrecht (2000)
7. Abercrombie, D.: Syllable Quantity and Enclitics in English. In: Abercrombie, D., Fry, D.B., MacCarthy, P.A.D., Scott, N.C., Trim, J.L.M. (eds.) In Honour of Daniel Jones, pp. 216–222. Longmans, London (1964)
8. Desain, P., Honing, H.: The Formation of Rhythmic Categories and Metric Priming. Perception 32, 341–365 (2003)
9. London, J.: Hearing in Time: Psychological Aspects of Musical Meter. Oxford University Press, New York (2004)
10. Povel, D.J.: Internal Representation of Simple Temporal Patterns. Journal of Experimental Psychology: Human Perception and Performance 7(1), 3–18 (1981)
11. ten Hoopen, G., Sasaki, T., Nakajima, Y., Remijn, G., Massier, B., Rhebergen, K., Holleman, W.: Time-Shrinking and Categorical Temporal Ratio Perception: Evidence for a 1:1 Temporal Category. Music Perception 24(1), 1–22 (2006)
12. Pinker, S.: The Language Instinct (Audiobook). Orion Publishing Group (2001)
13. Dauer, R.M.: Stress-Timing and Syllable-Timing Reanalyzed. Journal of Phonetics 11, 51–62 (1983)
14. Searle, J.: The Philosophy of Mind (Audiobook). The Teaching Company (1995)
15. Chomsky, N.: The Emerging Framework of World Power (CD). AK Press (2003)

Temporal Patterns in Polyphony

Mathieu Bergeron and Darrell Conklin

Department of Computing
City University London
{bergeron,conklin}@soi.city.ac.uk

Abstract. This paper formally characterizes the expressiveness of three approaches for polyphonic pattern representation and matching: $\mathcal{R}$ (relational patterns); $\mathcal{H}$ (Humdrum); and $\mathcal{SPP}$ (Structured Polyphonic Patterns). Relational networks have the highest expressiveness but $\mathcal{H}$ and $\mathcal{SPP}$ admit faster matching algorithms. It is shown how $\mathcal{H}$ and $\mathcal{SPP}$ can be cast as different restrictions of $\mathcal{R}$, both providing an expressive subset of full relational networks. In addition, the intersection of $\mathcal{H}$ and $\mathcal{SPP}$ yields yet another language: $\mathcal{SPP}_{\texttt{seq}}$, a restriction of $\mathcal{SPP}$ based on sequences of layered components. This new language is expressive enough to capture basic polyphonic patterns such as suspensions and parallel fifths and may be a new, more efficient approach to pattern extraction. The formal arguments contained in this paper are illustrated with musical examples extracted from J.S. Bach chorale harmonizations.

1 Motivation

Polyphony forms a large part of the western musical heritage and its essence — having multiple concurrent streams of musical events (with the temporal relations this implies) — is encountered in most kinds of modern music. However, there are few computational approaches for the expression and efficient matching of polyphonic patterns. This paper formally compares the expressiveness of three such languages and proposes a new one, establishing the hierarchy of Figure 1. To facilitate this presentation, arguments are restricted to patterns containing only two voices; results may however be generalized to denser polyphonic textures.

As a motivating example, consider the two-voice suspension of Figure 2. This typical polyphonic pattern is expressed in Figure 3 in the languages $\mathcal{R}$ (relational patterns); Humdrum; and $\mathcal{SPP}$ (Structured Polyphonic Patterns). As illustrated by the $\mathcal{R}$ expression (Figure 3i), even this simple pattern requires sophistication: variables to be instantiated by three events; inequality statements ensuring that the mapping from variables to events is injective; temporal relations between events (discussed below); and pitch relations such as consonance and dissonance.

This paper restricts its attention to the following binary temporal relations: i) $\texttt{m}(a,b)$ (a **m**eets b: a finishes when b begins), ii) the symmetric $\texttt{st}(a,b)$ (a and b **s**tart **t**ogether), iii) $\texttt{sw}(a,b)$ (a **s**tarts **w**hile b is sounding) and iv) the symmetric $\texttt{ov}(a,b)$ (a and b **ov**erlap: they sound together as some point in time). Figure 4 restates the relations in the notation of Allen [1] and Figure 2ii illustrates their musical relevance.

E. Chew, A. Childs, and C.-H. Chuan (Eds.): MCM 2009, CCIS 38, pp. 32–42, 2009.

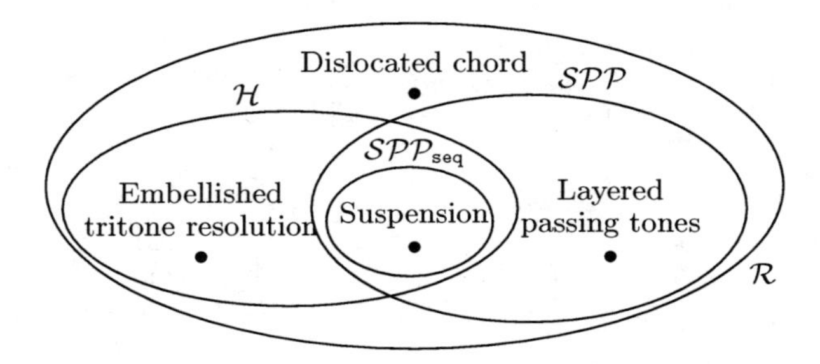

Fig. 1. Expressiveness of four polyphonic pattern languages: $\mathcal{R}$ (relational), $\mathcal{H}$ (Humdrum), $\mathcal{SPP}$ (Structured Polyphonic Patterns) and $\mathcal{SPP}_{\mathtt{seq}}$ ($\mathcal{SPP}$ restricted to sequences of layered components)

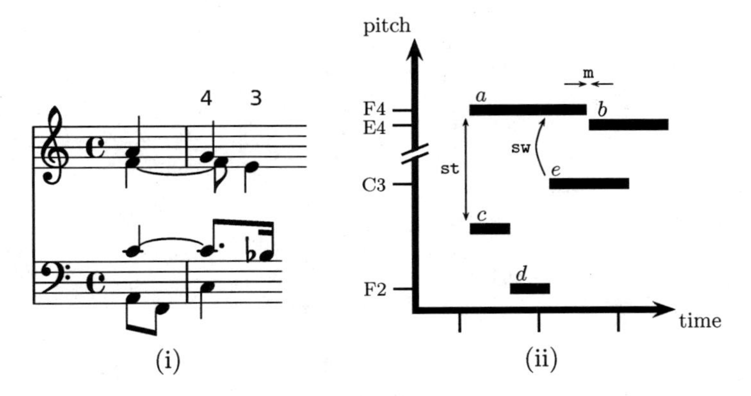

Fig. 2. (i) A 4-3 suspension between bass and alto voices in bars 16-17 of Bach's chorale BWV 283 and (ii) A piano-roll representation of the alto and bass voices

The Humdrum toolkit is widely-used for pattern matching in symbolic music data. Although Humdrum supports polyphony, it can be difficult to use for even simple patterns [5,2]. For example, Figure 3ii shows a Humdrum suspension pattern expressed with regular expressions. These typically do two things : i) match the beginning of events, the continuation of events or possibly no event at all (a "don't care" option) and ii) match features of those events by matching corresponding values in additional columns (this is the purpose of the "t" tokens at the end of the each lines, the first one matching a consonance feature and the second a dissonance feature).

In [4], the difficulties of Humdrum are circumvented by implementing a Prolog query that extracts all parallel fifths occurring in a corpus of J.S. Bach chorale harmonizations. The approach requires expert Prolog programming knowledge however, and even a slight reordering of Prolog clauses may have dramatic effects on pattern matching tractability. In general, the relational matching problem is

(i)

event(a)	voice(a, x)	m(a, b)
event(b)	voice(b, x)	sw(e, a)
event(e)	voice(e, y)	sw(b, e)
$a \neq b$	$x \neq y$	diss(a, e)
$a \neq e$		cons(b, e)
$b \neq e$		

(ii)

```
[a-g A-G]+[- # n]*[^)]*[(]  [a-g A-G]+[- # n]*.*.*t$
[a-g A-G]+[- # n]*[)]   [^(]*[a-g A-G]+[- # n]*.*t$
```

(iii)

$$\frac{-\{\}_x}{-\{\}_y} \; ; \; \frac{\{\texttt{sw\_cons}(y) : \texttt{t}\}_x}{-\{\texttt{sw\_diss}(x) : \texttt{t}\}_y}$$

Fig. 3. A suspension pattern using (i) relations; (ii) Humdrum and (iii) $\mathcal{SPP}$

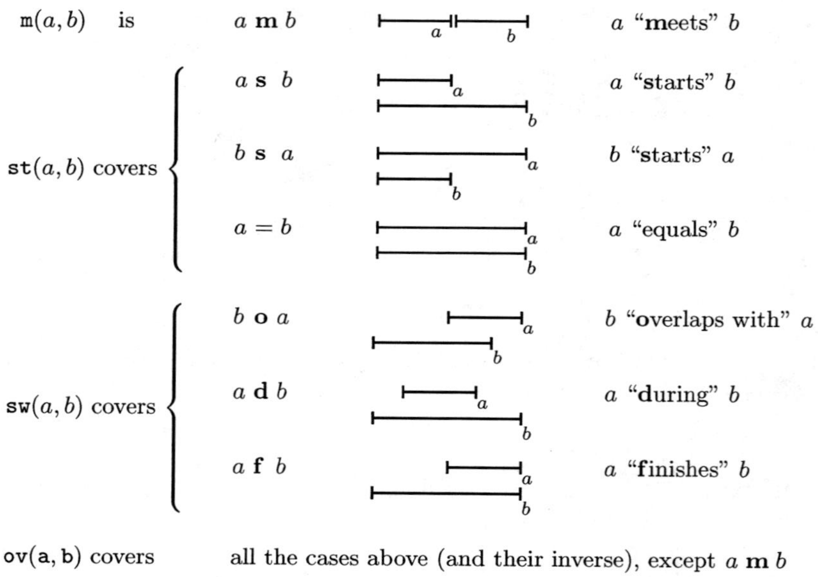

m(a, b) is	a **m** b		a "meets" b
st(a, b) covers	a **s** b		a "starts" b
	b **s** a		b "starts" a
	$a = b$		a "equals" b
sw(a, b) covers	b **o** a		b "overlaps with" a
	a **d** b		a "during" b
	a **f** b		a "finishes" b
ov(a, b) covers	all the cases above (and their inverse), except a **m** b		

Fig. 4. The temporal relation analyzed in this paper expressed in the notation of Allen [1]

a subgraph isomorphism problem, which is known to be NP-complete. Moreover, not all networks are satisfiable and it is also NP-complete to determine satisfiability. Hence, it is not practical to base a pattern extraction approach on relational patterns. In [2], the latter results are replicated with $\mathcal{SPP}$, an abstract polyphonic pattern representation based on sequencing and layering operators. In Figure 3iii for example, the sw_cons and sw_dis pattern components are layered to form a temporal relation. This layering holds for events forming a

Fig. 5. Dislocated V^7 chords captured by *Pattern 1*: BWV 284 bar 15 (i) and BWV 318 bar 13 (ii)

consonance and dissonance with opposite voices, and satisfying the `sw` temporal relation. $\mathcal{SPP}$ is further elaborated in Section 4.

2 Relational Patterns

A relational pattern r is simply a set of temporal relations over event variables ε:

Definition 1.

$$r \in \mathcal{R} ::= \omega, \ldots, \omega \quad \text{with} \quad \omega ::= \mathtt{m}(\varepsilon, \varepsilon) \mid \mathtt{ov}(\varepsilon, \varepsilon) \mid \mathtt{st}(\varepsilon, \varepsilon) \mid \mathtt{sw}(\varepsilon, \varepsilon)$$

With appropriate pitch relations, the pattern below could represent the "dislocated" V^7 chords shown in Figure 5. It enforces that chord tones eventually overlap with the root of the chord, but no other temporal relation is enforced:

Pattern 1.

$$\mathtt{ov}(a, b),\ \mathtt{ov}(a, c),\ \mathtt{ov}(a, d)$$

Alternatively, relational patterns can be represented as directed labelled graphs, where nodes represent event variables and edges represent relations (see Figure 6).

3 Humdrum

By contrast to $\mathcal{R}$, temporal relations in $\mathcal{H}$ are specified indirectly via a token matrix:

Definition 2.

$$h \in \mathcal{H} ::= \begin{bmatrix} h_{11}\ h_{12} \\ h_{21}\ h_{22} \\ \vdots \end{bmatrix} \quad \text{with} \quad h_{ij} ::= \varepsilon \mid (\varepsilon) \mid \star$$

The token ε refers to the beginning of a new event; the token (ε) is the continuation of the preceding event and the token $\star$ is the special "don't care" symbol that enforces no temporal relation. Note that time "flows" from top to bottom in Humdrum, e.g. the token h_{11} is followed by the token h_{21}. A $\mathcal{H}$ pattern is interpreted as follows with respect to the temporal relations it enforces:

Lines

$\mathcal{H}$	$\mathcal{R}$
$\begin{bmatrix} a\ b \end{bmatrix}$	$\rightsquigarrow \texttt{st}(a,b)$
$\begin{bmatrix} (a)\ b \end{bmatrix}$	$\rightsquigarrow \texttt{sw}(b,a)$
$\begin{bmatrix} a\ (b) \end{bmatrix}$	$\rightsquigarrow \texttt{sw}(a,b)$
$\begin{bmatrix} (a)\ (b) \end{bmatrix}$	$\rightsquigarrow \texttt{ov}(a,b)$
$\begin{bmatrix} a\ \star \end{bmatrix}, \begin{bmatrix} \star\ a \end{bmatrix}, \begin{bmatrix} \star\ (a) \end{bmatrix}, \ldots, \begin{bmatrix} \star\ \star \end{bmatrix}$	$\rightsquigarrow \emptyset$

Columns

$\mathcal{H}$	$\mathcal{R}$
$\begin{bmatrix} a \\ b \end{bmatrix}$	$\rightsquigarrow \texttt{m}(a,b)$
$\begin{bmatrix} (a) \\ b \end{bmatrix}$	$\rightsquigarrow \texttt{m}(a,b)$
$\begin{bmatrix} a \\ (a) \end{bmatrix}, \begin{bmatrix} \star \\ a \end{bmatrix}, \ldots, \begin{bmatrix} \star \\ \star \end{bmatrix}$	$\rightsquigarrow \emptyset$

The Humdrum pattern of Figure 3ii can be simplified to the following (variable names correspond to those of Figure 2ii):

Example 1.

$$\begin{bmatrix} e & (a) \\ (e) & b \end{bmatrix} \quad \rightsquigarrow \quad \begin{array}{l} \texttt{sw}(e,a), \texttt{sw}(b,e), \\ \texttt{m}(a,b) \end{array}$$

See Figure 6ii for the corresponding temporal network.

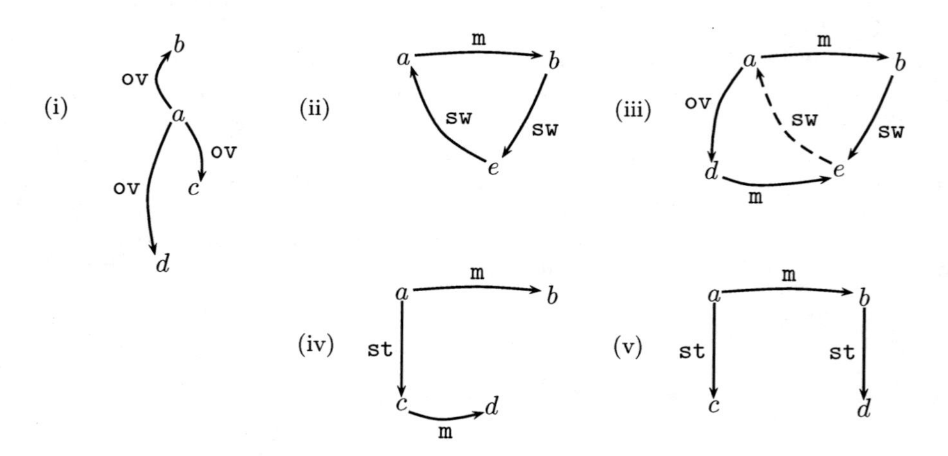

Fig. 6. Temporal networks enforced by (i) *Pattern 1*, (ii) *Example 1*, (iii) *Example 2* (the dashed edge is implied by the network), (iv) *Pattern 2* and (v) *Pattern 3*. The network (iv) can be represented in $\mathcal{SPP}$ but not in Humdrum. The network (v) can be represented in Humdrum but not in $\mathcal{SPP}$.

4 Structured Polyphonic Patterns

Patterns in $\mathcal{SPP}$ are defined according to the syntax below, where ε stands for an event and $-\varepsilon$ for a modified event (when layered, modified events start earlier than other events in the layer); the ";" operator joins two patterns in sequence (such that one finishes as the other starts) and the "$=$" operator layers two patterns (such that both start at the same time):

Definition 3.

$$\phi \in \mathcal{SPP} ::= \varepsilon \;\Big|\; -\varepsilon \;\Big|\; \phi \,;\, \phi \;\Big|\; \frac{\phi}{\phi}$$

A $\mathcal{SPP}$ pattern is interpreted as follows with respect to the temporal relations it enforces:

Layers

$\mathcal{SPP}$	$\mathcal{R}$
$\frac{a}{c}$	$\rightsquigarrow \mathtt{st}(a,c)$
$\frac{-a}{c}$	$\rightsquigarrow \mathtt{sw}(c,a)$
$\frac{a}{-c}$	$\rightsquigarrow \mathtt{sw}(a,c)$
$\frac{-a}{-c}$	$\rightsquigarrow \mathtt{ov}(a,c)$

Layers of sequences

$\mathcal{SPP}$	$\mathcal{R}$
$\frac{a\,;\,\ldots}{c\,;\,\ldots}$	$\rightsquigarrow \mathtt{st}(a,c)$
$\frac{-a\,;\,\ldots}{c\,;\,\ldots}$	$\rightsquigarrow \mathtt{sw}(c,a)$
$\frac{a\,;\,\ldots}{-c\,;\,\ldots}$	$\rightsquigarrow \mathtt{sw}(a,c)$
$\frac{-a\,;\,\ldots}{-c\,;\,\ldots}$	$\rightsquigarrow \mathtt{ov}(a,c)$

Sequences of layers

$\mathcal{SPP}$	$\mathcal{R}$
$\frac{a}{c}\,;\,\frac{b}{d}$	$\rightsquigarrow \mathtt{m}(a,b),\, \mathtt{m}(c,d)$
$\frac{-a}{c}\,;\,\frac{b}{d}$	$\rightsquigarrow \mathtt{m}(a,b),\, \mathtt{m}(c,d)$
$\frac{a}{-c}\,;\,\frac{b}{d}$	$\rightsquigarrow \mathtt{m}(a,b),\, \mathtt{m}(c,d)$
$\frac{a}{c}\,;\,\frac{-b}{d}$	$\rightsquigarrow \mathtt{m}(a,b),\, \mathtt{m}(c,d)$
$\vdots$	$\vdots$
$\frac{-a}{-c}\,;\,\frac{-b}{-d}$	$\rightsquigarrow \mathtt{m}(a,b),\, \mathtt{m}(c,d)$

Sequences

$\mathcal{SPP}$	$\mathcal{R}$
$a\,;\,b$	$\rightsquigarrow \mathtt{m}(a,b)$
$-a\,;\,b$	$\rightsquigarrow \mathtt{m}(a,b)$
$a\,;\,-b$	$\rightsquigarrow \mathtt{m}(a,b)$
$-a\,;\,-b$	$\rightsquigarrow \mathtt{m}(a,b)$

When ignoring pitch relations, the suspension example of Figure 3iii is simplified to the following pattern (also Figure 6iii):

Fig. 7. Layered passing tones captured by *Pattern* 2: BWV 255 bar 2 (i) and BWV 320 bar 19 (b)

Example 2.

$$\frac{-a}{-d} \; ; \; \frac{b}{-e} \quad \rightsquigarrow \quad \begin{array}{l} \texttt{ov}(a,d),\ \texttt{sw}(b,e), \\ \texttt{m}(a,b),\ \ \texttt{m}(d,e) \end{array}$$

One can verify that the temporal relations enforced by example *Example 2* are indeed consistent with those of a suspension.

5 $\mathcal{H}$ and $\mathcal{SPP}$ are Distinct

Claim 1. $\mathcal{SPP} \not\subseteq \mathcal{H}$: there exists at least one pattern in $\mathcal{SPP}$ that has no equivalent in $\mathcal{H}$.

Consider the following $\mathcal{SPP}$ pattern (also Figure 6iv):

Pattern 2.

$$\frac{a \; ; \; b}{c \; ; \; d} \quad \rightsquigarrow \quad \begin{array}{l} \texttt{st}(a,c), \texttt{m}(a,b), \\ \texttt{m}(c,d) \end{array}$$

With appropriate pitch relations, the pattern can capture layered passing tones (Figure 7), including pairs of passing tones that do not share the same rhythm: cases when b and d start together (Figure 7i) and cases when they are not synchronized (Figure 7ii).

Pattern 2 is not representable in Humdrum, due to the absence of a temporal relation between b and d. To capture the $\texttt{st}(a,c)$, $\texttt{m}(a,b)$ and $\texttt{st}(c,d)$ temporal relations enforced by the $\mathcal{SPP}$ pattern, the following three Humdrum patterns are possible:

$$\begin{bmatrix} a & c \\ b & d \end{bmatrix} \qquad \begin{bmatrix} a & c \\ (a) & d \\ b & \star \end{bmatrix} \qquad \begin{bmatrix} a & c \\ b & (c) \\ \star & d \end{bmatrix}$$

All of the above patterns enforce an additional temporal relation that is not enforced by the $\mathcal{SPP}$ pattern, respectively $\texttt{st}(b,d)$, $\texttt{sw}(d,a)$ and $\texttt{sw}(b,c)$. $\square$

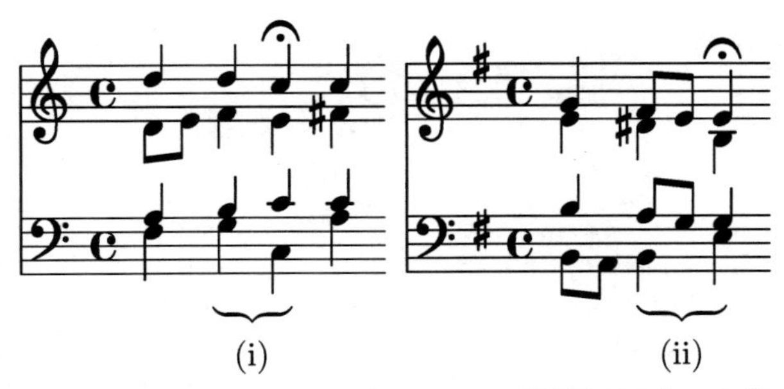

Fig. 8. Tritone resolutions captured by *Pattern* 3: BWV 257 bar 2 (i) and BWV 315 bar 13 (ii)

Claim 2. $\mathcal{H} \not\subseteq \mathcal{SPP}$: there exists at least one pattern in $\mathcal{H}$ that has no equivalent in $\mathcal{SPP}$.

Consider the following Humdrum pattern (also Figure 6v):

Pattern 3.

$$\begin{bmatrix} a & c \\ (a) & \star \\ b & d \end{bmatrix} \quad \rightsquigarrow \quad \begin{array}{l} \texttt{st}(a,c), \texttt{st}(b,d), \\ \texttt{m}(a,b) \end{array}$$

With appropriate pitch relations, this captures both embellished and unembellished tritone resolutions (Figure 8). Figure 8ii, for example, is matched by *Pattern* 3 even if the A forming the tritone in the tenor voice does not meet with the G of the final chord. Rather, there is an embellishment in the form of an anticipation.

Clearly, the Humdrum pattern enforces $\texttt{st}(a,c)$ and $\texttt{st}(b,d)$. The only way to do that in $\mathcal{SPP}$ is to join a,c and b,d with the "$=$"operator. As the Humdrum pattern also enforces $\texttt{m}(a,b)$, these two must be joined by the ";"operator:

$$\frac{a}{c} ; \frac{b}{d}$$

But then, the $\mathcal{SPP}$ pattern will also enforce the temporal relation $\texttt{m}(c,d)$ which is clearly not enforced by *Pattern* 3. □

By similar arguments, one can prove that the dislocated chord pattern (*Pattern 1* and Figure 5) cannot be represented in either Humdrum or $\mathcal{SPP}$. This explains its place in the language hierarchy of Figure 1. This is also why the figure shows that $\mathcal{R}$ properly subsumes $\mathcal{H}$ and $\mathcal{SPP}$.

6 The Common Denominator $\mathcal{SPP}_{\texttt{seq}}$

Characterizing the intersection between Humdrum and $\mathcal{SPP}$, the pattern language $\mathcal{SPP}_{\texttt{seq}}$ restricts $\mathcal{SPP}$ to sequences of layers:

Definition 4.

$$\varphi \in \mathcal{SPP}_{\texttt{seq}} ::= \psi \mid \varphi \,;\, \psi \qquad \text{with} \qquad \psi ::= \varepsilon \mid -\varepsilon \mid \frac{\psi}{\psi}$$

With the additional restriction that there can by only one "−" operator per layer, except for the first layer, in which any number of "−" may appear. One can easily check, for example, that the suspension pattern (e.g. *Example 2*) is in $\mathcal{SPP}_{\texttt{seq}}$. Also, as $\mathcal{SPP}_{\texttt{seq}}$ can be interpreted the same way as the unrestricted $\mathcal{SPP}$ (Section 4), it follows that $\mathcal{SPP}$ subsumes $\mathcal{SPP}_{\texttt{seq}}$.

Claim 3. $\mathcal{SPP}_{\texttt{seq}} \subseteq \mathcal{H}$: for every $\varphi \in \mathcal{SPP}_{\texttt{seq}}$ there exists a pattern $h \in \mathcal{H}$ enforcing the same temporal network.

The proof proceeds by structural induction over the ";" operator (i.e. the claim holds as the sequence grows). The base cases are:

$$\frac{a}{c} \qquad \frac{-a}{c} \qquad \frac{a}{-c} \qquad \frac{-a}{-c}$$

Those are clearly covered by the following Humdrum patterns:

$$[a\ c] \qquad [(a)\ c] \qquad [a\ (c)] \qquad [(a)\ (c)]$$

Now, suppose there exists a pattern $h \in \mathcal{H}$ that covers the $\mathcal{SPP}_{\texttt{seq}}$ pattern φ. The induction cases are as follows (the case with two modified events $-\varepsilon$ does not appear; by definition of $\mathcal{SPP}_{\texttt{seq}}$, this is only allowed in the first layer):

$$\varphi\,;\frac{a}{c} \qquad \varphi\,;\frac{-a}{c} \qquad \varphi\,;\frac{a}{-c}$$

Suppose h has n lines. The induction cases are covered by:

$$\begin{matrix} h \\ \cdot \\ [a\ c] \end{matrix} \qquad \begin{matrix} \left.\begin{matrix} \vdots \\ h_{n1}\ h_{n2} \end{matrix}\right] \\ \cdot \\ \begin{bmatrix} a & (h_{n2}) \\ (a) & c \end{bmatrix} \end{matrix} \qquad \begin{matrix} \left.\begin{matrix} \vdots \\ h_{n1}\ h_{n2} \end{matrix}\right] \\ \cdot \\ \begin{bmatrix} (h_{n1}) & c \\ a & (c) \end{bmatrix} \end{matrix}$$

The last two cases enforce an extra temporal relation (respectively $\texttt{sw}(a, h_{n2})$ and $\texttt{sw}(c, h_{n1})$) that the $\mathcal{SPP}_{\texttt{seq}}$ pattern does not enforce. However, that relation can be inferred by the temporal relations that the $\mathcal{SPP}_{\texttt{seq}}$ pattern do enforce. That is, referring back to Figure 2ii, whenever the relations $\texttt{m}(a, b)$, $\texttt{m}(d, e)$, $\texttt{sw}(b, e)$ and $\texttt{ov}(a, d)$ are present, then $\texttt{sw}(e, a)$ can be inferred. This inference is also indicated in Figure 6iii by a dashed edge. □

7 Discussion

This paper has presented three approaches that can accurately represent networks of temporal relations. Alternative approaches to polyphonic patterns often lack that accuracy. For example, vertical patterns [3] can only match polyphonic sources that have been expanded and sliced to yield a homophonic texture, hence not supporting the `sw` relation. A point set pattern representation [6] can only encode temporal relations with fixed duration ratios (capturing every instance of a `sw` relation would require a set of patterns, the size of which can grow quickly as many different ratios are likely to be found in the source). Techniques that rely on approximate matching to a source fragment [7] can confuse simultaneous notes with notes that overlap without being simultaneous, hence lacking precision with respect to the `st` relation.

With a little practice the musicologist should find it easy to write $\mathcal{SPP}$ patterns, in contrast to Humdrum, which requires extensive knowledge of Unix command line and regular expression tools. Relational patterns tend to be verbose and one quickly loses sight of the overall temporal structure of the pattern, where as the structure is syntactically expressed in $\mathcal{SPP}$. In Humdrum, this is readable when using the matrix form which this paper has developed. However, negations and disjunctions that can in principle appear in the regular expressions of a Humdrum pattern are not supported.

Finally, notice that $\mathcal{R}$ can express a great many temporal networks with unclear musical relevance (e.g. $\mathtt{sw}(a, b)$, $\mathtt{sw}(b, c)$) and even networks that are unsatisfiable (e.g. $\mathtt{m}(a, b)$, $\mathtt{st}(a, b)$). Perhaps there exists a restriction of $\mathcal{R}$ to "common sense" musical patterns. Ideally, such a restriction would preserve most of $\mathcal{R}$'s expressiveness, while being conducive to efficient pattern matching algorithms. Both Humdrum and $\mathcal{SPP}$ are candidate restrictions, yielding relational graphs that are always satisfiable. The graphs are also always connected and perhaps this connectedness is an interesting avenue to explore for future research. In parallel, a website with tools and tutorials is being developed in an effort to make the languages presented in this paper more easily applicable to musicological tasks.

References

1. Allen, J.F.: Maintaining knowledge about temporal intervals. Communications of the ACM 26(11), 832–843 (1983)
2. Bergeron, M., Conklin, D.: Structured polyphonic patterns. In: Ninth International Conference on Music Information Retrieval, Philadelphia, USA, pp. 69–74 (2008)
3. Conklin, D.: Representation and discovery of vertical patterns in music. In: Anagnostopoulou, C., Ferrand, M., Smaill, A. (eds.) ICMAI 2002. LNCS (LNAI), vol. 2445, pp. 32–42. Springer, Heidelberg (2002)
4. Fitsioris, G., Conklin, D.: Parallel successions of perfect fifths in the Bach chorales. In: Fourth Conference on Interdisciplinary Musicology, Thessaloniki, Greece (2008)

5. Jan, S.: Meme hunting with the Humdrum toolkit: Principles, problems, and prospects. Computer Music Journal 28(4), 68–84 (2004)
6. Meredith, D., Lemström, K., Wiggins, G.A.: Algorithms for discovering repeated patterns in multidimensional representations of polyphonic music. Journal of New Music Research 31(4), 321–345 (2002)
7. Typke, R., Veltkamp, R.C., Wiering, F.: Searching notated polyphonic music using transportation distances. In: ACM Multimedia Conference, New York, USA, October 2004, pp. 128–135 (2004)

Maximally Smooth Diatonic Trichord Cycles

Steven Cannon[*]

Abstract. In the usual seven-note diatonic scale, the maximally smooth cycle of triads contains a long section that uses only major and minor triads, the same triad that forms maximally smooth cycles within the twelve-note chromatic scale. Tonal music exploits this property of the scale to create sequences of similar chords. The goal of this study is to determine the extent to which such long chains containing inversionally related species exist in maximally smooth trichord cycles within microtonal scale systems that share certain properties with the diatonic. The study thus combines neo-Riemannian theory, especially Cohn's concept of maximally smooth cycles, with the diatonic scale theory developed by Clough and other authors. The patterns of maximally smooth trichord cycles depend on the type of scale within which they occur, and on the cardinalities of the scales. Among all scales, the usual diatonic supports the longest possible chain.

1 Introduction

An important feature of tonal music is its affinity for smooth motion from one harmony to another, maximizing common tones and minimizing melodic motion. In the diatonic scale, triads take part in such motion easily. The thrift of triadic chord progressions has led Richard Cohn to characterize the major and minor triads as *parsimonious trichords* [1]. Generally, theorists discuss the progressions in a chromatic context, but neo-Riemannian concepts are useful even within the limits of the diatonic set. To better understand how triads fit within the familiar diatonic, I will explain how they behave in scales from microtonal universes that use equal divisions of the octave other than the usual twelve. To my knowledge, this intersection of scale theory and neo-Riemannian theory has not yet been explored in detail in microtonal settings. While a number of essays in the recent collection *Music Theory and Mathematics: Chords, Collections, and Transformations* do bridge the gap between scale theory and neo-Riemannian theory, they mostly stay within the usual 12-edo [2].

2 Maximally Smooth Cycles and Parsimonious Triads

Cohn has shown that all scale systems, regardless of their cardinalities, contain a trichord capable of forming maximally smooth cycles [3]. The maximally smooth trichord cycle in the seven-note diatonic is given as Figure 1.[1] A striking feature of this cycle is the pattern with which the species of the chords change: the cycle contains a

[*] The author gratefully acknowledges the financial support of the Social Sciences and Humanities Research Council of Canada.

[1] To save space, I use common music-analytical symbols to indicate the qualities of chords: "+" or major, "–" for minor, and "o" for diminished.

E. Chew, A. Childs, and C.-H. Chuan (Eds.): MCM 2009, CCIS 38, pp. 43–56, 2009.

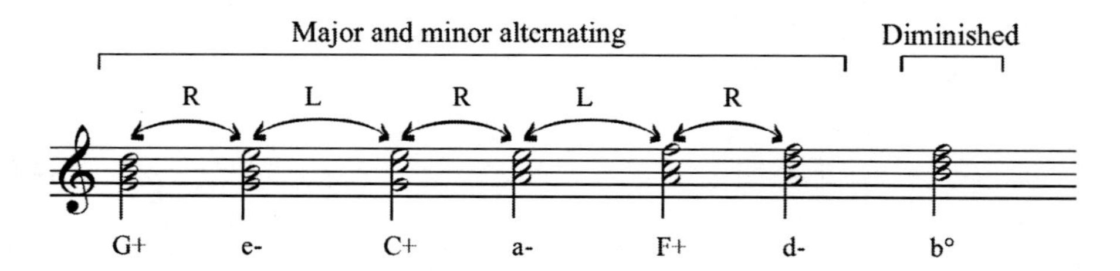

Fig. 1. Diatonic sequence of triads with root motion of falling thirds

long chain of alternating major and minor triads, two species that are related by inversion and thus, in a sense, equivalent. The major and minor triads are "parsimonious trichords" within the larger chromatic universe, but the diminished triad is not.

Cohn further explains how to find parsimonious sets within other chromatic universes when he notes that major and minor triads (as well as half-diminished and dominant sevenths) represent "minimal perturbations of a symmetrical division of the octave" (see [2], p. 39, n. 40). In scale systems where the cardinality of the chromatic universe is a multiple of three, the parsimonious trichord will be this kind of minimal perturbation. In systems where the cardinality of the chromatic set is not a multiple of three, the parsimonious trichord will simply be the closest approximation of an equal division of the chromatic scale. Since the generic triad is the closest approximation of an equal division of the diatonic scale into three, it is what Clough and Douthett call "maximally even," or "ME" within the diatonic. The diatonic scale is itself a maximally even distribution of seven notes within the twelve chromatic scale steps, so the triad is thus "second-order ME" [4]. In addition, it is a "generated" set, whose generator has a diatonic length (dlen) of two scale steps, or a third.[2]

3 Useful Scales

While it is clear enough that parsimonious trichords exist in all chromatic universes, how they fit in scales embedded within these universes has not yet been examined in much detail. Before proceeding any further, however, it is first necessary to decide which of the many possible scales we should examine. Clough and Myerson have described important features of the usual diatonic, including *cardinality equals variety* (CV) and *partitioning* [5]. In scales that share such features, the cardinalities of the chromatic universe (c for short) and the scale (d for short) relate in one of the following two ways, provided that d is odd:

$$c = 2d - 1 \tag{1}$$

or

$$c = 2d - 2\,. \tag{2}$$

[2] The abbreviations "dlen" and "clen" are also from Clough and Douthett.

Eytan Agmon was first to describe these classes of scales [6], which Clough and Douthett later named "family A" and "family B" respectively. Clough and Douthett call scales in family B "diatonic," setting the familiar 7-out-of-12 scale aside with the term "*usual* diatonic"; this usage has become standard in current literature. No useful adjective has been coined yet for the first class of scale, which most theorists (probably correctly) consider less interesting, so I adopt Clough and Douthett's term and say that these scales are members of "family A." I use the terms "family B scale" and "diatonic scale" interchangeably. I do, however, also occasionally use the term "diatonic" more loosely: the terms "maximally smooth *diatonic* cycle" and "*diatonic* length" (or "dlen") apply to both family A and family B scales. In scales of family A, c is always odd, and the scale generator (g_d for short) is always clen 2. The clen of g_d in diatonic scales is always equal to the value of d, and c is always a multiple of 4. Notably, chromatic universes in which c is even but not divisible by four (that is, 6, 10, 14, 18, 22, etc.) do not contain any useful scales. In Clough, Engebretsen, and Kochavi's taxonomy, members of *F-set 1* and *F-set 2* are both diatonic, and the complements of these scales fall into *F-set 5*; family A scales are members of what Clough and colleagues call *F-set 4*, which also includes the complements of family A scales [7]. I have omitted Clough and colleagues' *F-set 3*, which contains a type of scale first described by Gerald Balzano [8], because at higher cardinalities of d, trichords that are parsimonious within the full chromatic set are not subsets of the smaller scale.

I borrow several terms from the usual diatonic when describing scales in family A and family B. I call all clen 2 intervals "whole-tones" and all clen 1 intervals "semitones," regardless of the real size of these intervals. Family B scales have two semitones, separated as much as possible by whole tones. I call the highest note in the larger series of whole-tones the leading-tone, and the following note the tonic. I also number scale-degrees in ascending order starting with the tonic as $\hat{1}$. The lowest note in the larger series of whole-tones is the original note in the generating cycle—F in the usual diatonic with no key signature—which has a semitone below it. I call this

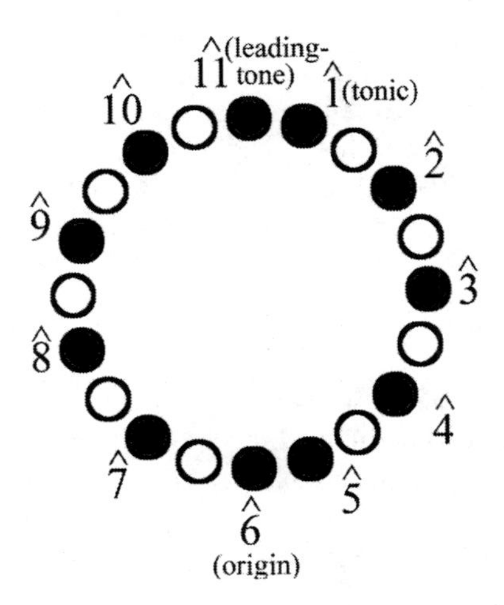

Fig. 2. Family B, or "diatonic" scale: 11/20

note the "origin." Figure 2 gives an example of a diatonic scale. Each circle in the clock-face represents a note in the chromatic scale, with the notes of the diatonic scale blacked in. Scales in family A contain but one scale-step of clen 1, with all other scale-steps having clen 2. I consider the note directly above this single semitone to be the tonic, and the note below it to be the leading-tone. See Figure 3 for an example of a family A scale. In family A, $\hat{1}$ is also the origin, but $\hat{1}$ is the first generated note after the origin in family B. To save space, I use a shorthand resembling a fraction to indicate the cardinalities of scales: the number before the slash indicates the value of *d*, and the number after the slash is the value of *c*. The scale is always a maximally even distribution of the smaller value within the larger.

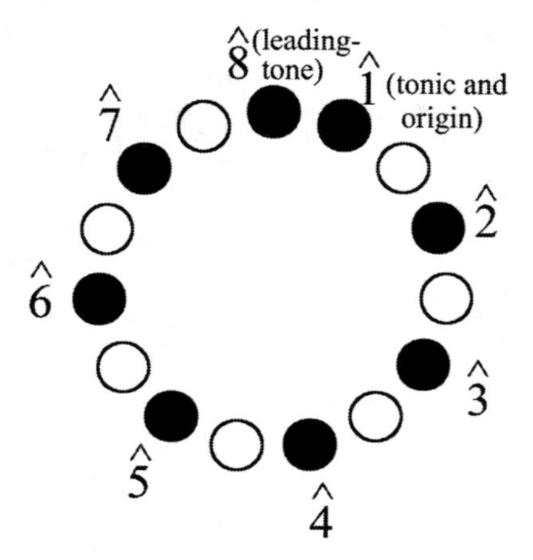

Fig. 3. Family A scale: 8/15

I have chosen to focus this study on family A and B in order to give these scale types thorough treatment, but the results are also applicable to the complements of such scales. Specifically, the pattern of the maximally smooth cycle for the second-order ME trichord is the same for a family A scale with a given value for *d* as it is for another set with the same value for *d* that is the complement of a family A scale. This is also true for family B scales and their complements. For example, the cycles in 8/15 and 8/17 share a common structure. Similarly, the structure of the cycle in the usual 7/12 diatonic, shown in Figure 1, also appears in the 7/16 "hyperpentatonic."

4 Trichord Species and Their Multiplicities

Species of second-order ME trichords in all universes may be categorized in the same way as they are in the usual diatonic, and much of the terminology from tonal theory is applicable. Since these trichords are generated, they have a "root," and as such, they can be arranged in "root position." According to MP, all generic intervals come in two specific sizes, which we can continue to call "major" (+) and "minor" (–). If the generating interval of the trichord (g_t for short) immediately above the root is major or minor, we can use this label for the trichord as a whole. Symmetrical trichords are augmented in some scales (two adjacent major g_t's) but diminished in others (two adjacent minor g_t's), so I use the term "symmetrical," (abbreviated as Ⓢ) rather than diminished or augmented.

To determine the structure of maximally smooth diatonic trichord cycles, we must first know how many chords of each species exist in the scale, and how they relate to each other. Fortunately, Clough and Myerson provide methods for finding this information quickly. An important feature implied by MP and CV is *structure yields multiplicity* (SM). According to Clough and Myerson, "within a particular genus, the number of chords in each species ... is directly inferable from the generic structure" (p. 250). This structure must be given in terms of the scale's generator (g_d), as measured in generic scale steps. That is, we can measure any interval not only in clen and dlen, but also by the number of g_d intervals it would take to generate it ($g_d$len for short).[3] When calculating $g_d$len, we must always count in generic, rather than specific g_d intervals, since specific intervals would give us different values for the two species of g_t. The notes of any diatonic scale form a g_d cycle, which we know as the cycle of fifths in the usual diatonic scale. In these diatonic scales, $g_d$len = 2(dlen); this doubling often forces us to octave-reduce the results, or at least to count shortest distance as a descending interval rather than ascending (this is equivalent to taking the complementary value within the modulus). In family A, $g_d$len is the same as dlen, so no translation is necessary, and the g_d cycle is identical to the scale itself.

To calculate the multiplicities of second-order ME trichord species we can start by finding their generators as measured in $g_d$len. These chords divide the diatonic scale as evenly as possible, so to find the generator g_e (measured in dlen) of any second-order ME chord of any cardinality *e,* we simply divide the cardinality of the scale by the cardinality of the chord, and round to the nearest integer:

$$g_e = \frac{d}{e} \quad . \tag{3}$$

$$\text{For trichords,} \quad g_t = \frac{d}{3} \quad . \tag{4}$$

Once we know the dlen chord generator, we must then translate it into $g_d$len (only necessary for family B scales) and generate the sets. Lastly we reduce the generated set to within one modular cycle (mod *d*), and find the multiplicity of each species by counting how many g_d intervals lie between each pair of adjacent notes.

For trichords in family A scales, the value of g_t in dlen also gives the multiplicities of the major and minor species. To find this value, divide *d* by three and round to the nearest integer, as in (4) above. The multiplicity of symmetrical species is what remains once the major and minor trichords are deducted from *d,* as in (6) below.

[3] Clough and Myerson have different ways of abbreviating the scale generator to show whether it is measured in clen or dlen. In clen it is d and in dlen it is c . They do not describe trichord or tetrachord generators.

Multiplicities of Trichord Species in Family A Scales

$$\text{no. of major or minor trichords} = g_t = \frac{d}{3}\ . \tag{5}$$

$$\text{no. of symmetrical trichords} = d - 2\ \frac{d}{3}\ . \tag{6}$$

To determine the multiplicities of major and minor trichords in diatonic scales, we have to alter the formulas to include a translation of dlen into $g_d$len, as in (7) below, and to measure descending rather than ascending intervals. The multiplicity of major and minor species is the complement of g_t, as measured in $g_d$len, as in (8). The multiplicity of symmetrical species is what remains once the major and minor are deducted from d, as in (9).

Multiplicities of Trichord Species in Family B Scales

$$g_t \text{ measured in } g_d\text{len} = 2\ \frac{d}{3}\ . \tag{7}$$

$$\text{no. of major or minor trichords} = d - 2\ \frac{d}{3}\ . \tag{8}$$

$$\text{no. of symmetrical trichords} = d - 2\ d - 2\ \frac{d}{3} = 4\ \frac{d}{3} - d\ . \tag{9}$$

Inputting different values for d, we can discard those that are multiples of 3, since scales of these cardinalities will not support maximally smooth diatonic cycles, and CV will not hold for harmonic trichords (although it will hold for three-note melodic lines).

5 Trichord Cycles

The patterns of the maximally smooth trichord cycles depend on the type of the scale, and on the cardinality of d, so we will consider each case in turn with one example of a specific scale. If d 2 (mod 3) in a family A scale, the trichords have the following properties:

- The parsimonious trichord is *asymmetrical*, and will give the *major* and *minor* species.
- The number of symmetrical trichords is *fewer* than the number of major or minor trichords.

See Figures 4 and 5 for illustrations of how trichords fit in this kind of scale. Each scale-degree number is accompanied by a symbol (+, –, or Ⓢ) indicating the species of trichord generated from that scale-degree. The arrows within the clock-face trace the minor trichord that has $\hat{8}$ as its root, which is parsimonious within the chromatic scale and is not inversionally symmetrical.[4] In these clock-face diagrams, the arrows always indicate parsimonious trichords, which are different species in different scales. We may now form a maximally smooth diatonic cycle in this universe, which is analogous to the cycle in Figure 1: pairs of adjacent trichords maintain two common tones, and the one voice that moves will only proceed by scale-step. The root motion along this cycle is by intervals of dlen g_t. The complete cycle is given in Figure 5. Note that inversionally related major and minor trichords always come in pairs, with the symmetrical trichords distributed as evenly as possible between these pairs. Since there are three pairs of major and minor chords, but only two symmetrical chords, two of the pairs must be adjacent. These adjacent pairs form a chain of four alternating major and minor trichords in a row, which is the longest such chain in the cycle.

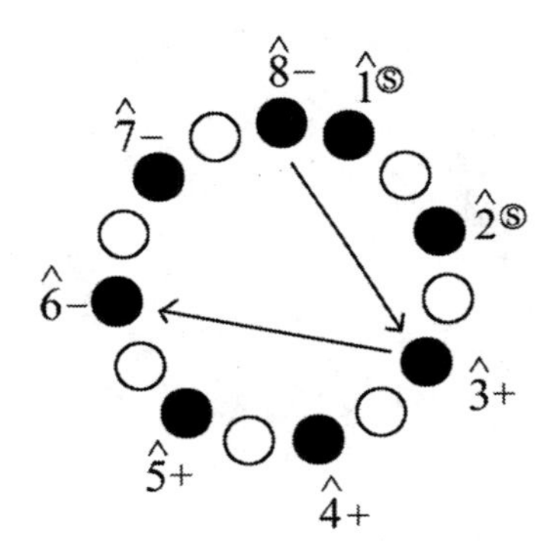

Fig. 4. Second-order ME trichord in 8/15
$c = 15$
$d = 8$
g_d = clen 2
g_t = dlen 3
major species prime form (0,5,9)
minor species prime form (0,4,9)
symmetrical species prime form (0,3,9)

longest chain

$\hat{5}$+, $\hat{8}$, $\hat{3}$+, $\hat{6}$, $\hat{1}$Ⓢ, $\hat{4}$+, $\hat{7}$, $\hat{2}$Ⓢ

Fig. 5. Maximally smooth diatonic trichord cycle in 8/15

In family B scales where d 1 (mod 3), the situation is similar except that the g_t intervals must be translated from dlen into $g_d$len. Figure 6 shows a thirteen-note diatonic

[4] Although $\hat{8}$ is the root in as much as it is the note from which the chord is generated, the sonority will sound to our ears as if $\hat{6}$ is the root because the chord's tuning is fairly similar to that of a minor triad in the usual diatonic.

scale in 24-edo. Figure 7 gives the cycle of g_d for this scale. The arc using a dotted line between $\hat{13}$ and $\hat{7}$ indicates the short g_d, equivalent to the diminished fifth between B and F in the usual diatonic.[5] The leading tone always occurs just before the short g_d, the origin immediately after, and proceeding clockwise the tonic appears next after the origin. The directions of the arrows appear to change from clockwise in Figure 6 to counter-clockwise in Figure 7 when the intervals are doubled to translate them from dlen to $g_d$len. Figure 7 also demonstrates SM well. The multiplicity of each species is given by the distance between the notes in the g_d cycle: moving counter-clockwise from the root of the chord, the distances are 5 , 5 , and 3 , which match the multiplicities of

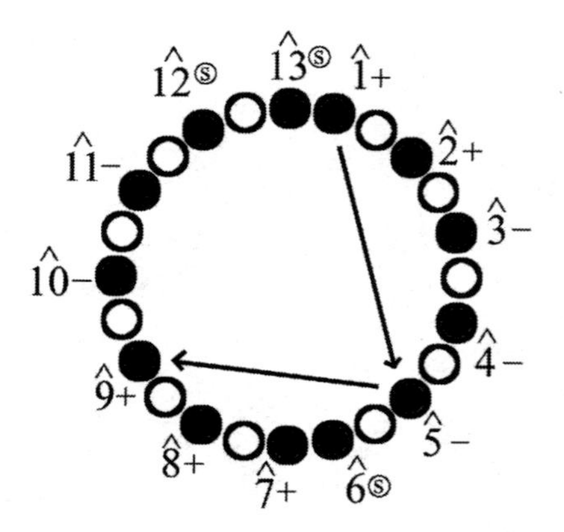

Fig. 6. Second-order ME trichord in 13/24, within scale cycle

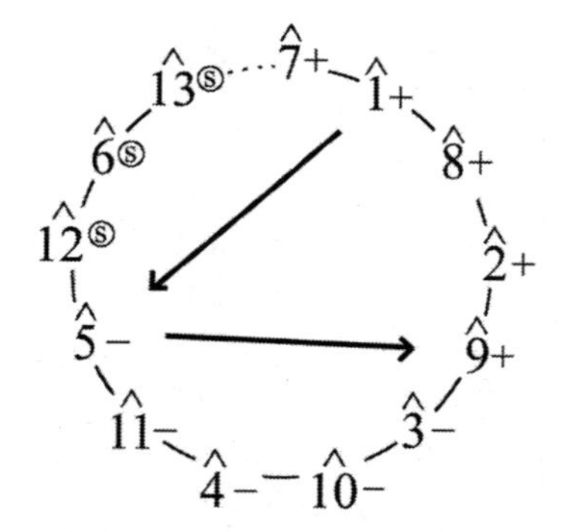

Fig. 7. Second-order ME trichord in 13/24, within g_d cycle

$c = 24$
$d = 13$
g_d = clen 13 = dlen 7
g_t = dlen 4 = $g_d$len 8
major species prime form (0,8,15)
minor species prime form (0,7,15)
symmetrical species prime form (0,7,14)

[5] I follow Clough and Myerson, who also indicate the short fifth in this way.

major, minor, and symmetrical trichords respectively. Looking at Figure 7, we notice that the trichord is not maximally even within the g_d cycle, so the multiplicities of species are not as balanced as they are in family A. The maximally smooth cycle is given in Figure 8. Notice again that the longest chains are four chords long.

$$\hat{10}\ ,\ \hat{1}+,\ \hat{5}\ ,\ \hat{9}+,\ \hat{13}\text{Ⓢ},\ \hat{4}\ ,\ \hat{8}+,\ \hat{12}\text{Ⓢ},\ \hat{3}\ ,\ \hat{7}+,\ \hat{11}\ ,\ \hat{2}+,\ \hat{6}\text{Ⓢ}$$

Fig. 8. Maximally smooth diatonic trichord cycle in 13/24

Four chords may be shorter than the series of six in the usual diatonic, but in other scales the chains are shorter still. In family A scales with *d* 1 (mod 3), and in family B scales with *d* 2 (mod 3), second-order ME trichords have the following properties:

- The parsimonious trichord is *symmetrical.*
- The number of symmetrical trichords is *greater* than the number of major or minor trichords.

Illustrations of second-order ME trichords in this kind of scale are given in Figures 9–13. Figures 9–10 illustrate a scale in family A, and Figures 11–13 illustrate a scale in family B. Note the lack of any long chains, as indicated by the short brackets above the cycles. At most, the chains are two chords long.

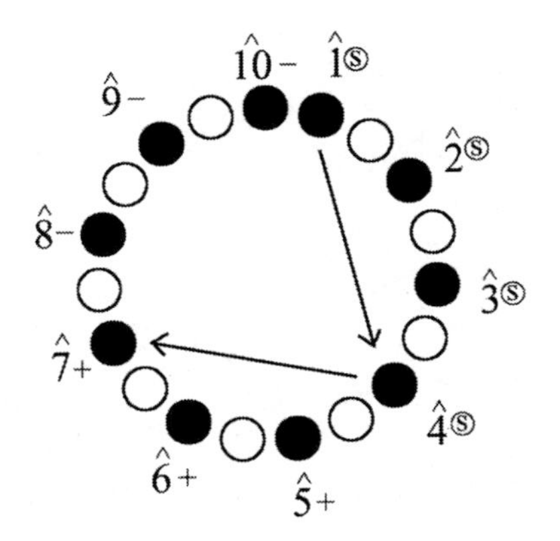

Fig. 9. Second-order ME trichord in 10/19
$c = 19$
$d = 10$
g_d = clen 2
g_t = dlen 3
major species prime form (0,6,11)
minor species prime form (0,5,11)
symmetrical species prime form (0,6,12)

$\hat{1}$Ⓢ, $\hat{4}$Ⓢ, $\hat{7}$+, $\hat{10}$, $\hat{3}$Ⓢ, $\hat{6}$+, $\hat{9}$, $\hat{2}$Ⓢ, $\hat{5}$+, $\hat{8}$

Fig. 10. Maximally smooth diatonic trichord cycle in 10/19

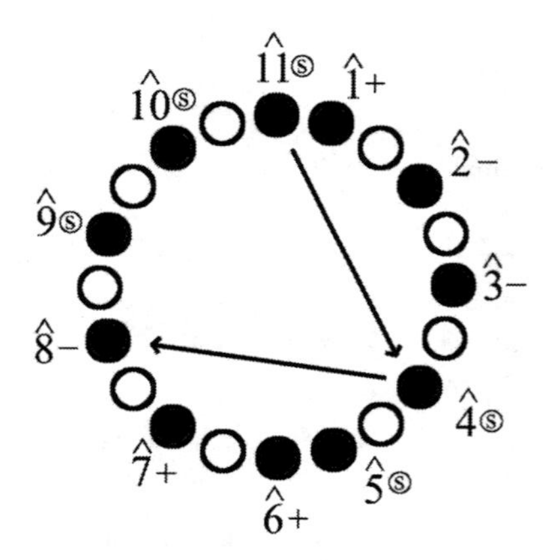

Fig. 11. Second-order ME trichord in 11/20, within scale cycle

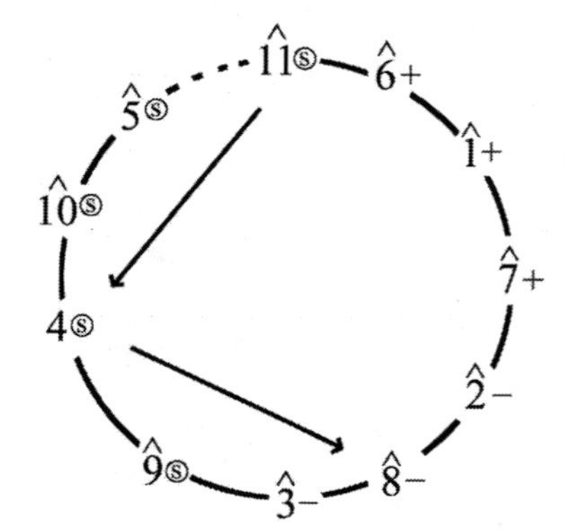

Fig. 12. Second-order ME trichord in 11/20, within g_d cycle

$c = 20$
$d = 11$
g_d = clen 11 = dlen 6
g_t = dlen 4 = $g_d$len 8
major species prime form (0,7,12)
minor species prime form (0,5,12)
symmetrical species prime form (0,6,13)

More abstract clock-faces appear in Figures 14–17. The circles represent an unspecified value of d. I do not indicate specific notes, but instead reckon the proportional number of notes using the lengths along the circumference. In family A scales,

$\widehat{11}$Ⓢ, $\hat{4}$Ⓢ, $\hat{8}$, $\hat{1}$+, $\hat{5}$Ⓢ, $\hat{9}$Ⓢ, $\hat{2}$, $\hat{6}$+, $\widehat{10}$Ⓢ, $\hat{3}$, $\hat{7}$+

Fig. 13. Maximally smooth diatonic trichord cycle in 11/20

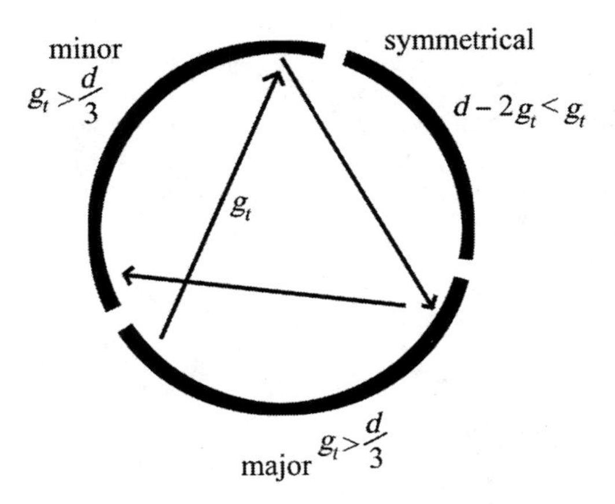

Fig. 14. Root progression of trichord cycles in family A scales with d 2 (mod 3)

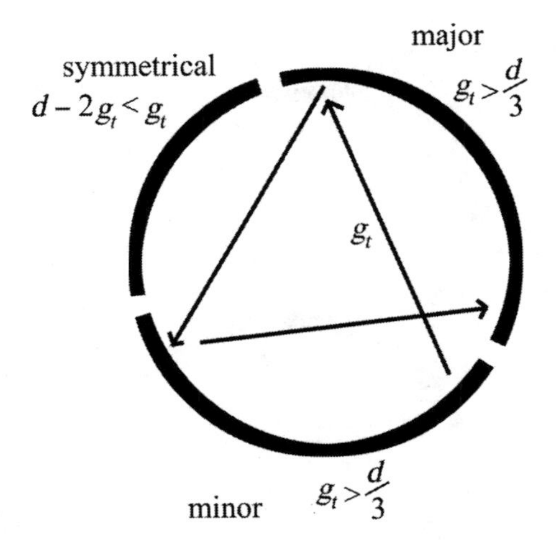

Fig. 15. Root progression of trichord cycles in family B scales with d 1 (mod 3)

distances along the circumference are measured in dlen; in family B scales, distances are measured in $g_d$len. The arrows inside the circle trace the root progressions in sequences of four trichords from maximally smooth cycles. The arcs for both major and minor species, as well as all of the arrows, all have the same value: g_t. Figures 14 and 15 illustrate scales with longer chains of alternating chords. Note that g_t is greater than

$d/3$, such that the third arrow places the root of the fourth chord past the root of the first chord, and the arrows cross. Since the number of symmetrical trichords is less than g_t, no symmetrical chords are adjacent in any maximally smooth diatonic cycles; since the numbers of major trichords and minor trichords are equal to g_t, they will always come in adjacent pairs. Moreover, at least one section of the cycle always has a chain of alternating major and minor trichords that is four chords long.

Figures 16 and 17 illustrate scales without long chains. Here g_t is less than $d/3$: the root of the fourth chord in the cycle does not pass the root of the first chord, and the arrows do not cross. Since the number of symmetrical trichords is greater than g_t, there will be at least one pair of adjacent symmetrical trichords in the maximally smooth diatonic cycle. Major and minor trichords still come in pairs, but at least one symmetrical trichord always sits between these pairs; no section of the cycle will have a chain longer than two chords.

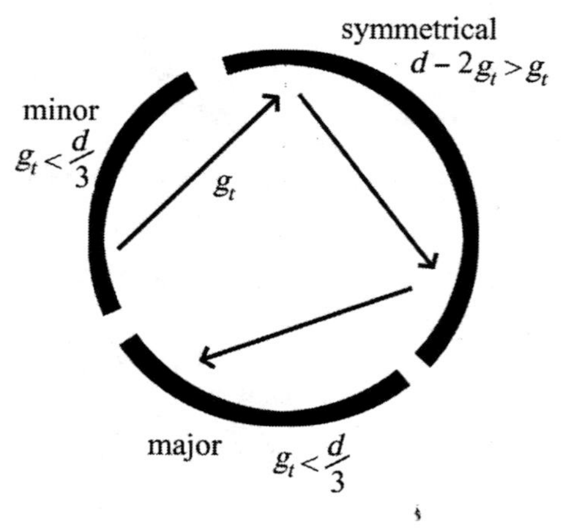

Fig. 16. Root progression of trichord cycles in family A scales with d 1 (mod 3)

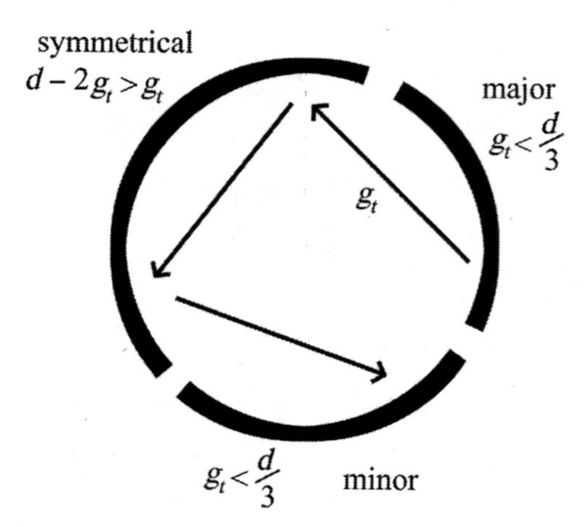

Fig. 17. Root progression of trichord cycles in family B scales with d 2 (mod 3)

6 Conclusions

No scale, in either family A or B can support a maximally smooth diatonic trichord cycle with as long segment of alternating major and minor trichords as the usual diatonic can. This segment includes six chords, as in Example 1, but the longest such segment in any other scale uses only four chords. The patterns of root progressions illustrated in Figures 14–17 show that the only way to get more than four chords in the segment is to have a very low number of symmetrical trichords, in proportion to the major and minor trichords. Specifically, the ratio of symmetrical to major (or minor) must be less than 0.5. Figures 18–22 illustrate. Major and minor trichords always come in pairs, as indicated by the brackets. The symmetrical trichords must be distributed as evenly as possible in the cycle without breaking up any pairs. If the ratio is greater than 1 (as it is when c is not a multiple of 3), some symmetrical trichords must be adjacent to one another, as in Figure 18. An extreme case is the five-note diatonic scale in eight-tone equal temperament with three adjacent symmetrical trichords, shown in Figure 19. If the ratio is less than 1 the pairs of major and minor trichords must be adjacent to each other, creating longer chains. In Figure 20, the cycle includes two segments with four adjacent major and minor trichords. Figures 21 and 22 demonstrate cycles in which the ratios are less than 0.5. The cycle in Figure 21 cannot exist in any real scale, since the cardinalities of major, minor, and symmetrical trichords cannot satisfy the formulas given above. Indeed, the only real scale that has a ratio less than 0.5 is the usual diatonic, so the longest possible chain of alternating major and minor trichords is the six-chord sequence depicted in Figure 1 and repeated below in Figure 22.

Fig. 18. Maximally smooth diatonic trichord cycle in 11/20

- Ratio of symmetrical to major: 1.667
- Major/minor pairs can not be adjacent, but two symmetrical trichords can
- Longest chains: 2 chords

Fig. 19. Maximally smooth diatonic trichord cycle in 5/8

- Ratio of symmetrical to major: 3
- Major/minor pairs cannot be adjacent but two symmetrical trichords can
- Longest chain: 3 chords

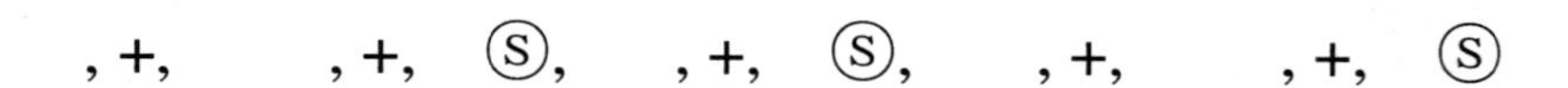

Fig. 20. Maximally smooth diatonic trichord cycle in 13/24

- Ratio of symmetrical to major: 0.6
- Major/minor pairs can be adjacent
- Longest chains: 4 chords

Fig. 21. Maximally smooth diatonic trichord cycle in fictional scale system

- Ratio of symmetrical to major: 0.286
- Major/minor pairs can be adjacent
- Longest chain: 8 chords

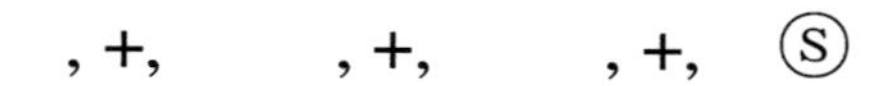

Fig. 22. Maximally smooth diatonic trichord cycle in 7/12 (usual diatonic)

- Ratio of symmetrical to major: 0.333
- Major/minor pairs can be adjacent
- Longest chain: 6 chords

References

1. Cohn, R.: Neo-Riemannian Operations, Parsimonious Trichords, and their *Tonnetz* Representations. Journal of Music Theory 41, 1–66 (1997)
2. Douthett, J., Hyde, M.M., Smith, C.J. (eds.): Music Theory and Mathematics: Chords, Collections, and Transformations. University of Rochester Press (2008)
3. Cohn, R.: Maximally Smooth Cycles, Hexatonic Systems, and the Analysis of Late-Romantic Triadic Progressions. Music Analysis 15, 9–40 (1996)
4. Clough, J., Douthett, J.: Maximally Even Sets. Journal of Music Theory 35, 93–173 (1991)
5. Clough, J., Myerson, G.: Variety and Multiplicity in Diatonic Systems. Journal of Music Theory 29, 249–270 (1985)
6. Agmon, E.: A Mathematical Model of the Diatonic System. Journal of Music Theory 33, 1–25 (1989)
7. Clough, J., Engebretsen, N., Kochavi, J.: Scales, Sets, and Interval Cycles: A Taxonomy. Music Theory Spectrum 21, 74–104 (1999)
8. Balzano, G.J.: The Group Theoretical Description of 12-Fold and Microtonal Pitch Systems. Computer Music Journal 4(4), 66–84 (1980)

Towards a Symbolic Approach to Sound Analysis

Carmine Emanuele Cella

Università degli studi di Bologna - Via Zamboni 38, 40126 Bologna, Italia

Abstract. In this article we will propose a new approach for music description, based on the connection between the symbolic (logic) level and the signal level. This approach relies on the possibility of representing sounds in terms of *types* inferred by some low-level descriptions of signals and subsequent learning stages. We will present *simple type theory* and we will introduce a twofold process to create *aggregate* representations with different degrees of abstraction thus making possible to describe and manipulate music at *variable* conceptual levels.

1 The Levels of Representation

Music, in its final stage of *performance*, can be described in many ways: it can be viewed as a time-varying *signal* and can be described by expressing the evolution of its physical properties over time. Music can be also viewed as a *symbolic system* exploiting relationships between sonic objects[1] and can be described with a formal language able to express these relationships over time. Common approaches for music description generally take into account the different points of view by selecting a particular degree of abstraction in the domain of the representation: either they rely on the *signal level*, either on the *symbolic level* or on a fixed mixture of both. The latter case is generally known as *mid-level* representation: this term is used in the computer audition community to indicate intermediate modelings of hearing usually based on perceptual criteria [2]. While signal-level representations are computationally efficient, invertible[2] and express some physical properties associated to the signal, they lack in abstraction and usually don't provide any kind of information about hierarchies, formal relationships between sonic-objects and so forth; they are unable to manipulate concepts other than the *basis* of the analysis itself (such as sinusoids, wavelets, etc.). The usual way to represent the signal-level decomposition of a signal $x[n]$ into expansion functions is a linear combination of the form:

$$x[n] = \sum_{k=1}^{K} \alpha_k g_k[n]. \quad (1)$$

[1] With this expression, here, we intuitively mean any kind of event that appears in the musical flow; a precise definition of sonic-objects is exactly the scope of any representation.

[2] With *invertibility*, here, we mean the possibility to go back to the signal domain from the representation itself.

E. Chew, A. Childs, and C.-H. Chuan (Eds.): MCM 2009, CCIS 38, pp. 57–64, 2009.

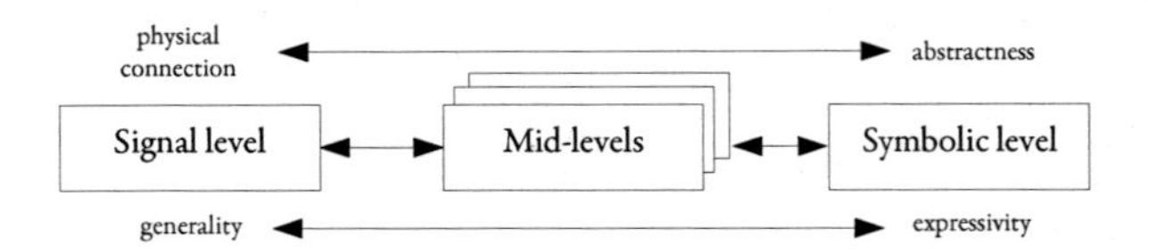

Fig. 1. The levels of representation

The coefficients α_k are derived from the analysis stage, while the functions $g_k[n]$ can or cannot be determined by the analysis stage and are used during the synthesis stage; both stages are related to a particular signal model [4].

On the other hand, symbolic-level representations can express complex relationships and hierarchies but are inefficient, non-invertible and are hardly related to the physical nature of sound: they are usually based on *logical rules* that cannot be verified by any model[3].

Mid-level representations, finally, try to address the issue related to the lack of generality by focusing on *relatively simple* concepts that are, however, more abstract than the basis of the analysis. These concepts are usually based on perceptual criteria related to the low-level hearing and are situated in between the constraints imposed on them by lower and higher levels. The power of this kind of representations stands in the fact that they are usually invertible and that the logical rules they involve are generally verifiable by some models related to perception.

All the representation levels discussed so far can be used to describe music; they are different because each one of them *captures* particular aspects of the sound. However, they share two common drawbacks: first, all of them have a *fixed degree of abstraction*. In other words, they are not *scalable*: once a representation level has been selected it is not possible to go smoothly to another level; while signal-level representations are very useful from numerical and computational points of view, higher level representations are essential to human reasoning. Second, all of them impose *their own* concepts onto the signal: each representation models the signal with it's own concepts, even if they are completely irrelevant to *that* particular signal; figure 1 roughly depicts the described ideas.

The main purpose of this article is to propose a representation method for music that, while being generic enough to be used for different signals, fulfills *by-design* the following requirements:

- **signal-dependent semantics**: the underlying logic and the involved concepts of the representation should be inferred from the signal, using learning techniques; this creates the possibility to describe concepts that are really *related* to the sound being analysed (adaptive);
- **scalability**: it should be possible to change the *degree of abstraction* in the representation, ranging from the signal level to the symbolic level in a

[3] In the context of this article, with symbolic-level representations we mean *highly formalized descriptions of music*, possibily based on a formal language and on its underlying logic [1]. First attempts to apply formal logic to music rely mainly on a deductive system called *first-order logic*. Later on, inspired by linguistic ideas, other extensions of logic have also been tested (temporal, modal, non-monotonic, etc.) [5], [6].

continuous manner; the degree of abstraction becomes a parameter of the representation;
- **weak invertibility**: the representation method should be able to generate the represented signal; this possibility does not imply, however, that the generated signal must be *waveform*-identical to the original one, but only that *relevant* parts of it can be reconstructed (that's why we call it *weak*); this also means that the representation has been able to capture *salient* aspects of a signal;
- **generativity**: it should be possibile to generate sounds *other* than the original one, according to some parameters in the domain of the representation that can be estimated from a given signal or deliberately created.

Defining such a kind of representation method is not an easy task; it involves, first of all, the selection of an underlying *logic* that is able to scale over abstraction and supports some special features like poly-valued semantics. Consequentially, it also involves the usage of specific signal processing techniques able to retrieve information from a signal (for example low-level features) and model that information statistically to find *salient* properties in order to provide a *validation model* for the underlying logic.

2 Sound Types

Symbolic-level and signal-level representations are complementary views of an underlying world: the former are expressive but don't relate easily with the modeled reality, the latter are physically-connected but lack in abstraction. The following sections will propose a connection between the signal and the simbolic level by suggesting a representation based on *types* inferred by some low-level descriptions of signals and subsequent learning stages. From a logical point of view, the concept of type is formalized in the so-called *theories of types*; from a computational point of view, low-level descriptions and statistical learning build to the so-called *audio indexing* theory.

2.1 Simple Type Theory

In order to avoid some set-theoretical paradoxes, Bertrand Russell proposed in 1908 a logic now known as *ramified theory of types*, subsequently expanded to the so-called *simple type theory* (STT) by Alonzo Church. There are many variants of STT; the presentation we give here is due to [3].

STT syntax is made of two principal objects: *types* and *expressions*. The former is a nonempty set of values used to build expressions. The latter denotes values including true and false. A type of STT is defined by the following formation rules:

T1. i is the type of individuals;
T2. $\star$ is the type of truth values;
T3. if α, β are types, then $\alpha \to \beta$ is the type of functions from elements of type α to elements of type β.

Rules T1 and T2 define he so-called *atomic* types while rule T3 defines *compound* types. The logical symbols of STT are defined as follows: function application: @, funcion abstraction: λ, equality: $=$, definite description: ι, an infinite set of symbols called variables: ν. We can now define a *language* of STT as the ordered pair $L = (C, \phi)$ where:

1. C is a set of symbols called *constants*;
2. $\nu \cap C = \oslash$ (the sets are disjoint);
3. $\phi : C \to \tau$ is a total function, where τ is a set of types of STT.

In other words, a language is a set of symbols with types that have been *assigned.* It is now possible to define an expression of the language L with another set of formation rules:

E1. if α is a type and $x \in \nu$, then $x : \alpha$ is an expression of type α (*variable*);
E2. if $c \in C$, then c is an expression of type $\phi(c)$ (*constant*);
E3. if A is an expression of type α and F is an expression of type $\alpha \to \beta$, then $F@A$ is an expression of type β (*function application*);
E4. if $x \in \nu$, α is a type and B is an expression of type β then $\lambda x : \alpha.B$ is an expression of type $\alpha \to \beta$ (*function abstraction*);
E5. if E_1 and E_2 are expressions of type α, then $E_1 = E_2$ is an expression of type $\star$ (*equality*);
E6. if $x \in \nu$, α is a type and A is an expression of type $\star$, then $\iota x : \alpha.A$ is an expression of type α (*definite description*).

2.2 Models for Simple Type Theory

For the languages of STT, like for first-order languages, it is possible to define a semantics based on models. A *standard model* for a language $L = (C, \phi)$ of STT is an ordered triple $M = (D, E, I)$ such as:

1. $D = \{D_\alpha : \alpha \in \tau\}$ is a set of nonempty domains;
2. $D_\star = \{true, false\}$;
3. for $\alpha, \beta \in \tau, D_{\alpha \to \beta}$ is the set of *all* functions from D_α to D_β;
4. $E = \{e_\alpha : \alpha \in \tau\}$ is a set of values such that $e_\alpha \in D_\alpha, \forall \alpha \in \tau$ (e_α is called the *canonical error* for α);
5. I maps each $c \in C$ to a member of $D_{\tau(c)}$.

Given a model $M = (D, E, I)$ for a language of STT, with *variable assignment* into M we mean a function ψ that maps each variable expression $x : \alpha$ to a member of D_α. Given a variable assignment ψ into M, an expression $x : \alpha$ and $d \in D_\alpha$, let $\psi(x : \alpha \to d)$ be a variable assignment $\psi\prime$ into M such that $\psi(x : \alpha) = d$ and $\psi\prime(v) = \psi(v), \forall v \neq x : \alpha$. Then, with *valuation function* we mean the binary function V^M that, for all variable assignments ψ and all expressions E of L, satisfies the following conditions:

1. if $E = x : \alpha$, then $V_\psi^M(E) = \psi(x : \alpha)$;

2. if $E = C$, then $V_{\psi}^{M}(E) = I(E)$;
3. if E is of the form $F@A$, then $V_{\psi}^{M}(E) = V_{\psi}^{M}(F)V_{\psi}^{M}(A)$;
4. if E is of the form $\lambda x : \alpha.B$ with B of type β, then $V_{\psi}^{M}(E)$ is the function $f : D_{\alpha} \to D_{\beta}$ such that $\forall d \in D_{\alpha}, f(d) = V_{\psi(x:\alpha \to d)}^{M}(B)$;
5. if E is of the form $E_1 = E_2$ and $V_{\psi}^{M}(E_1) = V_{\psi}^{M}(E_2)$, then $V_{\psi}^{M}(E) = true$; otherwise $V_{\psi}^{M}(E) = false$;
6. if E is of the form $Ix : \alpha.A$ with A of type α and there is a unique $d \in D_{\alpha}$ such that $V_{\psi(x:\alpha \to d)}^{M}(A) = true$, then $V_{\psi}^{M}(E) = d$; otherwise $V_{\psi}^{M}(E) = e_{\alpha}$.

Let E be an expression of type α of L and A be a formula of L. We will call $V_{\psi}^{M}(E)$ the *value* of E in M with respect of ψ. We also say that A is *valid* in M ($M \Vert\!= A$) if $V_{\psi}^{M}(A) = true$ for all variable assignments ψ into M. With *sentence* we mean a closed formula of L; A is a *semantic consequence* of a set of sentences Σ ($\Sigma \Vert\!= A$) if $M \Vert\!= A$ for every standard model M such that $M \Vert\!= B$ for all $B \in \Sigma$. With *theory* of STT we mean an ordered pair $T = (L, \Gamma)$ where L is a language of STT and Γ is a set of sentences called *axioms* of T; a *formula* A, therefore, is a semantic consequence of T ($T \Vert\!= A$) if $\Gamma \Vert\!= A$. Finally, a standard model of T is a standard model M for L such that $M \Vert\!= B, \forall B \in \Gamma$.

2.3 Low-Level Features and Audio-Indexing

Low-level features are numerical values describing the contents of an audio signal according to different kinds of inspection: temporal, spectral, perceptual, etc. The computation of the features is done by processing a given signal with specific algorithms on a small time scale (often called short-term analysis window) which is usually between 40 ms and 80 ms; different kinds of temporal modeling (like mean and variance computation) can then be applied to the features on larger time scales. A typical example of low-level feature is the so-called *spectral shape*, represented by the statistical moments of the spectrum: *mean, variance, skewness* and *kurtosis*. The probabilistic computation of the features takes the frequencies of the spectrum as the observed data and the amplitudes as the probabilites to observe the data; see [7] for more information.

A main field of application of low-level features is *audio indexing*: by combining different techniques it is possible to group together signals that share common properties. A typical approach to audio indexing is based on the projection of low-level features computed over a set of sounds (usually called *population*) in a multi-dimensional space (usually called *features space*); similar sounds, then, *tend* to project onto similar positions of the space, producing *clusters*. By analysing the space with some combined geometrical and statistical techniques (like Gaussian Mixture Models, Principal Component Analysis, etc.) it is possible to find the clusters of sounds present in the space.

2.4 The *Typed* Model

We are now ready to present a new approach to sound and music description based on the concept of *sound type*. The basic idea, is to **represent sound and music by means of a given theory of types** while providing a *verifiable model* for that theory through low-level descriptors plus statistical learning.

From a logical point of view, simple type theory provides a *variable abstraction level* by means of syntax constraints on functions: from a given set of elements it is possible to create a subset by applying a function only if that function *exists* for the original set. Mathematically, we want to be able to *translate* a signal-level representation into other forms involving different elements and operators (mid to symbolic-level representations) while providing real examples of such elements and operators. More formally:

$$
\begin{aligned}
x[n] &= \textstyle\sum_{k=1}^{K} \alpha_k g_k[n] \\
&= \alpha_1 g_1[n] + \ldots + \alpha_k g_k[n] \\
&= \beta_1 f_1[n] + \ldots + \beta_j f_j[n] \\
&\vdots \\
&= \omega_1 h_1[n] + \ldots + \omega_t h_t[n].
\end{aligned}
$$

In the equations above $\alpha, \beta, \ldots, \omega$ are weighting coefficients, $g_k, f_j, \ldots, h_t$ are variables belonging to different *types*, $+$ and $\cdot$ are relations defined for each type and $t < j < \ldots < k$ (i.e. last equation has less elements than first equation). Notice that $+$ and $\cdot$ are not algebraical sum and multiplication and are *not* required to be commutative: they can be any kind of binary relation defined over specific types. As long as it is possible to convert, say, from type g_k to type h_t and to define relations on both we can perform the translation. A model for the set of equations above is defined by providing real types, relations and functions that make all the equations *true*; in other words, we need some validation functions that clearly define whether a given variable belongs to a given type and whether a binary relation *exists* on that type and how it is possible to convert from a type to another. A possible way to achieve these requirements is by means of clusters of low-level descriptors in the features space. In order to provide a verifiable model for the proposed theory, we need a twofold process divided into different stages:

- **types inference:** during this stage the types involved in the representations are discovered;
- **rules inference:** a second stage is needed to discover the relations between the types and their conversions.

Since the $+$ relation is not the algebraical sum, we will suppose that our symbolic-level representation is a *sequence* of types and that $+$ is the *successor* function (i.e. $g + f$ means that variable f of type F follows variable g of type G[4]; remember that if F and G are types then $G \rightarrow_{+} F$ is a type).

[4] The *successor* relation is evidently non-commutative.

The following algorithm shows a possible implementation of the twofold process, using low-level descriptors plus statistical learning for types inference and Markov models for rules inference; a precomputational stage (atomic decomposition) is also performed to prepare the original signal for the analysis:

1. **(atomic decomposition):** subdivide a sound into small grains of approximately 40 ms called *atoms* or *0-types* overlapping in time and frequency (let's label them with integer numbers);
2. **(1-types inference):** compute a set of low-level descriptors on each atom obtained in the previous step, project the descriptors in a multi-dimensional space and compute the *clusters* by means of statistical techniques; each cluster will represent a *1-type* (let's label them $g_1, f_1, \ldots$);
3. **(1-rules inference):** implement a Markov model to describe the sequences of types present in the analysed sound (*1-rules)*;
4. **(n-types inference):** compute a set of low-level descriptors on the whole sequences found in the previous step (for example $g_1 + f_1$); project again the descriptors and compute again the clusters: each cluster will represent a *n-type* (let's label them $g_n, f_n, \ldots$);
5. **(n-rules inference):** repeat from step 3 until there are no more sequences (*n-rules*).

The number of iterations of the whole process are the *abstraction levels* of the representation. In terms of atomic decomposition, all the sets of the discovered types are time-frequency atoms with different time scales and spectral content; the higher the level of a type the less it is generic, the more expressive.

Figure 2 illustrates the approach.

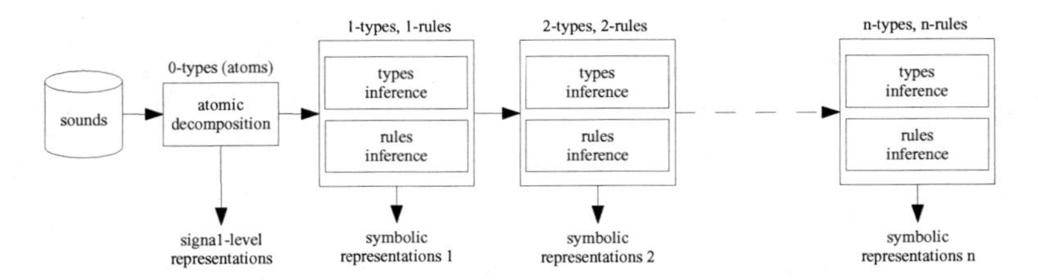

Fig. 2. An outline of the proposed algorithm for types and rules inference

Low-level descriptors and statistical techniques are not used to classify different sounds, but parts of a single sound; another approach could take into account a real population of sounds and compute sound types over a whole database; since different atoms and sequences (moleculae) belong to the same type as long as they share common properties (defined by the set of descriptors), they could theoretically be shared between different sounds. From an acoustical point of view, the information amount increases dramatically from level to level, ranging from the so-called *acoustical quanta* to segments of sounds that could be even recognized as *sections* of a musical composition.

2.5 Properties of Sound Types

The proposed model fulfills all the requirements of section 1:

- **signal-dependent semantics:** the atoms of the underlying signal-level representation and the + relations discovered by Markov models are derived from the signal;
- **scalability:** the possibility to scale over abstraction is implicit to theories of types; in the previous section we showed how it is possible to translate a representation to another;
- **weak invertibility** and **generativity:** there are many possibilities to create a signal back from sound types; simple ways could be to pick up randomly an element of each cluster or to generate a weighted sum of all the elements of a cluster.

3 Conclusions and Perspectives

The **theory of sound types** is still in an early stage and much work must be done: expansions and improvements of the theory should deal with both the logical level and the signal-processing level. Sound types seem to be promising entities to represent music because they are physically related to sound, are invertible and are also capable to represent formal relationships and hierarchies. In conclusion, we think that the strength of the proposed approach for music description stands in the *connection* between the underlying symbolic system (based on simple type theory) and the provided model (based on low-level descriptors plus statistical learning); a twofold process, divided into a *types inference* stage and a *rules inference* stage, represents the conceptual link between the two worlds. Since theories of types are able to scale over abstraction, the resulting method is a *scalable* representation, ranging from sound to music in the broadest sense.

References

1. Cella, C.E.: Sulla struttura logica della musica. Rivista umbra di musicologia 48, 1–82 (2004)
2. Ellis, D., Rosenthal, D.: Mid-level representations for Computational Auditory Scene Analysis. In: International joint conference on Artificial Intelligence (1995)
3. Farmer, W.M.: The seven virtues of simple type theory. Journal of Applied Logic 72 (2007)
4. Goodwin, M., Vetterli, M.: Atomic decompositions of audio signal. In: IEEE Audio Signal Processing Workshop (1997)
5. Leman, M.: Expressing coherence of musical perception in formal logic. Mathematics and music: a Diderot Mathematical Forum, pp. 184–198 (2002)
6. Marsden, A.: Timing in music and modal temporal logic. Journal of Mathematics and Music 1(3), 173–189 (2007)
7. Peeters, G.: A large set of audio features for sound description (similarity and classification) in the CUIDADO project. CUIDADO I.S.T. Project Report, 1–25 (2004)

Plain and Twisted Adjoints of Well-Formed Words

David Clampitt, Manuel Domínguez, and Thomas Noll

Abstract. This paper studies the mathematical basis for a new study of modes of well-formed (WF) scales, and presents a new characterization of special standard Sturmian morphisms.

We introduce WF words, which coincide with the step-interval patterns of modes of well-formed scales. WF words can be represented as conjugates of some Christoffel word (generalized Lydian mode).

To every WF word we may assign a pair of affine automorphisms f_w and g_w. These assignments induce a pair of involutions over the set of WF words: the *plain adjoint* and the *twisted adjoint*. We study the properties of these adjoints; in particular we show how the plain adjoint coincides with duality over the set of Christoffel words and also that the twisted adjoint extends Sturmian involution to the set of WF words. Thomas Noll's *divider incidence* result holds, *inter alia*, that w is special standard if and only if $f_w(1) = 1$.

1 Geometrical Motivations

In [7], two topics were connected: step-interval patterns of WF scales (see [4] and [5]) and Christoffel words (see [9], [1] and [10]). This connection initiated the possibility of further integration of the algebraic combinatorial theory of words into mathematical scale theory, and possibly reciprocally. Further music-theoretical interpretation of the results herein are explored in [6].

We consider the monoid $\{x, y\}^*$ of words on a two-letter alphabet $A = \{x, y\}$, $\{x, y\}^* = \{w | w = w_1 \dots w_N;\ w_i \in A\}$. The empty word, denoted ϵ, belongs to $\{x, y\}^*$ and the monoid operation is concatenation of words. If F is an endomorphism of $\{x, y\}^*$, $F(w) = F(w_1 \dots w_N) = F(w_1) \dots F(w_N)$, so F is entirely determined by the images of x and y. The monoid St of *Sturmian morphisms* on $\{x, y\}^*$ can be generated by the following morphisms:

$$G: \begin{array}{l} x \to x \\ y \to xy \end{array} \quad \tilde{D}: \begin{array}{l} x \to xy \\ y \to y \end{array} \quad \tilde{G}: \begin{array}{l} x \to x \\ y \to yx \end{array} \quad D: \begin{array}{l} x \to yx \\ y \to y \end{array} \quad E: \begin{array}{l} x \to y \\ y \to x. \end{array}$$

We define the monoid of *special Sturmian morphisms*, denoted by St_0, as the monoid $St_0 = \langle G, D, \tilde{G}, \tilde{D} \rangle$. We use the word theory notations $|w|$ to refer to the length of the word w, i.e., the total number of letters it contains, while $|w|_x$ means the number of appearances of the letter x in the word w, and similarly for $|w|_y$. We make use also of the rotation operator γ which transforms $w = w_1 \dots w_N$ into $\gamma(w_1 \dots w_{N-1} w_N) = w_N w_1 \dots w_{N-1}$.

E. Chew, A. Childs, and C.-H. Chuan (Eds.): MCM 2009, CCIS 38, pp. 65–80, 2009.

Given an irrational number $0 < \theta < 1$, we know that the scale of N notes generated by θ is well-formed if and only if N is the denominator of a (semi-) convergent of θ. The succession of (semi-)convergents of θ produces an ordered infinite family of well-formed scales called a *hierarchy of WF scales generated by* θ. If the two distinct step intervals are of sizes a and b with $a < b$, then the larger step interval splits into two intervals of sizes a and $b - a$ in the next level of the hierarchy. We can start from the two-note scale, $\{0, \theta\}$, which has step-interval pattern xy, then, every subsequent step-interval pattern of the hierarchy can be obtained as the image of xy by an element of the monoid $\langle G, \tilde{D} \rangle$. These morphisms, G and $\tilde{D}$ connect two consecutive cases in the hierarchy, (see Figure 1, (a) and (b)).

Let us consider **generalized generated sets** of the type

$$\Sigma_k = \{\{-k \cdot \theta\}, \ldots, \{(N-1-k) \cdot \theta\}\}, \quad \text{with } k \in \mathbb{Z}_N,$$

where $\{x\}$ means decimal part of x. These sets are rotations of the scale of N notes generated by θ. Recall that the generator θ coincides with the arc length swept clockwise from every note on the circle σ_i to the next one in generation order σ_{i+1}. Thus, modes of a scale can be represented by generalized generated sets Σ_k. It follows from the circular representation of the scale that morphisms D and $\widetilde{G}$ connect consecutive cases of WF modes when notes are added to the scale *in the negative direction* (see Figure 1, (c) and (d)).

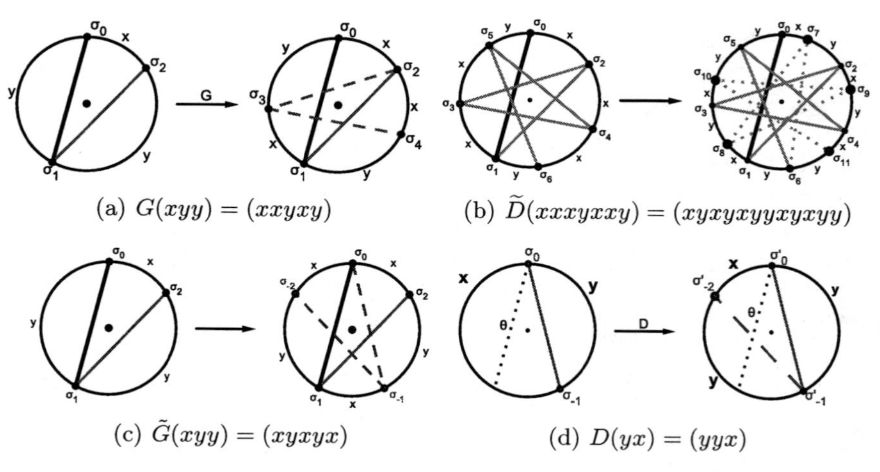

(a) $G(xyy) = (xxyxy)$ (b) $\widetilde{D}(xxxyxxy) = (xyxyxyyxyxyy)$

(c) $\tilde{G}(xyy) = (xyxyx)$ (d) $D(yx) = (yyx)$

Fig. 1. Geometrical interpretation of the monoid $\Big\langle G, D, \tilde{G}, \tilde{D}$

The proposition below relates rotations of the step pattern of a WF scale with step patterns of generalized generated sets.

Proposition 1. *Given a WF scale Σ with step pattern w, $\Sigma_{|w|_x \cdot k}$ has as step pattern $\gamma^k w$.*

Proof. One has just to recall that the homothecy $h_{|w|_y}(k) = |w|_y \cdot k \mod N$ transforms scale order into generation order, and thus $h_{|w|_x}$ transforms scale order into generation order in *negative direction.* □

We say that two words u, v are conjugated if $u = xy$ and $v = yx$ for some words x, y. Notice that conjugation can be thought of as an equivalence relation via circle rotations, if we write the words around a circle. Therefore the last proposition asserts that generalized generated scales have conjugated step-interval patterns. This geometric interpretation suggests that the modes of WF scales may be presented as words that encode rotations of generalized WF sets. The next section will define such words as WF words, which will be shown to be conjugates of Christoffel words.

2 Well-Formed Words

If we take as point of departure the geometrical interpretations of generalized generated scales and their step-interval patterns, we can define in a purely word-theoretical context a **well-formed (WF) word**.

Definition 1. *Let $w \in \{x, y\}^*$ be a word formed by letters x and y.*

1. *The **balance map** of w is the map $\beta_w : \{1, \ldots, |w|\} \longrightarrow \mathbb{Z}$ with*

$$\beta_w(k) = \begin{cases} |w|_y \text{ for } w_k = x \\ -|w|_x \text{ for } w_k = y \end{cases}.$$

2. *The **accumulation map** of w is the map $\alpha_w : \mathbb{Z}_{|w|} \longrightarrow \mathbb{Z}$ with*

$$\alpha_w(0) := 0 \; \alpha_w(k) := \sum_{r=1}^{k} \beta_w(r).$$

Let $\{n \cdot \theta\}_{n=0,\ldots,N-1}$ be a WF scale with step pattern $w \in \{x, y\}^*$. The size of the jumps between consecutive notes in generation order is $\{\theta\}$ and $\{1 - \theta\}$. On the other hand, these jumps will be $|w|_y$ steps clockwise or $N - |w|_y = |w|_x$ counterclockwise, since $|w|_y$ is the diatonic length of the generator (see [5], [3]). Thus, the balance map can be seen as the number of step intervals of the generator in each appearance, with positive sign for clockwise and negative for counterclockwise, whereas the accumulation map transforms scale order into (generalized) generation order.

Definition 2. *A word w is called **well-formed** if there exists an integer $\mu_w \in \{0, \ldots, |w| - 1\}$ such that $\{\alpha_w(0) + \mu_w, \ldots, \alpha_w(|w| - 1) + \mu_w\} = \{0, \ldots, |w| - 1\}$. μ_w is called the mode of w.*

The notion of a well-formed word was introduced by Thomas Noll in [11] and it is a generalization of a Christoffel word, as the following proposition states.

Proposition 2. *A word w of length $|w| = N$ is well-formed with mode 0 if and only if it is a Christoffel word.*

Proof. The accumulation map of a given word w can be represented geometrically as a path that starts from $(0,0)$, ends in $(-p \cdot q, -p \cdot q)$ and consists of $|w|_x = p$ horizontal segments of length $|w|_y = q$ and q vertical segments of length $-p$ (see Figure 2 (b)). If we apply a pair of dilatations, namely $x' = \frac{1}{q}x$ and $y' = \frac{-1}{p}x$ we will have transformed the previous path into another one made of vertical and horizontal segments of length 1 (see Figure 2 (a)). This last path represents a mechanical word of slope $\frac{p}{q}$ if and only if the $N-1$ points $(k, \alpha_w(k))$ for $k = 1, \ldots, N-1$ in Figure 2 (b) are above the segment that joins $(0,0)$ with $(-p \cdot q, -p \cdot q) \Leftrightarrow$ the values of the accumulation map are non-negative. □

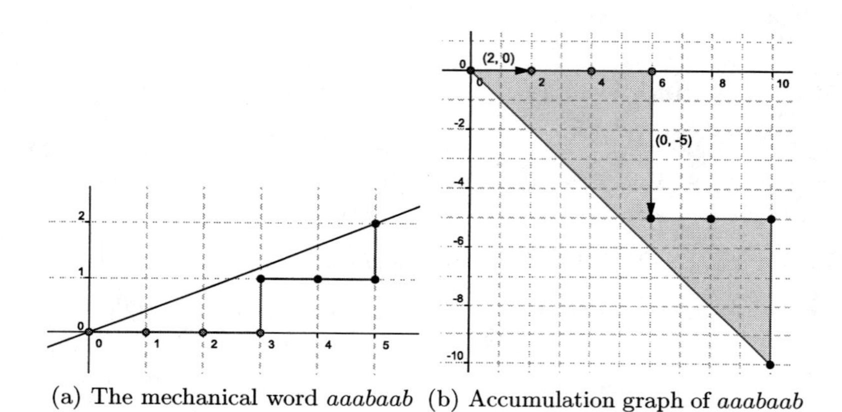

(a) The mechanical word *aaabaab* (b) Accumulation graph of *aaabaab*

Fig. 2. The accumulation map in the Christoffel case

The next step is the characterization of the set of well-formed words.

Lemma 1. *If w is well-formed, $\gamma^k w$ is also well-formed, furthermore:*

$$\mu_{\gamma^k w} = \mu_w + k \cdot |w|_y \mod |w|.$$

Proof. We just show that γw is well-formed whenever w is so. By the definition of the balance map one has:

$$\begin{aligned} &\beta_{\gamma w}(k) = \beta_w(k+1) \ \forall k = 1, \ldots, |w| - 1 \\ &\beta_{\gamma w}(|w|) = \beta_w(1). \end{aligned}$$

Therefore, we have that

$$\alpha_{\gamma w}(k) = \sum_{r=1}^{k} \beta_{\gamma w}(r) = \sum_{r=1}^{k} \beta_w(r+1) = \sum_{r=2}^{k+1} \beta_w(r) = \alpha_w(k+1) - \beta_w(1),$$

and thus $\mu_{\gamma w} = -\inf \alpha_{\gamma w}(k) = -(\inf \alpha_w(k+1) - \beta_w(1)) = \mu_w + \beta_w(1) \equiv \mu_w + |w|_y$. □

This last result, together with proposition 2 and the fact that a morphism of words F is Sturmian if and only if the word $F(xy)$ is conjugated with some Christoffel word, yields the central result of this section:

Theorem 1 (Characterization of well-formed words). *A word is well-formed $\Longleftrightarrow$ it is conjugated with some Christoffel word. The set of all well-formed words coincides with the set of all words of the type $w = F(xy)$ where $F \in St$.*

We conclude the section with a description of special standard words in terms of well-formed words of mode $|w|_y - 1$.

Lemma 2. *The special standard word $w = uxy$ and the Christoffel word xuy are conjugated and one has*

$$\gamma^{|w|_y^{-1}-1}w = xuy.$$

Proof. A word w is special standard $\Leftrightarrow w = w_1 \cdot w_2$ with (w_1, w_2) a standard pair. Following [10, Lemma 2.2.8] we have

$$\text{either} \begin{cases} w_1 = pyx = qr \\ w_2 = qxy \end{cases} \text{or} \begin{cases} w_1 = qyx \\ w_2 = pxy = qr \end{cases}.$$

where p, q and r are in PAL (the set of palindromes). In the first case we have that

$$\gamma^{|w_1|-1}w = x \cdot qxyp \cdot y = x \cdot qrq \cdot y.$$

Therefore $\gamma^{|w_1|-1}w = xuy$ with $u \in PAL \cap PALxyPAL$ that coincides, by [10, Corollary 2.2.9], with the set of central words. xuy is thus a Christoffel word. The second case is completely analogous. Notice finally that $|w_1| = |w_1 \cdot w_2|_y^{-1}$ mod $|w|$ and $|w_2| = |w_1 \cdot w_2|_x^{-1}$ mod $|w|$. □

Corollary 1. *A well-formed word w is a special standard word $\Longleftrightarrow \mu_w = |w|_y - 1$*

Proof. Proof. We have just to compute the mode of the special standard word w depending on the mode of the Christoffel word $\gamma^{-(|w|_y^{-1}-1)}w$, which is zero:

$$\mu_w = 0 - (|w|_y^{-1} - 1) \cdot |w|_y = |w|_y - 1.$$

□

Let us finally relate the step-interval pattern of generalized generated sets Σ_k to WF words. With that purpose we denote by $\sigma_i = \{i \cdot \theta\}$ the i-th note in generation order (where i may be negative), and by $\Sigma_0 = \{0 = \rho_0 < \rho_1 < \ldots < \rho_{N-1}\}$ the scale order of the associated WF scale.

Proposition 3. *If $\Sigma = \Sigma_0$ is a WF scale with N notes, the scale pattern of Σ_k is a WF word w of mode k for every $k \in \mathbb{Z}_N$ and one has:*

$$\rho_{(i+k|w|_x^{-1}) mod N} = \sigma_{(|w|_y \cdot i) mod N-k}, \quad \forall i \in \mathbb{Z}_N.$$

3 Plain and Twisted Adjoints

Given a WF word w of mode μ, we introduce two affine automorphisms of $\mathbb{Z}_N$, f_w (plain affinity) and g_w (twisted affinity), associated with the word w and defined by the formulas below.

$$f_w(k) = |w|_y \cdot k - \mu \bmod N \qquad g_w(k) = |w|_x \cdot k + N - 1 - \mu \bmod N.$$

If w is the step pattern of a WF mode Σ_μ, $-\mu$ and $N-1-\mu$ are the minimum and the maximum, respectively, of the accumulation map α_w and therefore the morphism f_w transforms scale order (starting from the first note in generation order $\sigma_{-k} = \{-k \cdot \theta\}$) into generation order. On the other hand, the morphism g_w starts from the last generated note σ_{N-1-k} and it covers the scale in a negative direction (see Figure 3).

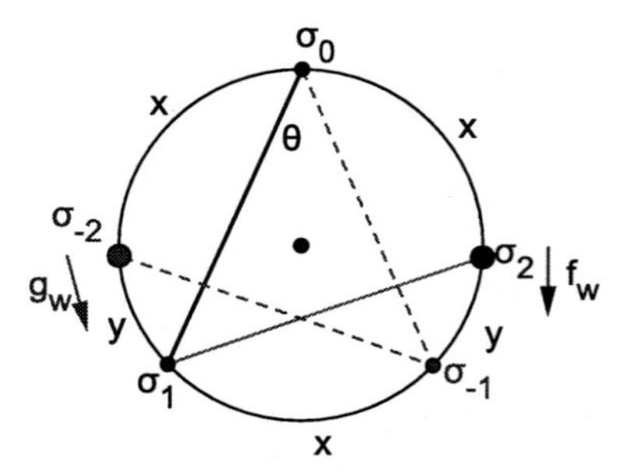

Fig. 3. Generated scale of pattern $(xyx, yx) = \widetilde{D}\widetilde{G}(x, y)$

Definition 3. *Given a WF word w, we call the* **plain adjoint** *of w, denoted by $w^\square$, the unique word whose associated affinity coincides with the inverse affinity of w. In other words, the plain adjoint $w^\square$ is defined by the equation:*

$$f_{w^\square} = (f_w)^{-1}$$

Example 1. We show in the following table the conjugation class of the diatonic step pattern $xxxyxxy$ in relation with the conjugation class of the dual pattern (plain adjoint).

w	$xxxyxxy$	$xxyxxyx$	$xyxxyxx$	$yxxyxxx$	$xxyxxxy$	$xyxxxyx$	$yxxxyxx$
$f_w(k)$	$2k$	$2k-2$	$2k-4$	$2k-6$	$2k-1$	$2k-3$	$2k-5$
$f_{w^\square}(k)$	$4k$	$4k-6$	$4k-5$	$4k-4$	$4k-3$	$4k-2$	$4k-1$
$w^\square$	$xyxyxyy$	$yyxyxyx$	$yxyyxyx$	$yxyxyyx$	$yxyxyxy$	$xyyxyxy$	$xyxyyxy$

Observe that the inverse of an affinity $g(k) = a \cdot k + b$ is the affinity $g^{-1}(k) = a^{-1} \mod n \cdot k + (-a^{-1} \cdot b) \mod n$. Thereby one has that

$$\beta_{w^\square}(k) = \alpha_{w^\square}(k) - \alpha_{w^\square}(k-1) = |w|_y^{-1} \mod |w|.$$

Therefore the balance of the word $w^\square$ is equal to

$$\beta_{w^\square}(k) = \begin{cases} |w|_y^{-1} & \text{for } w_k^\square = x \\ -|w|_x^{-1} & \text{for } w_k^\square = y. \end{cases}$$

If w is a Christoffel word, its plain adjoint $w^\square$ is also a Christoffel word because its mode is 0. In fact, it coincides with the Christoffel word of slope $\frac{|w|_y^{-1}}{|w|}$. Recall that this last word coincides by definition (see [2]) with the dual word of w. Thus, we have shown:

Proposition 4. *For Christoffel words, the plain adjoint $w^\square$ coincides with the dual word w^*.*

The canonical injection $St_0 \hookrightarrow WF$ associates to every Sturmian morphism F the WF word $F(xy)$. If we focus on the monoid of Christoffel morphisms $\langle G, \widetilde{D} \rangle$ we have the following result:

Proposition 5. *The diagram:*

$$\begin{array}{ccc} \langle G, \widetilde{D} \rangle & \hookrightarrow & WF \\ {\scriptstyle rev}\downarrow & \circlearrowleft & \downarrow{\scriptstyle \square} \\ \langle G, \widetilde{D} \rangle & \hookrightarrow & WF \end{array}$$

is commutative or, in other words:

$$f(xy)^\square = f^{rev}(xy)$$

where F^{rev} denotes the retrogradation of F as a word in $\langle G, \widetilde{D} \rangle$.

Proof. Observe that $w^\square$ is the dual of w and the Christoffel morphisms related to dual words are retrograde (see [7, Proposition 10]). □

One has also the same result for the standard monoid:

Proposition 6. *The following diagram is commutative (where F^{rev} denotes the retrogradation of F as a word in $\langle G, D \rangle$):*

$$\begin{array}{ccc} \langle G, D \rangle & \hookrightarrow & WF \ , \\ {\scriptstyle rev}\downarrow & \circlearrowleft & \downarrow{\scriptstyle \square} \\ \langle G, D \rangle & \hookrightarrow & WF \end{array}$$

Proof. The square in the proposition may be decomposed in the following way:

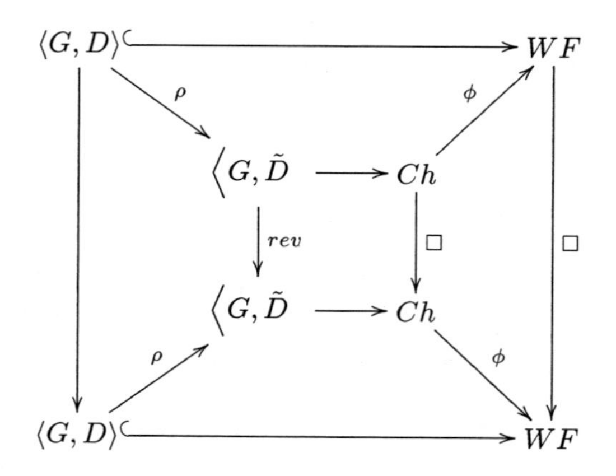

where $\rho : \begin{array}{c} G \to G \\ D \to \tilde{D} \end{array}$ and $\phi(xuy) = uxy$. One has just to show that every smaller square commutes. The left square trivially commutes and the square in the center is commutative by proposition 5. If the characteristic morphism of a Christoffel word xuy is F, then the characteristic morphism of uxy is $\rho(F)$. Thus upper and lower squares also commute. The commutativity of the square on the right, finally, is equivalent to the formula

$$uxy^{\Box} = u'xy,$$

where u' is the central word of the dual of xuy. $\Box$

The problem is that this nice formula $F(xy)^{\Box} = F^{rev}(xy)$ does not extend to the whole special Sturmian monoid St_0. That is, the following diagram *is not* commutative:

$$\begin{array}{ccc} St_0 & \hookrightarrow & WF \\ {\scriptstyle rev}\downarrow & \circlearrowleft & \downarrow{\scriptstyle \Box} \\ St_0 & \hookrightarrow & WF \end{array}$$

Notice that $\tilde{G}(xy)^{\Box} = (y, xx)$ is not a morphic WF word. That is, a so-called *bad conjugate*, not the image of a morphism.

We can solve this problem by giving a definition in a parallel way to the plain adjoint, namely the *twisted adjoint*:

Definition 4. *Given a well-formed word w, we define the twisted adjoint of the word w by the same formula as the plain adjoint, but using the morphism g_w instead:*

$$g_{w^{\boxtimes}} = g_w^{-1}.$$

In other words, twisted maps associated to words which are twisted adjoints of each other are each other's inverses.

Example 2. The following table shows the twisted adjoints of all words conjugated with the diatonic step pattern $xxxyxxy$.

w	$xxxyxxy$	$xxyxxyx$	$xyxxyxx$	$yxxyxxx$	$xxyxxxy$	$xyxxxyx$	$yxxxyxx$
$g_w(k)$	$5k+6$	$5k+4$	$5k+2$	$5k$	$5k+5$	$5k+3$	$5k+1$
$f_{w^\boxtimes}(k)$	$3k+3$	$3k+2$	$3k+1$	$3k$	$3k+6$	$3k+5$	$3k+4$
$w^\boxtimes$	$yxyxyxy$	$yxyxyyx$	$yxyyxyx$	$yyxyxyx$	$xyxyxyy$	$xyxyyxy$	$xyyxyxy$

We define the Sturmian involution as the anti-automorphism $*$ over the monoid of special Sturmian morphisms $St_0 \xrightarrow{*} St_0$, which fixes G and $\tilde{G}$ and exchanges D and $\tilde{D}$. Sturmian involution extends Christoffel duality to the monoid of Sturmian morphisms (see [2] Proposition 4.1). In a parallel way, the twisted adjoint extends the involution of special Sturmian morphisms to the set of WF words:

Proposition 7. *The following diagram is commutative:*

$$\begin{array}{ccc} St_0 & \hookrightarrow & WF \\ {\scriptstyle *}\downarrow & \circlearrowleft & \downarrow{\scriptstyle \boxtimes} \\ St_0 & \hookrightarrow & WF \end{array}$$

To prove this result we need to recall the description of the set of special Sturmian morphisms F such that $F(xy)$ is conjugated with a given Christoffel word w. There are exactly $N-1$ such morphisms, and they can be ordered as $F_1, F_2, \dots, F_{N-1}$ with F_1 a special standard morphism (generated by G and D) and $\gamma F_i(xy) = F_{i+1}(xy)$ for all $i = 1, \dots, N-2$ (see [2]).

Lemma 3. *Given a WF word w and let $F_1, F_2, \dots, F_{N-1}$ be the set of special Sturmian morphisms related to w. Then, the twisted affinity associated with the special Sturmian morphism F_i is $g_{F_i}(k) = |w|_x \cdot k + |w|_x \cdot i$.*

Proof. From [2], Lemma 4.2 we have that $F_{|w|_y^{-1}}(xy) = u$ is a Christoffel word, and thus its twisted affinity is $g_u(k) = |w|_x \cdot k + N - 1$. By definition of twisted affinity one has that $g_{\gamma^j w} \equiv g_w + j \cdot |w|_x \mod N$ and thus, we have that $F_1 = \gamma^{(|w|_x^{-1}+1)} F_{|w|_y^{-1}}$ has as twisted affinity

$$g_{F_1(xy)}(k) = |w|_x \cdot k + N - 1 + (|w|_x^{-1}+1) \cdot |w|_x = |w|_x \cdot k + N + |w|_x \equiv |w|_x \cdot k + |w|_x$$

We close the argument by recursion. □

Proof. (Proposition 7) Notice that $g_{w^\boxtimes}$ is a homothecy $\Leftrightarrow \nu_w = 0 \Leftrightarrow \alpha_w(k) \leq 0 \ \forall k = 0, \dots N-1 \Leftrightarrow w$ is an amorphic word ($w = yux$ with u a central word), that is, w is a negative Christoffel word. Then the twisted adjoint sends amorphic words to amorphic words and therefore it sends morphic words to morphic words.

Recall a last result from [2]: if $F_1, F_2, \dots, F_{N-1}$ (resp. $F_1', F_2', \dots, F_{N-1}'$) is the succession of special Sturmian morphisms related to a Christoffel word w (resp.

to $w^{\boxtimes}$) then one has $F_i^* = F'_{i|w|_y}$. It is enough to show that the twisted affinities associated to F_i and $F'_{i\cdot|w|_y}$ are inverses one of each other. By the previous lemma one has:

$$g_{F_i^*} = g_{F'_{i\cdot|w|_y}} = |w|_x^{-1} \cdot k + |w|_x^{-1} \cdot |w|_y \cdot i = |w|_x^{-1} \cdot k + |w|_x^{-1} \cdot (N - |w|_x) \cdot i \equiv$$
$$\equiv |w|_x^{-1} \cdot k - i = (|w|_x \cdot k + |w|_x \cdot i)^{-1} = g_{F_i}^{-1},$$

which completes the proof. □

We conclude the section with a result that analyzes the concatenation of plain and twisted adjoints. Let WF_N denote the set of WF words w of length N in $\{x, y\}^*$ and let $Aff^*(\mathbb{Z}_N)$ denote the group of affine automorphisms of the cyclic group $\mathbb{Z}_N$ of order N. Let furthermore $j_{\square}, j_{\boxtimes} : \mathbb{Z}_N \to \mathbb{Z}_N$ denote the affine automorphisms given by the formulas $j_{\square}(k) = -k+1 \mod N$ and $j_{\boxtimes}(k) = -k - 1 \mod N$, respectively.

Proposition 8. *The concatenation $\square \circ \boxtimes$ is an involution over the set WF_N. Furthermore, we have:*

$$\square \circ \boxtimes = \boxtimes \circ \square : WF_N \to WF_N,$$

and the application induced over $Aff * (\mathbb{Z}_N)$ is an inner automorphism:

$$f_{w^{\boxtimes\square}} = j_{\square} \circ f_w \circ j_{\square}^{-1} \quad \textit{and} \quad g_{w^{\square\boxtimes}} = j_{\boxtimes} \circ g_w \circ j_{\boxtimes}^{-1}.$$

Proof. Notice that $\Delta(f_{w^{\boxtimes\square}}) = f_{w^{\boxtimes\square}}(k+1) - f_{w^{\boxtimes\square}}(k) = \Delta(j_{\boxtimes} \circ g_w \circ j_{\boxtimes}^{-1}) = |w|_y$ and therefore one has just to check that $j_{\boxtimes} \circ g_w \circ j_{\boxtimes}^{-1}(0) = f_{w^{\boxtimes\square}}(0)$ to conclude that the affinities $f_{w^{\boxtimes\square}}$ and $j_{\boxtimes} \circ g_w \circ j_{\boxtimes}^{-1}$ coincide. An analogous argument yields that $g_{w^{\square\boxtimes}} = j_{\boxtimes} \circ g_w \circ j_{\boxtimes}^{-1}$. □

4 Divider Incidence

In this section we extend Noll's result for plain adjointness (see [12]), Divider Incidence, which holds, *inter alia*, that a word w is positive standard Sturmian if and only if $f_w(1) = 1$. (See Figure 4). Before we can explain, why $f_w(1) = 1$ (and likewise in the twisted case $g_w(1) = 1$) are expressions of divider incidence, we shall first understand the difference between plain and twisted adjoint in terms of height and width trajectories. First recall the plain case (see left graph in Figure 4 and [12] for details), where the height trajectory of a word w is built as a point sequence $\Phi : \{0, \dots, N\} \to \mathbb{Z}^2$, whose difference vectors $\Phi(k) - \Phi(k-1)$ are either $v_a = (|w|_b, 1)$ or $v_b = (-|w|_a, 1)$. The sequence of indices (i.e., of a's and b's), which is associated with the sequence of vectors $(\Phi(1) - \Phi(0)), (\Phi(2) - \Phi(1)), ..., (\Phi(N) - \Phi(N-1))$, coincides with the order of the letters a and b in the word w. The width trajectory of the plain adjoint word $w^{\square}$ is the point sequence $\Psi : \{0, \dots, N\} \to \mathbb{Z}^2$, whose difference vectors $\Psi(k) - \Psi(k-1)$ are either $v_x = (1, |w^{\square}|_y)$ or $v_y = (1, -|w^{\square}|_x)$. Here again the sequence of indices

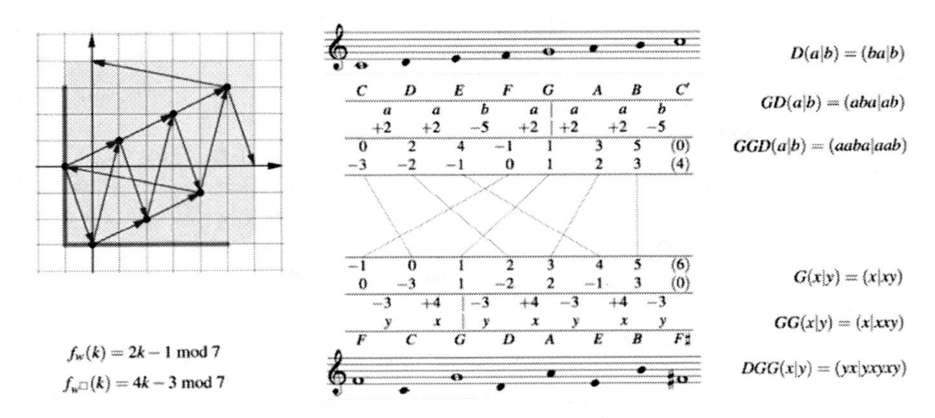

Fig. 4. Divider Incidence for plain adjoints

x or y, which is associated with the sequence of vectors $(\Psi(1) - \Psi(0)), (\Psi(2) - \Psi(1)), ..., (\Psi(N) - \Psi(N-1))$, coincides with the order of the letters x and y in the word $w^{\square}$. Thereby we have an equality of point sets: $\{\Phi(0), \ldots, \Phi(N-1)\} = \{\Psi(0), \ldots, \Psi(N-1)\}$. What characterizes the construction of the plain adjoint is the fact that in the primary vectors $v_a = (1, |w|_b)$ as well as in $v_x = (|w^{\square}|_y, 1)$ both coordinates have the same sign. In the twisted case the signs differ from one another. In the present paper we choose the primary vectors $v_a = (-1, |w|_b)$ and $v_x = (|w^{\boxtimes}|_y, -1)$, i.e., we switch the direction of the trajectories $\Phi', \Psi' : \{0, \ldots, N\} \to \mathbb{Z}^2$, while keeping the positive coordinates $|w|_b$ and $|w^{\boxtimes}|_y$ (see also Figure 5). Thereby we have again the defining equality of point sets: $\{\Phi'(0), \ldots, \Phi'(N-1)\} = \{\Psi'(0), \ldots, \Psi'(N-1)\}$.

In terms of the height and width trajectories both equations $f_w(1) = 1$ and $g_w(1) = 1$ mean that the point $(1, 1)$ represents the divider in both trajectories. In the plain case we have $\Phi(|w^{\square}|_y) = \Psi(|w|_b) = (1, 1)$ and in the twisted

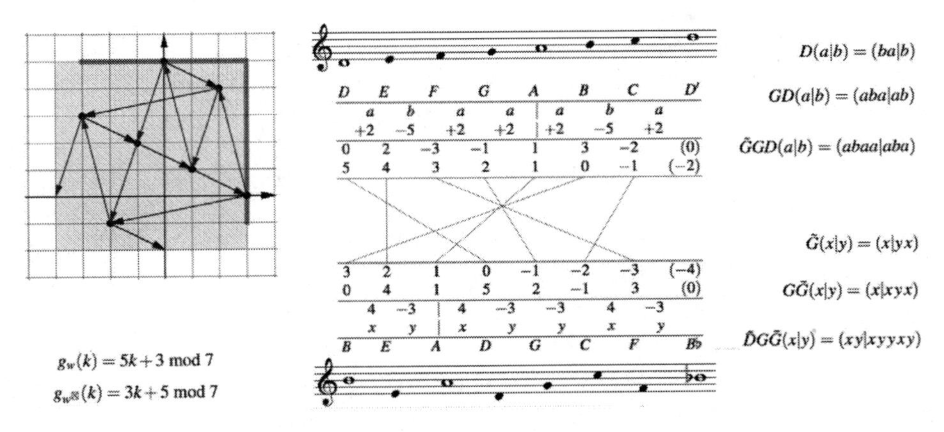

Fig. 5. Divider Incidence for twisted adjoints

case it is $\Phi'(|w^{\boxtimes}|_y) = \Psi'(|w|_b) = (1,1)$. Why is it the point $(1,1)$? For any special Sturmian morphism F the accumulation at the divider of $w = F(x)|F(y)$ is $\alpha_w(|F(x)|) = 1$. This is a consequence of the known fact that the incidence matrix $M_F = \begin{pmatrix} |F(x)|_x & |F(y)|_x \\ |F(y)|_x & |F(y)|_y \end{pmatrix}$ is an element of $SL_2(\mathbb{Z})$ (see [10]): $\alpha_w(|F(x)|) = |w|_y|F(x)|_x - |w|_x|F(x)|_y = (|F(x)|_y + |F(y)|_y)|F(x)|_x - (|F(x))|_x + |F(y)|_x)|F(x)|_y) = |F(y)|_y|F(x)|_x - |F(y)|_)|F(x)|_y = Det(M_F) = 1$. Thus the main point in the equations $f_w(1) = 1$ and $g_w(1) = 1$ is the assertion that it is precisely the *argument* 1, where the affine morphisms f_w and g_w take this accumulation value 1.

Theorem 2 (Divider Incidence). *Given a WF word w of length N, which is the step-interval pattern of a generated mode $\Sigma_k = \{\sigma_{-k}, \ldots, \sigma_{N-1-k}\}$, then:*

1. *The following assertions are equivalent (D.I. for plain adjoint):*
   (a) *$w = F(xy)$ with $F \in \langle G, D \rangle$ a special standard morphism.*
   (b) *$f_w(1) = 1$ where f_w is the plain affinity associated to w.*
   (c) *The notes σ_{-k} and σ_1 are consecutive in scale order, i.e., the origin of the scale folding is the divider predecessor in scale order.*
2. *The following assertions are also equivalent (D.I. for twisted adjoint):*
   (a) *$w = F(xy)$ with $F \in \langle \widetilde{G}, \widetilde{D} \cdot G \cdot \langle \widetilde{G}, D$.*
   (b) *$g_w(1) = (1)$ with g_w the twisted affinity associated with w.*
   (c) *The notes σ_1 and σ_{N-1-k} are consecutive in scale order, i.e., the origin of the scale folding is the divider successor in scale order.*

Proof. We shall first show that the conditions (b) and (c) are equivalent in both cases (1) and (2). Remember that in the plain case the value $f_w(0)$ is the reduction modulo N of the minimal accumulation. Condition (1b) is therefore equivalent to the condition (1c), namely that the index with minimal accumulation is the divider predecessor (simply because 0 proceeds 1 in the ascending order of the arguments). In the twisted case, the value $g_w(0)$ is the reduction modulo N of the maximal accumulation, and condition (2b) is therefore equivalent to the condition (2c), namely that the index with maximal accumulation is the divider successor (because 0 follows 1 in the descending order of the arguments). The main part of the proof is therefore to show the equivalence of either (b) or (c) with condition (a).

For the plain case this is done in [12] by means of structural induction down the free monoid $\langle G, D \rangle$ of special standard morphisms. The same technique applies to the twisted case. But some comments should be made in the beginning, as the set $\mathcal{T} = \langle \widetilde{G}, \widetilde{D} \cdot G \cdot \langle \widetilde{G}, D$ is redundantly presented here. To begin with, the redundant presentation perfectly shows that $\mathcal{T}$ is invariant under Sturmian involution: If we have a special Sturmian morphism $F = F_1 \cdot G \cdot F_2$ with anti-standard morphism $F_1 \in \langle \widetilde{G}, \widetilde{D}$ and anti-Christoffel morphism $F_2 \in \langle \widetilde{G}, D$ we obtain $F^* = F_2^* \cdot G \cdot F_1^*$, which is also in $\mathcal{T}$ because F_2^* is anti-standard and F_1^* is anti-Christoffel. In other words, the formulation of condition (2a) is consistent with Proposition 7. To remove the redundancy in the presentation of $\mathcal{T}$, recall that the special Sturmian monoid $St_0 = \langle G, D, \tilde{G}, \tilde{D}$ is known to satisfy a number of relations. Among these we find:

$$\tilde{G}\tilde{D}^k G = GD^k\tilde{G}, \text{ for } k = 0, 1, 2, \ldots.$$

This leads to the (non-redundant) presentation $\mathcal{T} = \langle \tilde{D} \; \cdot G \cdot \langle \tilde{G}, D$.

We are interested in the orbit $\mathcal{T}(x|y)$ of the pair (or divided word) $x|y$. The same set of pairs is also the orbit of $x|y$ under the following monoid of transformations (see [10] or [12]): $\mathfrak{T} = \langle \tilde{\Gamma}, \Delta \; \cdot \Gamma \cdot \langle \tilde{\Delta}$, where for any ordered pair $w_1|w_2$ of words

$$\begin{array}{ll} \Gamma(w_1|w_2) = w_1|w_1w_2 & \Delta(w_1|w_2) = w_2w_1|w_2 \\ \tilde{\Gamma}(w_1|w_2) = w_1|w_2w_1 & \tilde{\Delta}(w_1|w_2) = w_1w_2|w_2. \end{array}$$

We need to understand an important aspect of the behavior of the accumulation maps $\alpha_{\Gamma(w_1|w_2)}(t)$, $\alpha_{\Delta(w_1|w_2)}(t)$, $\alpha_{\tilde{\Gamma}(w_1|w_2)}(t)$, and $\alpha_{\tilde{\Delta}(w_1|w_2)}(t)$ in comparison to the accumulation map $\alpha_{w_1|w_2}(t)$. For each factor w_1 or w_2 within the four words $w_1w_1w_2$, $w_2w_1w_2$, $w_1w_2w_1$, $w_1w_2w_2$ we may consider maps such as $\omega : \alpha_{w_1w_1w_2}(\{0, 1, ..., |w_1|\}) \to \alpha_{w_1w_2}(\{0, 1, ..., |w_1|\})$ sending the local accumulation values to those of the original factors in w_1w_2, e.g. $\omega : \alpha_{w_1w_1w_2}(t) \mapsto \alpha_{w_1w_2}(t)$ for $t \in \{0, 1, ..., |w_1|\}$. Each of these maps ω sends a set of integers to a set of integers bijectively. The relevant aspect for our proof is the fact that these maps are completely order-preserving. Although the concrete accumulation values differ from factor to factor they strictly preserve their relative order relations. In particular, if in w_2 the highest accumulation $\alpha_{w_1|w_2}(t)$ occurs after the second letter of w_2, i.e., at $t = |w_1|+1$, this remains locally the case for every occurrence of $w2$ in the four words $w_1w_1w_2$, $w_2w_1w_2$, $w_1w_2w_1$, $w_1w_2w_2$.

For the structural induction we go down the binary $\langle \tilde{\Gamma}, \Delta$ tree, with an initial map $\Gamma \cdot \tilde{\Delta}^k$ prepended (from the right). Therefore, as the initial part for structural induction we inspect the special type of pair

$$w = w_1|w_2 = \Gamma \cdot \tilde{\Delta}^k(x|y) = \Gamma(xy^k|y) = xy^k|xy^ky = \tilde{D}^k(x|xy) = \tilde{D}^k \cdot G(x|y).$$

The balance and accumulation maps are given as

$$\begin{array}{l} \beta_w(1) = \beta_w(k+2) = 2k+1, \\ \beta_w(2) = \beta_w(3) = \cdots = \beta_w(k+1) = \beta_w(k+3) = \cdots = \beta_w(2k+3) = 2, \\ \alpha_w(0) = 0, \alpha_w(1) = 2k+1, \alpha_w(2) = 2k-2+1, \ldots, \alpha_w(k+1) = 1, \\ \alpha_w(k+2) = 2k+2, \alpha_w(k+3) = 2k, \alpha_w(k+4) = 2k-2, \ldots, \alpha_w(2k+3) = 0. \end{array}$$

The associated affine morphism satisfies (with $\mu = 0$ and $N = 2k+3$)

$$g_w(t) = |w|_x \cdot k + N - 1 - \mu = 2t + (2k+3) - 1 - \mu = 2t - 1 \mod (2k+3).$$

Thus, $g_w(1) = 1$ and the maximal value $g_w(0) = 2k+2$ is the successor of value 1 in scale order. Now suppose, we have a pair $w = w_1|w_2$, for which the maximum value of the accumulation is reached after the first letter of w_2. Furthermore — for technical reasons — we suppose that either w_1 is a prefix of w_2 (which is actually the case in our initial situation, where $w_1|w_2 = xy^k|xy^ky$) or that w_2 is a prefix of w_1 in such a way that the factorization $w_1 = w_2^n w_3$ satisfies $w_2 = w_3 w_4$ for some non-empty word w_4 and some $n \geq 1$. We need to study two cases:

1. $\tilde{\Gamma}(w_1|w_2) = w_1|w_2 w_1$. If w_1 is a prefix of the divider suffix w_2 (in $w_1|w_2$) it is also a prefix of the longer divider suffix $w_2 w_1$ in $\tilde{\Gamma}(w_1|w_2)$. If, however, w_2 is a prefix of the divider prefix $w_1 = w_2^n w_3$ (in $w_1|w_2$) such that there is a factor w_4, satisfying $w_2 = w_3 w_4$ and thus $w_1 = (w_3 w_4)^n w_3$, then w_1 is a prefix of the new divider suffix $w_2 w_1 = w_2(w_2)^n w_3 = (w_3 w_4)^{n+1} w_3 = ((w_3 w_4)^n w_3) w_4 w_3$ in $\tilde{\Gamma}(w_1|w_2)$. This means that the second (technical) part of the induction hypothesis is satisfied for $\tilde{\Gamma}(w_1|w_2)$.
   We already noticed that highest accumulation value $\alpha_{w_1|w_2 w_1}(t)$ within the factor w_2 is still at the index $t = |w_1| + 1$. We must exclude the possibility that a higher value is being reached in one of the two w_1 factors. There are three accumulation values, which we immediately know: $\alpha_{w_1|w_2 w_1}(0) = 0$, $\alpha_{w_1|w_2 w_1}(|w_1|) = 1$ and $\alpha_{w_1|w_2 w_1}(|w_1 w_2 w_1|) = 0$. We conclude that $\alpha_{w_1|w_2 w_1}(|w_1 w_2|) = -1$. From this we may conclude that, if the global maximum is being reached in one of the two w_1 factors, it needs to be the first one. But as w_1 is a prefix of the new divider suffix $w_2 w_1$ its accumulation must behave analogously to this prefix, up to a shift by some number. But we know this shift explicitly. The accumulation at the divider is 1, while the accumulation at the beginning of the word is 0, by definition. So conclude that the global maximum of the accumulation is reached after the first letter after the divider.
2. $\Delta(w_1|w_2) = w_2 w_1|w_2$. If w_1 is a prefix of the divider suffix $w_2 = w_1 w_3$ (in $w_1|w_2$), we obtain $\Delta(w_1|w_2) = w_1 w_3 w_1|w_1 w_3$, which satisfies the second (technical) part of the induction hypothesis in this case. If, however, w_2 is a prefix of the divider prefix $w_1 = w_2^n w_3$ (in $w_1|w_2$) such that there is a factor w_4, satisfying $w_2 = w_3 w_4$ and thus $w_1 = (w_3 w_4)^n w_3$, then, of course, w_2 is again a prefix of the new divider prefix $w_2 w_1$. But it also satisfies the factorization property with the same factor f_4 and power $n+1$ instead of n: $w_2 w_1 = (w_3 w_4)(w_3 w_4)^n w_3 = (w_3 w_4)^{n+1} w_3$ in $\Delta(w_1|w_2)$. This means that also in this case the second (technical) part of the induction hypothesis is satisfied. With the maximal accumulation we may argue as in the previous case. The first factor w_2 starts with accumulation 0, while divider successor w_2 starts with accumulation 1. The possibility to reach a maximum within w_1 can be excluded because the order of the accumulation values of the factor $w_1 w_2$ within $w_2 w_1 w_2$ needs to be the same as in the original word $w_1 w_2$, where the maximum was reached after the first letter of w_2 by the induction hypothesis. □

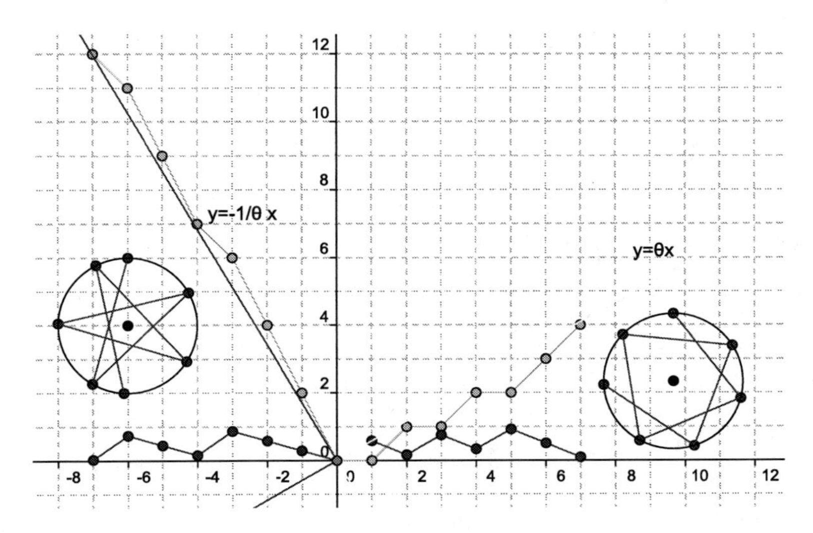

Fig. 6. $xyxyxyy$ and $xxxyxxy$: prefixes of the Sturmian words of slopes $log_2(\frac{3}{2})$ and $-\frac{1}{log_2(\frac{3}{2})}$

5 Final Remarks

1. David Clampitt objected to the musical interpretation (that is to say, to the graphical representation) of twisted adjoints in the present formulation. When our treatment is extended to the free group of two letters (that is, when inverse letters are admitted), the proper representations may be given for the twisted cases, which for St_0 are connected with upward step-interval patterns/backward scale-folding patterns (i.e., negative generation orders), and downward step-interval patterns/forward scale-folding patterns (i.e., positive generation orders). This extension is proposed in [6].
2. For the diatonic scale of 7 notes, the step-interval pattern and scale-folding pattern can be generated as prefixes of two Sturmian words of perpendicular slopes (see Figure 6). One can check that this is not always the case, but there is a tight connection between WF duality, reverse of morphisms and retrogradation of the continued fraction representation of the generator. A further investigation of this subject is presented in [8], where the description of modes in terms of hight and width coordinates determined by generator vector $(1, g)$ and its perpendicular $(-g, 1)$ is taken into account.

References

1. Berstel, J., de Luca, A.: Sturmian words, Lyndon words, and trees. Theoretical Computer Science 178, 171–203 (1997)
2. Berthé, V., de Luca, A., Reutenauer, C.: On an involution of Christoffel words and Sturmian morphisms. European Journal of Combinatorics 29(2), 535–553 (2008)

3. Carey, N.: Distribution modulo 1 and musical scales. Ph.D. diss., University of Rochester (1998)
4. Carey, N., Clampitt, D.: Aspects of well-formed scales. Music Theory Spectrum 11(2), 187–206 (1989)
5. Carey, N., Clampitt, D.: Self-Similar Pitch Structures, Their Duals, and Rhythmic Analogues. Perspectives of New Music 34(2), 62–87 (1996)
6. Clampitt, D., Noll, T.: Modes, the height-width duality, and divider incidence. Paper presented at Society for Music Theory national conference, Nashville, TN (2008)
7. Domínguez, M., Clampitt, D., Noll, T.: Well-formed scales, maximally even sets and Christoffel words. In: Proceedings of the MCM 2007, Berlin, Staatliches Institut für Musikforschung (2007)
8. Domínguez, M., Noll, T.: A specific extension of Christoffel duality to a certain class of Sturm numbers and their characteristic words. In: WORDS 2009 Conference (submitted to, 2009)
9. Lothaire, M.: Combinatorics on Words. Cambridge University Press, Cambridge (1983)
10. Lothaire, M.: Algebraic Combinatorics on Words. Cambridge University Press, Cambridge (2002)
11. Noll, T.: Sturmian sequences and morphisms: a music-theoretical application. SMF, Journée Annuelle, 79–102 (2008)
12. Noll, T.: Ionian theorem. Journal of Mathematics and Music 3(3) (to appear, 2009)

Regions and Standard Modes

David Clampitt and Thomas Noll

[1] Ohio State University, School of Music, USA
clampitt.4@osu.edu
[2] Escola Superior de Musica de Catalunya, Barcelona, Spain
noll@cs.tu-berlin.de

Abstract. Norman Carey and David Clampitt observed in [4] that each region has two well-formed scales as its prefixes. If one looks at this finding from the viewpoint of word theory, one observes that regions are central words and the two prefixes are their independent periods. More precisely, each region, understood as a word in a two-letter alphabet, contains two distinct prefixes, both of which represent well-formed scales. One period is a special standard word, and the other period is a non-special standard word. Thomas Noll proposed in [13] to generalize the authentic Ionian mode through special standard words. He showed that the property of divider incidence characterizes these words among their conjugates. Thus there are two parallel lines of generalization which can be further enriched by observations from [7], [8], as well as by further combinatorial connections between central and standard words.

Two independent lines of research turn out to have so many conceptual cross-links, that a productive synergy emerges immediately from their contact (see [10], [6], [13], [14]). In the past two decades Norman Carey and David Clampitt developed mathematical music theory for the study of scales, regions and related concepts. At the same time mathematicians such as Aldo de Luca, Jean Berstel, Valérie Berthé, Christian Kassel, and Christophe Reutenauer investigated a certain branch of algebraic combinatorics on two-letter words, which includes the study of central words, standard words, Christoffel words. We refer the reader to chapter 2 in [11], as well as to [3], [12], [1], [2].

1 Regions

In [4] Carey and Clampitt attempted a rational reconstruction of certain pitch-space diagrams found in medieval treatises, such as the *heptactys* in *Scolica enchiriadis* and elsewhere and the diamond-shaped diagrams in the *Micrologus* of Guido of Arezzo. They were motivated by a care to understand aspects of diatonicism that were perhaps more available to medieval theorists, who were not so in the thrall of the notion of pitch class, i.e., octave equivalence. The notion of a *region* arises from an alternative path suggested at the outset of their earlier article, [5]: “At the ... purely mathematical level, the octave and the fifth play perfectly symmetrical roles: they are simply numbers which generate other

E. Chew, A. Childs, and C.-H. Chuan (Eds.): MCM 2009, CCIS 38, pp. 81–92, 2009.

numbers. At the level of the formal theory presented here, however, octave and fifth are presumed to play fundamentally dissimilar roles: the octave establishes a primary equivalence relation — octave equivalence — while the fifth generates the different pitch and interval classes. The fifth generates material which fills the frame provided by the octave." What are the implications of taking seriously the symmetry announced at the beginning of this quotation? It implies first of all that the imposed asymmetry may be reversed: as opposed to the procedure in [5] where, for example, the diatonic scale is generated by the perfect fifth with periodicity at the octave, we may understand the perfect fifth as the frame of a scale with diatonic step intervals that is generated by the octave. The latter case is precisely the *dasian* scale of the *Enchiriadis* treatises. In both cases, the scales satisfy the well-formedness condition: generated sets where the generator is everywhere spanned by the same number of step intervals. Abolishing the asymmetry, rather than reversing it, leads to the notion of a region, a pitch-space construction within which modes of the two alternative well-formed scales are enclosed.

We will not need the very concrete instantiations of regions that the definition in [4] provides, but for definiteness we consider a small region, the *heptactys*. If we consider the perfect octave and perfect twelfth as co-generators, either one is potentially the frame of a non-degenerate well-formed scale with step intervals perfect fourth (a), and whole step (b): C F G C' yields the step-interval pattern aba, while C F G C' F' G' yields the step-interval pattern $abaab$. The largest pitch space within which modes of both well-formed scales may coexist is the region C F G C' F' G' C": above the region if the note F" were chosen, that would confirm the octave above F' but would contradict the twelfth above G, D", and conversely were D" to be chosen. Similarly, below the region a choice is forced between the B-flat a twelfth below F' and the octave below G. The *heptactys* is the maximal space within which both well-formed scales remain in balance, in the above sense.

The *heptactys* region corresponds to the palindromic word $abaaba$, and the driving conception behind the paper — that always two well-formed scales of different periods have modes within an enveloping region — can be rephrased in terms of the mathematical fact that the two prefixes of a central word that are the fundamental patterns for its two periodicities are standard words: namely, one positive standard word and one negative standard word. On the other hand, the driving idea of an enveloping region was already subverted by another example in [4], Guido's hexachord: a region enclosed within a well-formed scale *par excellence*, the diatonic (major) scale. This countervailing conception is similarly realized in word theory in the mathematical fact that every central word may be extended in two ways to standard words; every region extends into two standard modes.

1.1 *Ut-Re-Mi-Fa-Sol-La*

Let us consider that prominent music-theoretical object, the *Guidonian hexachord.* Figure 1 displays two arrangements of its six notes, namely (Ut, Re, Mi, Fa, Sol, La) (step order) and (Fa, Ut, Sol, Re, La, Mi) (generation order folded into an octave). Both arrangements deploy binary interval patterns, namely ascending major and minor seconds in the step pattern and ascending fifths and descending fourths in the folding of the chain of fifths into the ambit of one octave. We represent these binary patterns in terms of two two-letter words, namely $u = aabaa$ (for the step-interval pattern with letters a and b representing the ascending major and minor seconds respectively) and $u' = yxyxy$ (for the folding pattern with letters x and y representing the ascending fifth and descending fourth, respectively). Let $q = aaba$ and $p = aab$ denote the two prefixes of u of lengths 4 and 3, respectively. When we write $u = qa$, we see that u has a periodic continuation as $qq = (aaba)(aaba)$. When we write $u = paa$, we see that u has also a periodic continuation as $pp = (aab)(aab)$. An analogous observation can be made with the folding pattern $u' = yxyxy$ and its two prefixes $q' = yx$ of length 2 and $p' = yxyxy$ of length 5: u' has the potential periodic continuation $qqq = (yx)(yx)(yx)$, and $p = (yxyxy)$ is exactly a complete period of p, and can still be extended to $pp = (yxyxy)(yxyxy)$.

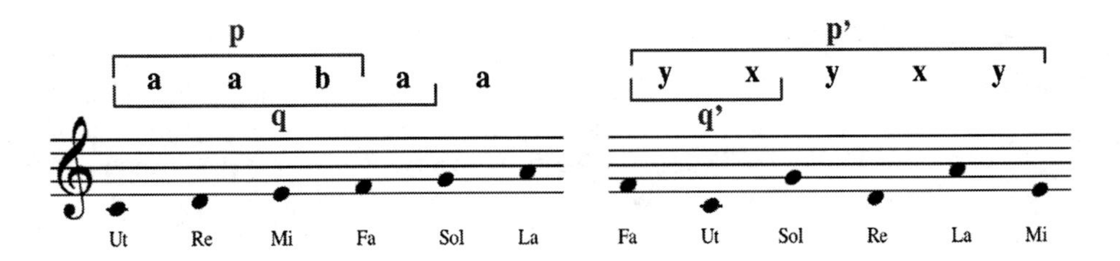

Fig. 1. Step pattern and fifth/fourth folding of the Guidonian hexachord as instances of central palindromes

The possibility for such a potential double periodicity is only given for words which are short enough. It turns out that for periods 4 and 3, the length $5 = 3 + 4 - 2$ is already the maximum. The same length $5 = 2 + 5 - 2$ is also the maximum for the periods 2 and 5. The following proposition is a well-known fact in the algebraic combinatorics on words. It provides the minimal word length, for which such a potential double-periodicity becomes impossible.

Proposition 1 (Theorem of Fine and Wilf). *(c.f. [11], prop. 1.2.1)* *Let q and p be words of lengths $n = |q|$ and $m = |p|$, respectively. Let $d = gcd(n, m)$ denote the greatest common divider of n and m. If two powers $q^k = qq...q$ and $p^l = pp...p$ of q and p have a common prefix v of length at least $n + m - d$, then q and p are powers of the same word.*

We inspect again the words $q = aaba$ of length $n = 4$ and $p = aab$ of length $m = 3$ in the light of this theorem. The longest common prefix of the powers $qq = aabaaaba$ and $pp = aabaab$ is indeed the word $aabaa$ of length $5 = 4 + 3 - 2$, the step-interval pattern of the Guidonian hexachord. 5 is the largest number below the threshold $m + n - gcd(m, n) = 6$ in the presupposition of the theorem. And

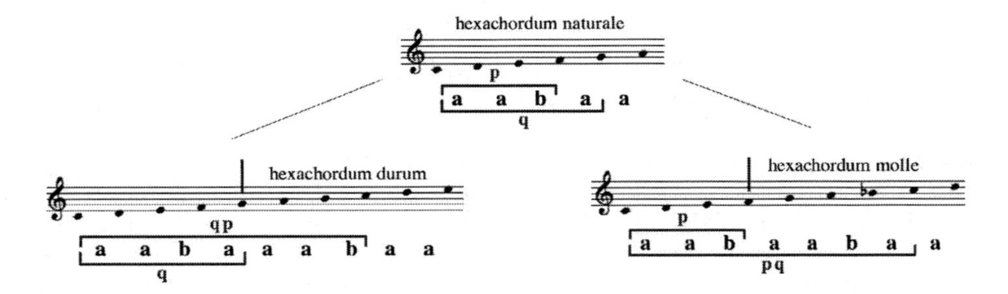

Fig. 2. Construction of two new central words *aabaaabaa* and *aabaabaa* from the central word *aabaa*

therefore it is still possible to have double-periodicity without the consequence of the theorem, namely that the words q and p would necessarily be powers of one and the same word.

1.2 Central Words

The present paper focuses exactly on the generalization of this limiting case, where the word length of u undercuts the critical length $n + m - gcd(n, m)$ of the theorem of Fine and Wilf by 1. We also restrict ourselves to the case where the two periods m and n are mutually co-prime, i.e., where $gcd(n, m) = 1$, and where $|u| = m + n - 2$. The double-periodic two-letter words of this type are known as *central words* or *central palindromes* in the literature. There are two ways to arrange the central words in a strict binary tree structure and the relation between these two trees can be interpreted as a *duality*.

We look again at the example $aabaa$ in order to explain how the *central tree* is organized. Figure 2 shows the two successor nodes of the node $u = aabaa$. We write $P_a(u) = qu = aabaaabaa$ for the successor to the left and $P_b(u) = pu = aabaabaa$ for the successor to the right. Each successor is defined as the concatenation of one of the two prefixes q or p with the word u itself. We shall check that these successors are both double-periodic. Recall, that the concrete node $u = aabaa$ is double-periodic with respect to its two prefixes $q = aaba$ and $p = aab$. Each of the two successors inherits one of the two periodicities and abandons the other: $qu = aabaaabaa$ inherits period $n = 4$ with pattern q and abandons period $m = 3$, while $pu = aabaabaa$ inherits period $m = 3$ with prefix p and abandons period $n = 4$ with pattern q. In both cases one new period comes into play, namely $n+m = 7$. The associated prefixes are the two concatenations qp and pq of the original prefixes q and p. The other prefix of $P_a(u) = qu$ is $qp = uab = aabaaab$, and the other prefix of $P_b(u) = pu$ is $pq = uba = aabaaba$. Note that the sum $7 = 3 + 4$ is co-prime with both 3 and 4, and therefore both new pairs of periods, 4 and 7, as well as 3 and 7 are co-prime.

The example can be easily generalized. For every double-periodic word u in letters a and b with co-prime periods n and m and associated prefixes q of length $n = |q|$ and p of length $m = |p|$ we construct

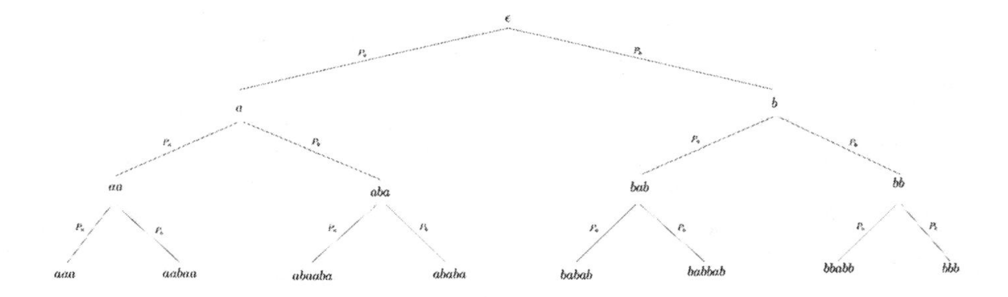

Fig. 3. Beginning of the infinite binary tree of central words

1. the left successor $P_a(u) = qu$ with co-prime periods n and $n + m$ and associated prefixes q and qp,
2. the right successor $P_b(u) = pu$ with co-prime periods m and $m + n$ and associated prefixes p and pq.

The entire tree (see Figure 3) starts from the empty word ϵ, which — somewhat counterintuitively, but for good systematic reasons — is supposed to have the periods $n = m = 1$ and associated "ghost-prefixes" $q = a$ and $p = b$. This exotic behavior continues along the outermost branches of the tree, where either only a's or b's occur.

1.3 Duality for Central Words

In this subsection we want to generalize and thereby explain the nature of the relation between the step-interval pattern $u = aabaa$ and the associated folding pattern $u' = yxyxy$ of the hexachord. Figure 1 helps to acknowledge these patterns on the basis of music-theoretical background knowledge. Likewise, we may use our music-theoretical intuition in order eventually to find an octave/twelfth-folding pattern in association with the step-interval pattern $abaaba$ of the *heptactys* region. In this case the tones C F G C' F' G' C" (in ascending height order) are re-ordered as G G' C C' C" F F' with the folding pattern $xyxxyx$. What we are looking for is a word-theoretic explanation for the associations below:

$$aabaa \leftrightarrow yxyxy \qquad abaaba \leftrightarrow xyxxyx$$

The tree in Figure 3 provides an alternative reading, which opens an additional perspective on central words: Central words are palindromes of a very special kind, (which is why they are also called *central palindromes* in the literature). Every node u on the tree can be constructed as an *iterated right palindromic extension* of its associated *directive word*, which encodes the path on the tree from the root to the node u. The (simple) *right palindromic extension* of a word w is the unique shortest palindrome w^+ having w as a prefix. This means: one writes $w = w_1 w_2$ with w_2 being the longest palindromic suffix of w. If $\tilde{w_1}$ denotes the reversal of the word w_1 one obtains $w^+ = w_1 w_2 \tilde{w_1}$. The iterated

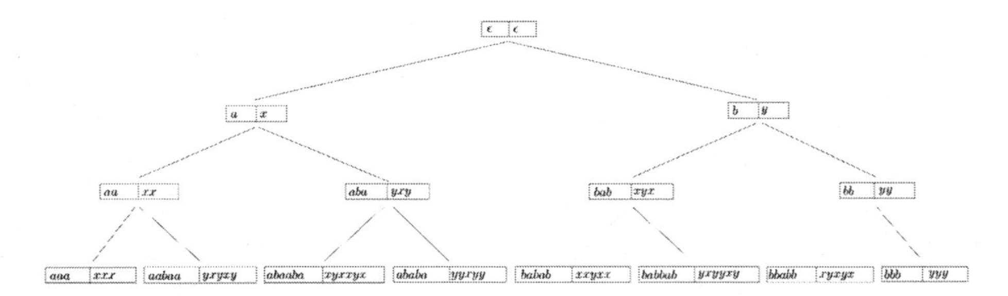

Fig. 4. Compilation of the binary tree of central words with its dual tree. In each box the word $u \in \{a, b\}^*$ (on left side) denotes a node of the central tree, while the word $u' \in \{x, y\}^*$ (on the right side) denotes the associated dual central word, which belongs to the associated node of the dual tree.

right palindromic closure $Pal(w)$ of a word $w = w'w_n$ is defined recursively as $(Pal(w')w_n)^+$, where w_n denotes the last letter of w. The recursion terminates with $Pal(\epsilon) := \epsilon$ for the empty word ϵ. One can easily prove that the map P_a associates any central word u with the right palindromic extension $(ua)^+$ of ua, and likewise P_b maps u to the right palindromic extension $(ub)^+$ of ub. In our hexachord example the central palindrome $u = aabaa = P_b(P_a(P_a(\epsilon)))$ is the iterated palindromic extension $u = Pal(aab)$ of the *directive word* $dir(u) = aab$, because $Pal(aab) = (((a^+)a)^+b)^+ = (aab)^+ = aabaa$.

The central palindrome u' is the iterated right palindromic extension $u' = Pal(yxx)$ of the *directive word* $dir(u') = yxx$, because $Pal(yxx) = (((y^+)x)^+x)^+ = ((yx)^+x)^+ = (yxyx)^+ = yxyxy$. Observe that — up to a renaming of the letters ($x \rightsquigarrow a$, $y \rightsquigarrow b$) — the directive words aab and yxx are retrogrades of each other. Without proof we mention that the retrogradation of the directive words provides an appropriate duality for central words. An expression of this duality is the conversion of periods into letter-frequencies and vice versa. More precisely: Let q' and p' denote the periodic prefixes of the dual word u' of a central word u with periods $n = |q|$ and $m = |p|$, and let $n' = |q'|$ and $m' = |p'|$ denote the periods of u'. Then the letter frequencies $|u|_a$, $|u|_b$, $|u'|_x$, $u'|_y$ satisfy

$$|u|_a + 1 = n', |u|_b + 1 = m' \quad \text{and} \quad |u'|_x + 1 = n, |u'|_y + 1 = m.$$

Systematically adjoining every node u in the tree of Figure 3 with its dual central word u' (written in letters a and b instead of x and y), we construct a dual binary tree (c.f. 4).

The dual of the central tree can also be defined directly, i.e., in terms of two maps specifying the left and right successors for each node of the dual central tree. We postpone their definitions to the subsequent section.

2 Standard Modes

In the previous section we studied central words with two prefixes q and p. Thereby the genealogy of central words down the tree goes along with the iterated concatenations of the two prefixes. The ordered pair (q, p) (of prefixes for u) has the two successors (q, qp) (of prefixes for $P_a(u) = qu$) and (pq, p) (of prefixes for $P_b(u) = pu$). The first prefix q in the ordered pair (q, p) is identified by the property that it ends with letter a, while the second one, p, ends with letter b. The present section is dedicated to the study of these pairs.

2.1 *Do-Re-Mi-Fa-Sol-La-Ti-(Do′)*

We return to our prominent music-theoretical example, but now in relation to an historically much later, equally if not more prominent example: the *diatonic major (Ionian) mode*. In consideration of this historical disjunction, we will use modern solmization syllables, replacing *Ut* with *Do* and adjoining *Ti*. The main subject of this section is the interdependence of central words and (positive) standard words as well as the close connection between the two dualities (i.e., between the duality for central words on the one hand and the duality for positive standard words on the other). The adjunction of the note *Ti* and of *Do′* (the repetition of the *finalis Do* one octave higher) to the hexachord is a quite natural procedure from the viewpoint of word theory.

As mentioned in the introduction to Section 1, every region encompasses particular modes of two well-formed scales. Closer inspection of [4] under the perspective of central words reveals that these two modes coincide with the periodic prefixes q and p. Well-formed scales are mostly known as scales with a periodicity at the interval of the octave. In this case, however, the tones *Do, Re, Mi, Fa, (Sol)* with step-interval pattern $q = aaba$ form a well-formed scale modulo perfect fifth, which is generated by the major second, starting from *Fa*. Analogously, the tones *Do, Re, Mi, (Fa)* with step-interval pattern $p = aab$ form a well-formed scale modulo perfect fourth, which is also generated by the major second, starting from *Do*.

When we concatenate these two scales, we obtain the fifth-generated diatonic scale *Do, Re, Mi, Fa, Sol, La, Ti, (Do′)* modulo octave. More precisely, we obtain the well-formed Ionian mode with step-interval pattern *aabaaab* from the concatenation of two shorter well-formed modes with step-interval patterns *aaba* and *aab*. Note that under this *concatenative* construction of the Ionian mode from the two shorter modes *aaba* and *aab* the music-theoretical interpretation of the letters a and b does not change: a stands for (ascending) major second and b stands for (ascending) minor second. What does change is the modulus, the interval of periodicity. We see below, that this concatenation qp is an instance of a transformation on word pairs. Henceforth we use the notation $w_1|w_2$ for word pairs (w_1, w_2) with the additional meaning that we regard this pair as two factors of the concatenation w_1w_2.

For any such pair $w_1|w_2$ we define two successors, namely

$$\Gamma(w_1|w_2) := w_1|w_1w_2 \quad \text{and} \quad \Delta(w_1|w_2) := w_2w_1|w_2.$$

Starting from the pair $a|b$ we obtain $aaba|aab$ as

$$\Delta(\Gamma(\Gamma(a|b))) = \Delta(\Gamma(a|ab)) = \Delta(a|aab) = aaba|aab.$$

There is an entirely different way to construct the Ionian mode of the diatonic scale of the basis of transformations. This is based on *substitutions* rather than concatenations. For the free monoid $\{a,b\}^*$ of finite words with letters a and b let $G_{a,b}$ and $D_{a,b}$ denote the following monoid morphisms. For single letters we define:

$$G_{a,b}(a) = a, G_{a,b}(b) = ab \quad \text{and} \quad D_{a,b}(a) = ba, D_{a,b}(b) = b.$$

For words with more than two letters the transformations are applied to each letter and the images are concatenated. For pairs $w_1|w_2$ we may trace the images separately, i.e., we write $G_{a,b}(w_1|w_2) = G_{a,b}(w_1)|G_{a,b}(w_2)$ as well as $D_{a,b}(w_1|w_2) = G_{a,b}(w_1)|G_{a,b}(w_2)$.

Again, starting from the pair $a|b$ we obtain $aaba|aab$ as

$$G_{a,b}(G_{a,b}(D_{a,b}(a|b))) = G_{a,b}(G_{a,b}((ba|b)) = G_{a,b}(aba|ab) = aaba|aab.$$

Note, that with respect to the substitutive transformations it is music-theoretically meaningful to assume a common modulus before and after the transformation and to change the interpretation of the letters a and b instead:

authentic division:	$a\|b$	$= P5 \mid P4$
tetractys:	$ba\|b$	$= P4\,M2 \mid P4$
pentatonic:	$aba\|ab$	$= M2\,m3\,M2 \mid m3\,M2$
diatonic:	$aaba\|aab$	$= M2\,M2\,m2\,M2 \mid M2\,M2\,m2$

What remains constant in the meaning of the letters a and b under substitution notwithstanding their change in size, is the interpretation of a as a *primary step-interval* and of b as a *secondary step-interval*. The substitutive transformations are the heart of the hierarchy of well-formed scales.

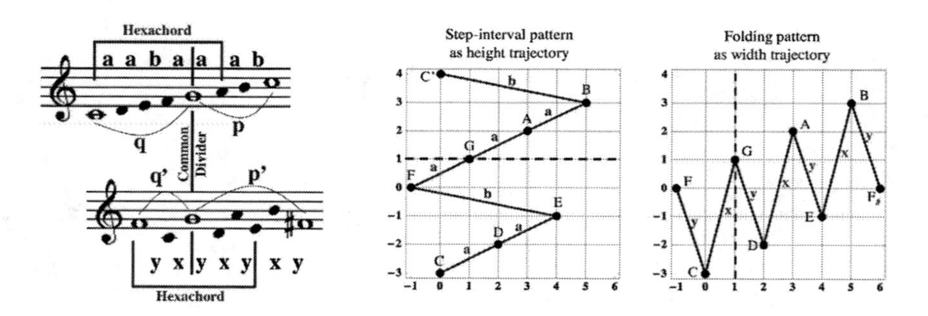

Fig. 5. Authentic Division of the Ionian Mode. The figures to the right display height- and width-trajectories in association with the step-interval pattern and the folding pattern of the Ionian mode.

For a moment we leave the two transformational constructions of the step-interval pattern $aaba|aab$ aside and consider their common result simply as an extension of the hexachord step-interval pattern $aabaa$ by the suffix ab. As Figure 5 suggests, we may analogously append the suffix xy to the folding pattern $yxyxy$ of the hexachord. The penultimate note achieved by both patterns is Ti, whereas the ultimate notes Do' (in the step-interval scale) and $Fa\sharp$ (in the folding) are not equal. It can be shown however (see [13]) that the intervals $P8$ (the octave between Do and Do' and $A1$ (the augmented prime between Fa and $Fa\sharp$) are duals of each other in a music-theoretically very convincing way. Observe, for example, that the augmented prime A1, i.e., the ambit interval of the folding, is also the difference $a-b$ between the ascending major and minor steps. Dually, the octave P8, i.e., the ambit interval of the step-interval scale, is also the difference $x-y$ between the ascending fifth and the descending(!) fourth.

Figure 5 includes a division $yx|yxyxy$ of the folding pattern of the Ionian mode. This is consistent with the two periods $2 = |q'|$ and $5 = |p'|$ of the folding pattern $yxyxy$ of the hexachord. We return now to the two constructions of $aaba|aab$ by concatenative and substitutive transformations. We may find two strictly dual constructions of the pair $yx|yxyxy$:

$$\Gamma(\Gamma(\Delta(x|y))) = \Gamma(\Gamma(yx|y)) = \Gamma(yx|yxy) = yx|yxyxy.$$

$$D_{x,y}(G_{x,y}(G_{x,y}(x|y))) = D_{x,y}(G_{x,y}(x|xy)) = D_{x,y}(x|xxy) = yx|yxyxy$$

Closer inspection shows that the chain of pairs $x|y \mapsto yx|y \mapsto yx|yxy \mapsto yx|yxyxy$, which we obtain by the concatenative transformations Δ, $\Gamma\Delta$ and $\Gamma\Gamma\Delta$ are precisely the folding patterns to the chain of pairs $a|b \mapsto ba|b \mapsto aba|ab \mapsto aaba|aab$, which are the result of the iterated substitutional transformations $G_{a,b}$, $G_{a,b}G_{a,b}$, and $D_{a,b}G_{a,b}G_{a,b}$. Under this view the interpretation of x as ascending fifth and y as descending fourth remains constant. Conversely, the intermediate stages $x|y \mapsto x|xy \mapsto x|xxy \mapsto yx|yxyxy$ of the substitutive construction of the folding pattern $yx|yxyxy$ via $G_{x,y}$, $G_{x,y}G_{x,y}$ and $D_{x,y}G_{x,y}G_{x,y}$ are precisely the folding patterns of the intermediate stages of the concatenative construction of $aaba|aab$ along $a|b \mapsto ba|b \mapsto aba|ab \mapsto aaba|aab$ via $\Gamma, \Gamma\Gamma$ and $\Delta\Gamma\Gamma$. In this case the more abstract music-theoretical meaning of x and y (namely, *primary* vs. *secondary folding interval*) remains unchanged under transformation, while the actual sizes of these intervals change.

In figure Figure 5 there is a vertical line, connecting the dividing note Sol of the step-interval pattern with the dividing note Sol of the folding. For ascending scales and forward foldings (ascending fifths) this is characteristic for the Ionian mode among all other modes (c.f. [13]).

2.2 Standard Pairs and Their Duality

The concrete example may easily be turned into a definition for the general case. Every central word u has two periodic prefixes q and p and we can write $qp = uab$. Its dual central word u' has periodic prefixes q' and p' and again we

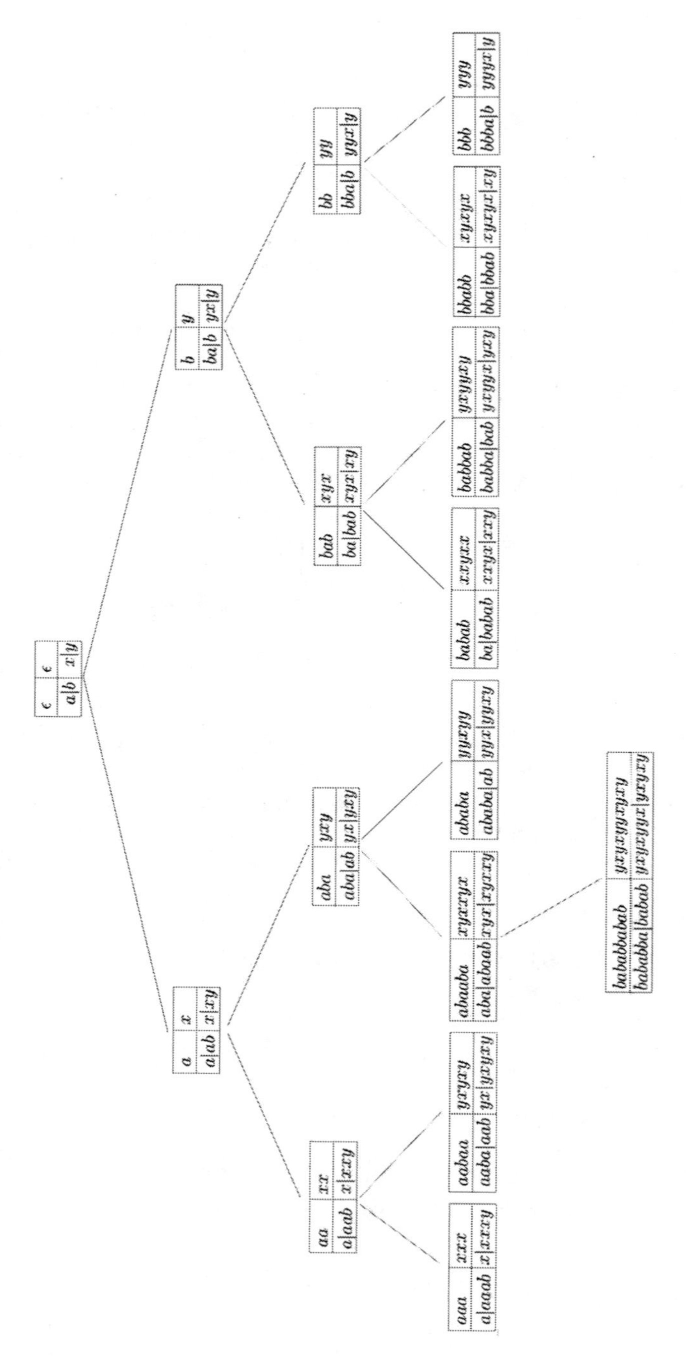

Fig. 6. Complilation of four binary trees: central words and positive standard words. Observe that the duality is manifest through the operation of path reversal.

can write $q'p' = u'xy$. Figure 6 extends Figure 4. In each of the 2×2-boxes there are now four nodes from four different — by highly related — binary trees:

1. The tree of central words (displayed in the left upper fields of each node-box). The left and right successors of node u are the nodes $P_a(u) = qu = (ua)^+$ and $P_b(u) = qu = (ub)^+$, respectively.
2. The dual tree of central words (displayed in the right upper fields of each node-box). The left and right successors of node u' are the nodes $C_a(u') = G_{x,y}(u')a$ and $C_b(u') = D_{x,y}(u')b$, respectively.
3. The tree of standard pairs (displayed in the left lower fields of each node-box). The left and right successors of node $q|p$ are the nodes $\Gamma(q|p) = q|qp$ and $\Delta(q|p) = pq|p$, respectively.
4. The dual of this tree of standard pairs (displayed in the right lower fields of each node-box). The left and right successors of node $q|p$ are the nodes $G_{x,y}(q|p)$ and $D_{x,y}(q|p)$, respectively.

Concluding Remark: In [9] we take a closer look at the way in which the central word u (with period-prefixes q and p) interacts with the authentic standard mode $q|p$ and its conjugate modes. This paper revisits observations by the first author (see [7]) in connection with Guido's and Hermannus's concept of *affinity* in the light of the double periodicity of the central words. It also investigates the *sensitive interval* between divider and leading tone by means of generalization.

Acknowledgments. The authors thank Emmanuel Amiot and Norman Carey for stimulating discussions. We also thank Fabian Singler for dedicating his thesis ([15]) to this research subject.

References

[1] Berstel, J., de Luca, A.: Sturmian words, Lyndon words and trees. Theoretical Computer Science 178, 171–203 (1997)
[2] Berstel, J., Lauve, A., Reutenauer, C., Saliola, F.V.: Combinatorics on words: Christoffel words and repetitions in words. CRM Monograph Series, vol. 27. American Mathematical Society, Providence (2009)
[3] Berthé, V., de Luca, A.: Christophe Reutenauer. On an involution of Christoffel words and Sturmian morphisms. European Journal of Combinatorics 29(2), 535–553 (2008)
[4] Carey, N., Clampitt, D., Regions: A theory of tonal spaces in early medieval treatises. Journal of Music Theory 40(1), 113–147 (1996)
[5] Carey, N., Clampitt, D.: Aspects of well-formed scales. Music Theory Spectrum 11(2), 187–206 (1989)
[6] Clampitt, D., Noll, T.: Modes, the height-width duality, and divider incidence. In: Society for Music Theory national conference, Nashville, TN (2008)
[7] Clampitt, D.: Double neighbor polarity (unpublished paper) (2008)
[8] Clampitt, D.: Sensitive intervals: major third analogues in standard well-formed words. In: Workshop on Mathematical Music Theory, South Bristol, ME (2008)
[9] Clampitt, D., Noll, T.: Regions within enveloping standard modes (unpublished paper) (2009)

[10] Domínguez, M., Clampitt, D., Noll, T.: Well-formed scales, maximally even sets and Christoffel words. In: Proceedings of the MCM 2007, Berlin, Staatliches Institut für Musikforschung (2007)
[11] Lothaire, M.: Algebraic Combinatorics on Words. Cambridge University Press, Cambridge (2002)
[12] Kassel, C.: From Sturmian morphisms to the braid group B_4. In: The Workshop on Braid Groups and Applications, Banff International Research Station (2004)
[13] Noll, T.: Ionian theorem. Journal of Mathematics and Music 3(3) (to appear, 2009)
[14] Noll, T., Clampitt, D., Domínguez, M.: What Sturmian morphisms reveal about musical scales and tonality. In: Proceedings of the WORDS 2007 Conference, Marseille (2007), http://iml.univ-mrs.fr/words2007/
[15] Singler, F.: Zur Dualität zwischen doppelter Periodizität und binärer Intervall-Struktur in der Theorie der Tonregionen. Thesis (final paper). Hochschule für Musik und Theater Felix Mendelssohn Bartholdy, Leipzig (2008)

Compatibility of the Different Tuning Systems in an Orchestra

Alfonso del Corral[1], Teresa León[2], and Vicente Liern[3]

[1] Taller de Música Jove, Mistral, 22, 46020 - Valencia – Spain
[2] Dep. Estadística e Investigación Operativa, Universidad de Valencia, Vicente Andrés Estellés, s/n, 46100-Burjassot, Valencia – Spain
[3] Dep. Matemáticas para la Economía y la Empresa, Universidad de Valencia, Avda. Tarongers, s/n, 46022,Valencia – Spain
fons.vila@gmail.com, {teresa.leon,vicente.liern}@uv.es

Abstract. Focusing on the daily practice of musicians, we give flexibility to the mathematical treatment of musical notes, tuning systems and the relations between them. This allows us to connect the theory and the practice of music. Using the techniques of fuzzy logic, we describe the concepts with fuzzy sets and introduce the α-compatibility as a degree of interchangeability between tuning systems. To show how our proposal works, we use a fragment of Haydn and analyze the compatibility of the notes taken from 48 recordings for the tuning systems of Pythagoras, Zarlino and Equal Temperament of 12 notes.

Keywords: Tuning Systems, Fuzzy Sets, Fuzzy Numbers.

1 Introduction

Different criteria have been used to select the sounds that music uses. A set containing these sounds (musical notes) is called a tuning system. Most of them have been obtained through mathematical arguments. The numerical nature of these systems facilitates their transmission and the manufacture of instruments, etc. However, the harshness of the mathematical arguments relegated these tuning systems to theoretical studies while in practice musicians tuned in a more flexible way. They implicitly deal with complex mathematical processes involving some uncertainty in the concepts. In fact, most of the musicians in a classical orchestra must adjust their instruments to tune well. For instance, wind instrument players modify the air pressure or the finger positions to fit their notes to the ensemble. Because of this, many musicians feel that the mathematical arguments that justify tuning systems are impractical.

Sometimes, probability distributions are useful for handling uncertainty (*stochastic uncertainty*) [7],[8], but in other cases it cannot be justified that the given concepts follow a predetermined distribution (*fuzzy uncertainty*). As musicians need flexibility in their reasoning, the use of fuzzy logic to connect music and uncertainty is appropriate (see [6],[7],[10]). Therefore, we propose to model the notes as fuzzy sets and analyze the compatibility between them. With this idea, we can extend the concept of tuning systems, connecting theory and practice, and understand how musicians work

E. Chew, A. Childs, and C.-H. Chuan (Eds.): MCM 2009, CCIS 38, pp. 93–103, 2009.

in real-life. In order to compare the notes that constitute theoretical tuning systems and those performed by musicians, we have studied the compatibility of a set of notes (recorded by a professional musician) with the corresponding notes of the Pythagorean, Zarlinean and Equal Temperament of 12 notes systems.

2 Some Concepts and Notation

We will identify each musical note with the frequency of its fundamental harmonic (the frequency that chromatic tuners measure). The usual way to relate two frequencies is through their ratio; this number is called the interval. It is well known that, in the middle zone of the audible field, the "pitch sensation" changes somewhat according to the logarithm of the frequency, so the distance between two sounds whose frequencies are f_1 and f_2 can be estimated by means of the expression

$$d(f_1, f_2) = 1200 \times \left| \log_2 \frac{f_1}{f_2} \right|, \tag{1}$$

where the logarithm in base 2 and the factor 1200 have been used in order to express d in cents [5], [6]. Let us mathematically define the well-known concept of an octave:

Definition 1. Given two sounds with frequencies f_1 and f_2, we say that f_2 is one octave higher than f_1 if f_2 is twice f_1.

Two notes one octave apart from each other have the same letter-names. This naming corresponds to the fact that notes an octave apart sound like the same note produced at different pitches and not like entirely different notes. Based on this idea, we can define in $\mathbb{R}^+$ (the subset of all the frequencies of all the sounds) a binary equivalence relation, denoted by $\mathcal{R}$, as follows:

$$f_1 \, \mathcal{R} \, f_2 \text{ if and only if } \exists\, n \in Z \text{ such that } f_1 = 2^n \times f_2.$$

Therefore, instead of dealing with $\mathbb{R}^+$, we can analyze the quotient set $\mathbb{R}^+/\mathcal{R}$, which for a given fixed note f_0 (diapason) can be identified with the interval $[f_0, 2f_0[$. However, for the sake of simplicity, we will assume that $f_0 = 1$ and work in the interval [1, 2].

Tuning systems based on a unique interval (like the Pythagorean) admit a direct mathematical construction. However, the definition of systems generated by more than one interval requires specifying when and how many times each interval appears. Next, we give a give a general definition of a tuning system (see [1], [10]):

Definition 2. Let $\Lambda = \{\lambda_i\}_{i=1}^k \subset [0,1[$ be a family of functions $F = \{h_i : Z \to Z\}_{i=1}^k$. We call the tuning system generated by the intervals $\{2^{\lambda_i}\}_{i=1}^k$ and F the set

$$S_\Lambda^F = \left\{ 2^{c_n} : c_n = \sum_{i=1}^k \lambda_i h_i(n) - \left\lfloor \sum_{i=1}^k \lambda_i h_i(n) \right\rfloor,\ n \in Z \right\} \tag{2}$$

where $\lfloor x \rfloor$ is the integer part[1] of x.

[1] Note that the integer part in (2) is added to gain octave equivalence.

If every element in the tuning system is a rational number, we say that it is a tuned system, whereas if some element is an irrational number then the system is a temperament. The advantage of expressing the tuned notes as 2^{c_n} is that if our reference note is 2^0, in accordance with (1) the exponent c_n provides the pitch sensation. Let us mention that the family of integer-valued functions F mark the "interval locations". In those systems generated by one interval (for instance the Pythagorean) they are not really necessary. However, in the other systems they are. For instance, in the Just Intonation h_1(n) and h_2(n) indicate the position of the t fifths and the thirds considered as tuned. Table 1 displays some examples of tuning systems.

Table 1. Examples of generators of some tuning systems

S	Λ	*F*
Pythagorean	$\lambda_1 = \log_2(3/2)$	$h_1(n) = n$
Equal Temperament	$\lambda_1 = 7/12$	$h_1(n) = n$
Zarlinean (Just Intonation)	$\lambda_1 = \log_2(3/2)$ $\lambda_2 = \log_2(5/4)$	$h_1(n) = n - 4h_2(n)$ $h_2(n) = \frac{n+1}{7} + \frac{n+4}{7}$
Neidhart's temperament (1/2 & 1/6 comma)	$\lambda_1 = \log_2(3/2)$ $\lambda_2 = 1/6 \cdot \log_2(2^{12}/3^6)$ $\lambda_3 = 7/12$	$a_n = n - 12\, n/12$ $h_1(n) = \frac{a_n+2}{12} + \frac{n+3}{12} + \frac{a_n+10}{12} + \frac{n+11}{12}$ $h_2(n) = \frac{a_n+6}{12} + \frac{n+7}{12} + \frac{a_n+8}{12} + \frac{n+9}{12}$ $h_3(n) = \frac{a_n+1}{12} + \frac{n+4}{12} + \frac{a_n+5}{12}$

In this article we only analyze Pythagorean, Zarlinean and Equal Temperament (with 12 notes) systems [5]. However, note that the study of other tuning systems would be similar.

3 Introducing Fuzzy Logic

The main idea in Fuzzy Logic is to substitute the characteristic function of set A, which takes the value 1 when the element belongs to A and 0 otherwise, with a membership function $\mu_{\tilde{A}}(x)$ which takes values in the interval [0, 1] (see [11]). The value $\mu_{\tilde{A}}(x)$ is understood as the membership degree of element x to the set. A nule membership degree is understood as non-membership, 1 as membership in the Boolean sense, and intermediate numbers imply that membership is uncertain, which will be understood in different ways, depending on the case [3]. If the initial reference set is X, the most usual way of denoting the fuzzy sets is the following:

$$\tilde{A} = \{(x, \mu_{\tilde{A}}(x)),\ x \in X\}. \tag{3}$$

In our context, if we take the note A = 440Hz (*diapason*) as our fixed note, a frequency of 442Hz, from the point of view of the Boolean logic, would be out of tune.

However, for every musician, or anybody that hears it, that note is slightly more out of tune than another note of 450Hz. The step between tuned or not tuned is represented as a fuzzy set in which a tolerance level has been fixed (see Fig. 1).

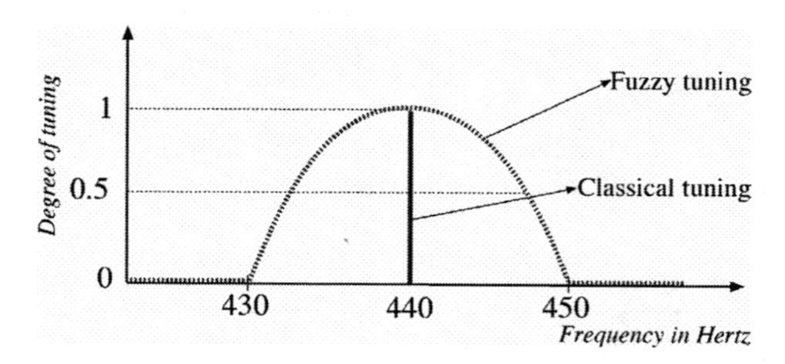

Fig. 1. Membership and characteristic functions for the fuzzy and classical membership functions

Before explaining how the tolerance levels are fixed, we will introduce the idea of a fuzzy number [4]:

Definition 3. A fuzzy number is a fuzzy set whose membership function is piecewise continuous and convex and there is a value whose membership degree is equal to one.

In this paper we will use a particular case of fuzzy numbers, the *LR*-fuzzy numbers [3], [4] and the relationship between them.

Definition 4. A fuzzy number $\tilde{M}$ is said to be an LR-fuzzy number, $\tilde{M} = (m^L, m^R, \alpha^L, \alpha^R)$, if its membership function has the following form:

$$\mu_{\tilde{M}}(x) = \begin{cases} L\left(\dfrac{m^L - x}{\alpha^L}\right), & x < m^L \\ 1 & m^L \le x \le m^R \\ R\left(\dfrac{x - m^R}{\alpha^R}\right), & x > m^R \end{cases}$$

where L and R are reference functions, i.e., $L, R : [0, +\infty[\to [0, 1]$ are strictly decreasing in $\operatorname{supp} \tilde{M} = \{x : \mu_{\tilde{M}}(x) > 0\}$ and upper semi-continuous functions such that $L(0) = R(0) = 1$. If $\operatorname{supp} \tilde{M}$ is a bounded set, L and R are defined on $[0, 1]$ and satisfy $L(1)=R(1)=0$. Moreover, if L and R are linear functions, the fuzzy number is called *trapezoidal* (see Fig. 2) and their arithmetic is easy to perform. They are defined by four real numbers, $\tilde{A} = (a^L, a^R, \alpha^L, \alpha^R)$, and values below $a^L - \alpha^L$ and above $a^R + \alpha^R$ are not acceptable.

4 Fuzzy Musical Notes

It is well known that the human ear perceives notes with very close frequencies as if they were the same note [2]. In 1948, N. A. Garbuzov made thousands of experiments and used them to assign band frequencies to every musical interval, called the Garbuzov zones (see [7], [8]). In this study, he showed that we perceive as the same note, unison, two frequencies that are 12 cents apart. Other authors reduce this interval to 6 cents. For the purpose of our work, we will define this band of unison as:

$$\left]f2^{-\varepsilon}, f2^{\varepsilon}\right[\cong f \tag{4}$$

where $\varepsilon > 0$, and 1200ε expresses, in cents, the accuracy of the human ear to the perception of the unison.

Moreover, if the amount of notes per octave is q, we can divide the octave into q intervals with a length of 1200/q cents. So, we can express the interval of the note f as:

$$\left]f2^{-\delta}, f2^{\delta}\right[, \tag{5}$$

where $\delta = 1/(2q)$, the quantity $\Delta = 1200\delta$ expresses, in cents, the tolerance that we admit for every note. Actually, this is what chromatic tuners do: they assign 12 divisions per octave, $\delta = 1/(2\times 12)$, and then, the tolerance corresponding to every note is $\Delta = 1200\frac{1}{2\times 12} = 50$ cents.

Remark 1. It is necessary that $\varepsilon < q$ since a tuning system with more notes than the human ear can distinguish would have no practical sense.

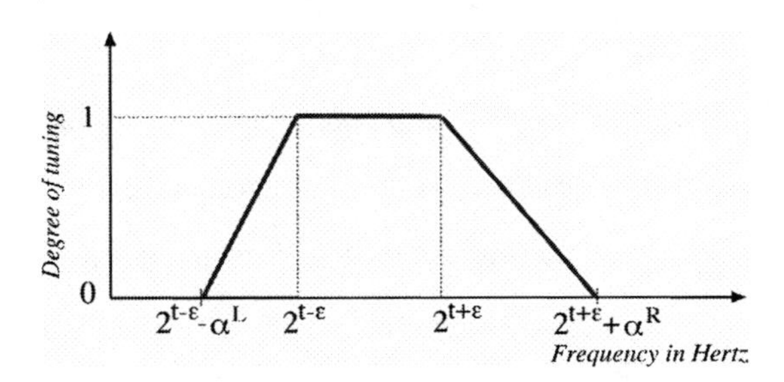

Fig. 2. Membership function of a musical note

Taking into account (4) and (5), we can express a musical note as a trapezoidal fuzzy number with peak $[f2^{-\varepsilon}, f2^{\varepsilon}]$ and support $[f2^{-\delta}, f2^{\delta}]$. Besides, according to Definition 2, the notes are expressed as powers of two, 2^{c_n}, so it is more practical to express the fuzzy musical notes using their exponent, c_n.

Definition 5. Let $\tilde{t} = (t-\varepsilon, t+\varepsilon, \delta, \delta)$ be a symmetric trapezoidal fuzzy number, where $t, \delta \in [0,1]$. A fuzzy musical note can be defined as the following trapezoidal fuzzy number:

$$2^{\tilde{t}} = (2^{t-\varepsilon}, 2^{t+\varepsilon}, \alpha^L, \alpha^R),$$

where $\alpha^L = 2^{t-\varepsilon}(1-2^{-\delta})$, $\alpha^R = 2^{t+\varepsilon}(2^{\delta}-1)$ and its membership function is

$$\mu_{2^t}(x) = \begin{cases} 1 - \dfrac{2^{t-\varepsilon} - x}{2^{t-\varepsilon} - 2^{t-\varepsilon-\delta}}, & 2^{t-\varepsilon-\delta} < x \le 2^{t-\varepsilon} \\ 1, & 2^{t-\varepsilon} < x \le 2^{t+\varepsilon} \\ 1 - \dfrac{x - 2^{t+\varepsilon}}{2^{t+\varepsilon+\delta} - 2^{t+\varepsilon}} & 2^{t+\varepsilon} < x \le 2^{t+\varepsilon+\delta} \\ 0, & \text{otherwise.} \end{cases}$$

Let us point out that two trapezoidal fuzzy numbers are involved in the definition of a fuzzy musical note: one for the exponent, $\tilde{t}$, which reflects the pitch sensation and therefore is a symmetric fuzzy number, and the other for the fuzzy note which is non-symmetric, as the expression of its membership function shows.

5 Measuring Compatibility

Let us recall the definition of intersection or similitude between two fuzzy sets [3].

Definition 6. The fuzzy intersection of the fuzzy sets $\tilde{A}$ and $\tilde{B}$ is a new fuzzy set $\tilde{A} \cap \tilde{B}$ with membership function

$$\mu_{\tilde{A}\cap\tilde{B}}(x) = \min\{\mu_{\tilde{t}}(x), \mu_{\tilde{s}}(x)\}. \tag{6}$$

The concept of compatibility between two notes can be derived from this definition [10].

Definition 7. Let $2^{\tilde{t}}$ and $2^{\tilde{s}}$ be two musical notes, where $\tilde{t} = (t-\varepsilon, t+\varepsilon, \delta, \delta)$ and $\tilde{s} = (s-\varepsilon, s+\varepsilon, \delta, \delta)$. We define the degree of compatibility between $2^{\tilde{t}}$ and $2^{\tilde{s}}$ as

$$\text{Comp}[2^{\tilde{t}}, 2^{\tilde{s}}] = \max_x \mu_{\tilde{s}\cap\tilde{t}}(x), \tag{7}$$

and we say that $2^{\tilde{t}}$ and $2^{\tilde{s}}$ are α-compatible, $\alpha \in [0,1]$, if $\text{Comp}[2^{\tilde{t}}, 2^{\tilde{s}}] \ge \alpha$.

Although the compatibility between notes could have been defined for notes with different degrees of tolerance δ and δ' (with $\delta \ne \delta'$), in practice we are equally tolerant of all the notes in an octave.

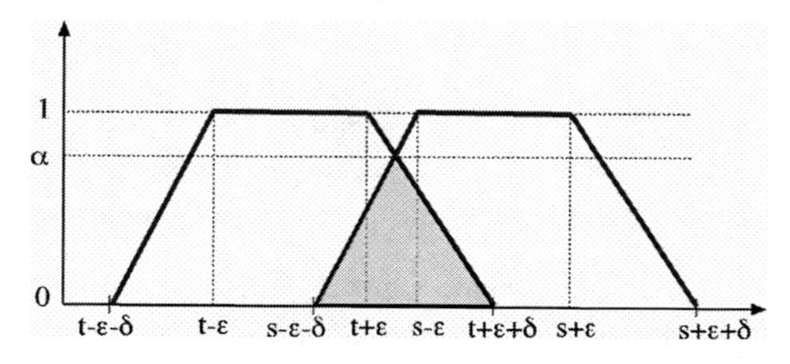

Fig. 3. Graph showing the concept of α-compatibility between two notes

Figure 3 illustrates Definition 7 and shows that the intersection of two trapezoidal numbers is not necessarily trapezoidal.

By a direct calculus we can obtain the next result that allows us to calculate the compatibility between notes.

Proposition 1: Two musical notes $2^{\tilde{t}}$ and $2^{\tilde{s}}$, where $\tilde{t}=(t-\varepsilon,t+\varepsilon,\delta,\delta)$ and $\tilde{s}=(s-\varepsilon,s+\varepsilon,\delta,\delta)$, are α-compatible, $\alpha\in[0,1]$ if and only if $|t-s|\leq 2\delta(1-\alpha)+2\varepsilon$.

Taking into account (7) and Proposition 1, the compatibility can be expressed as:

$$\mathrm{Comp}[2^{\tilde{t}},2^{\tilde{s}}]=\max\left\{0,1-\frac{|t-s|-2\varepsilon}{2\delta}\right\}. \tag{8}$$

However, for musicians it is usually more convenient to calculate the compatibility between two notes in terms of their frequencies. Hence, given two notes with frequencies f_1 and f_2, for which we admit a tolerance of Δ cents, according to (8), the compatibility between f_1 and f_2 is given by

$$\mathrm{Comp}[f_1,f_2]=\max\left\{0,1-\frac{d(f_1,f_2)-2E}{2\Delta}\right\}, \tag{9}$$

where d is the distance defined in (1), $\Delta=1200\times\delta$ and $E=1200\times\varepsilon$.

In order to extend the concept of compatibility to tuning systems, we need to define the fuzzy tuning systems.

Definition 8. Let $\delta\in[0,1]$, $\Lambda=\{\lambda_i\}_{i=1}^{k}\subset[0,1[$ and a family of functions $F=\{h_i:Z\to Z\}_{i=1}^{k}$. We call a fuzzy tuning system generated by the intervals $\{2^{\lambda_i}\}_{i=1}^{k}$ and F the set

$$\tilde{S}_{\Lambda}^{F}(\delta)=\left\{2^{\tilde{c}_n}:\tilde{c}_n=\left(\sum_{i=1}^{k}\lambda_i h_i(n)-\left\lfloor\sum_{i=1}^{k}\lambda_i h_i(n)\right\rfloor,\delta\right), n\in Z\right\}. \tag{10}$$

Next, we introduce the concept of compatibility between two systems which reflects both the idea of proximity between their notes and also whether their configuration is similar.

Definition 9. Let $\tilde{S}_q(\delta)$, $\tilde{T}_q(\delta)$ be two tuning systems with q notes. We say that $\tilde{S}_q(\delta)$ and $\tilde{T}_q(\delta)$ are α-compatible, if for each $2^{\tilde{s}_i}\in\tilde{S}_q(\delta)$ there is a unique $2^{\tilde{t}_i}\in\tilde{T}_q(\delta)$ such that

$$\mathrm{Comp}[2^{\tilde{s}_i},2^{\tilde{t}_j}]\geq\alpha.$$

The quantity α is the degree of interchangeability between $\tilde{S}_q(\delta)$ and $\tilde{T}_q(\delta)$ and the uniqueness required in Definition 9 guarantees that these systems have a similar distribution in the cycle of fifths.

Remark 2. It is important to note that the α-compatibility does not define a binary relation of equivalence in the set of tuning systems, because the transitive property is not verified.

6 Computational Results

We have made 48 recordings of 23 bars[2] of the Concert in E flat major Hob. VIIe, N. 1 for trumpet and orchestra of Franz Joseph Haydn (1732-1809) (see Fig. 4). All the recordings were played by the same player with a Bach® trumpet in B^b.

Fig. 4. Piece of the concert in E^b Major of Haydn

Our aim is to analyze the compatibility between the notes recorded and the tuned notes in the Pythagorean, Zarlinean and Equal Temperament (with 12 notes) systems. In our approach both kinds of notes are fuzzy.

The frequencies corresponding to a practical note (which is performed slightly differently at each recording) belong to the interval

$$[f_{min}, f_{max}],$$

where f_{min} (resp. f_{max}) is its lowest (resp. its highest) interpreted frequency (see Columns 1—2 in Table 2). The exact frequencies of some tuning systems[3] appear in Columns 3—6. The logarithms in base 2 of the frequencies in the table provide the values for t in Definition 5 and we also take $\Delta = 50$ cents (as the chromatic tuners) and $E = 6$ cents as tolerances (see [8], [10]).

---

[2] The different notes that appear in the fragment are $E^b{}_3$, F_3, G_3, $B^b{}_3$, $A^b{}_3$, $B^b{}_3$, C_4, D_4, $E^b{}_4$, F_4, G_4, $B^b{}_4$, B_3 and A_3.

[3] To be more precise, the Just Intonation system is a family of tuning systems and we have selected the Zarlinean [5].

Table 2. Exact frequencies of the notes that appear in the piece of Haydn

Note	Lower	Upper	Pythagor.	Zarlinean	Temp. (12)
E^b	305	316	309,0261	310,8272 311	,127
F	346	356	347,6543	350,3555 349	,2282
G	376	398	391,1111 396		391,9954
B^b	226	239	231,7695	233,5703 233	,0819
A^b	406	414	412,0347	414,4362 415	,3047
B^b	453	479	463,5391	467,1406 466	,1638
C	502	533	521,4815 528		523,253
D	571	598	586,6667 594		587,3296
E^b	587	636	618,0521	621,6544 622	,254
F	668	713	695,3086	700,7109 698	,4564
G	781	795	782,2222 792		783,9908
B^b	548	570	549,3797	559,4889 554	,3652
B	487	498	495	497,3235 493	,8833
A	440	443 440		440	440

Making use of (9) we construct the compatibility intervals obtained as

$$[\text{Comp(lower, tuning system)}, \text{Comp(upper, tuning system)}]$$

where the tuning systems considered are those in Table 2.

Table 3. Compatibility of the notes that appear in the piece by Haydn

Note	Interval of comp. Temp(12)	Interval of comp. Pythag.	Interval of comp. Zar.
E^b	[0,6556, 0,7309]	[0,6136, 0,7729]	[0,6723, 0,7142]
F	[0,6675, 0,8392]	[0,5893, 0,9174]	[0,7834, 0,7233]
G	[0,2787, 0,7368]	[0,3178, 0,6977]	[0,1027, 0,9127]
B^b	[0,4658, 0,5659]	[0,4681, 0,5635]	[0,4295, 0,6021]
A^b	[0,6077, 0,9455]	[0,7445, 0,9176]	[0,6439, 0,9817]
B^b	[0,5041, 0,5297]	[0,4319, 0,6018]	[0,4678, 0,5659]
C	[0,2821, 0,6804]	[0,3408, 0,6217]	[0,1257, 0,8368]
D	[0,5118, 0,6882]	[0,5313, 0,6687]	[0,3163, 0,8838]
E^b	[0, 0,6217]	[0,1075, 0,5044]	[0,0069, 0,6050]
F	[0,2281, 0,6432]	[0,3063, 0,5650]	[0,1723, 0,6990]
G	[0,7585, 0,9338]	[0,7194, 0,9729]	[0,7578, 0,9345]
B^b	[0,5184, 0,8001]	[0,3621, 0,9564]	[0,6407, 0,6777]
B	[0,7570, 0,8562]	[0,7179, 0,8953]	[0,6368, 0,9764]
A	[0,8823, 1]	[0,8823, 1]	[0,8823, 1]

In Table 3, we have underlined the compatibility intervals that are lower than 0.6. This means that its compatibility is lower than 60%. If we fix this percentage as the minimum to accept that the notes are compatible, we can see in the table that 50% of the notes are not acceptable in all the tuning systems.

These results are very useful to the player that needs to modify the pressure and/or the position to raise the level of compatibility.

7 Conclusions

Many musicians think that the technical treatment of musical concepts comes into conflict with their daily practice. However, their methods are used in this paper to make the concept of a musical note more flexible. In this framework, fuzzy mathematical rules and practice are the same thing. In fact, the adjustments that the musicians make constitute a method for increasing the compatibility level among systems. In this way, describing the tuning systems as fuzzy sets allows us to include the daily reality of musicians and their theoretical instruction in a mathematical structure. In our opinion, this constitutes a good model of reality.

From the idea of compatibility, the possibility of substituting a tuning system with another one arises. Therefore, when a tuning system presents many harmonic difficulties, such as not allowing certain transpositions, we can use a compatible system to avoid these disadvantages. On the other hand, knowing the compatibility between notes allows musicians to improve their performance by choosing between different tune positions, increasing lip pressure, etc.

Finally, we find it necessary to warn that many players and composers have difficulties when they work with proposals from the area of science. So, we should continue our attempts to make them feel more comfortable with technical arguments.

Acknowledgments. The authors acknowledge the kind collaboration of Dr. J. Ibañez in making the recordings used in computational tests and would also like to thank the financial support of research projects TIN2008-06872-C04-02 and TIN2006-10134 from the Science and Innovation Department of the Spanish government.

References

1. Benson, D.: Music: a Mathematical Offering. Cambridge University Press, Cambridge (2006)
2. Borup, H.: A History of String Intonation, `http://www.hasseborup.com/ahistoryofintonationfinal1.pdf`
3. Carlsson, C., Fullér, R.: Fuzzy Reasoning in Decision Making and Optimization. Physica-Verlag, Heidelberg (2002)
4. Dubois, D., Prade, H.: Fuzzy Sets and Systems:Theory and Applications. Academic Press, New York (1980)
5. Goldáraz Gaínza, J.J.: Afinación y temperamento en la música occidental, Alianza Editorial, Madrid (1992)

6. Hall, R.W., Josíc, K.: The mathematics of musical instruments. Amer. Math. Monthly 108, 347–357 (2001)
7. Halusca, J.: Equal temperament and pythagorean tuning:a geometrical interpretation in the plane. Fuzzy Sets and Systems 114, 261–269 (2000)
8. Haluska, J.: The Mathematical Theory of Tone Systems. Marcel Dekker, Inc., Bratislava (2005)
9. Lattard, J.: Gammes et tempéraments musicaux, Masson Éditions, Paris (1988)
10. Liern, V.: Fuzzy tuning systems: the mathematics of the musicians. Fuzzy Sets and Systems 150, 35–52 (2005)
11. Zadeh, L.A.: Fuzzy sets as a basis for a theory of possibility. Fuzzy Sets and Systems 1, 3–28 (1978)

Formal Diatonic Intervallic Notation

Jack Douthett[1] and Julian Hook[2]

[1] University of New Mexico, Albuquerque, NM, USA
douthett@comcast.net
[2] Indiana University, Bloomington, IN, USA
juhook@indiana.edu

Abstract. Numbers called *quality modifiers* are used to identify interval qualities: 0 numerically represents perfect, ½ represents major, –½ represents minor, and so on. These modifiers are linked with *diatonic class intervals* as ordered pairs that mimic common interval notation. For example, a minor third is represented by (–½, 2). A binary operator is constructed that allows these ordered pairs to be added consistent with our expectations. Similarly, *accidental modifiers* numerically identify the number of sharps or flats attached to a given note: 0 indicates no attached accidentals, negative integers indicate the number of flats attached, and positive integers indicate the number of sharps attached. These modifiers are linked with *diatonic classes* as ordered pairs that mimic common note names. For example, the note G♭ is represented by (–1,4) and G𝄪 by (2,4). Intervals and notes represented by these ordered pairs are said to be in *MD-notation* (MD for *modifier-diatonic*). A group action and generalized interval system are defined for intervals and notes in MD-notation. An implied quarter-tone system is also discussed.

Keywords: quality modifier, accidental modifier, diatonic class, diatonic class interval, PD-notation, MD-notation, enharmonic system, maximally even set, generalized interval system.

1 Introduction

While current numerical diatonic intervallic notations reveal much about diatonic intervals, these notations in general do not mimic what we call *common diatonic notation*; that is, the interval qualities (major, minor, perfect, ...) are not immediately apparent in the notation. As it turns out, to mimic common diatonic notation in a mathematically consistent way is surprisingly complicated. For example, sometimes the sum of two major intervals yields an augmented interval (M2 + M3 = A4), and other times the sum yields another major interval (M2 + M2 = M3). How would one construct a group of intervals that mimic common intervals and overcome this seeming ambiguity? This problem is addressed in Sections 2 and 3. In Section 4 Hook's [1] notation that models common notes (e.g., A♭, F♯, ...) by linking *accidental modifiers* and diatonic classes is coupled with Clough and Douthett's [2] algorithm for maximally even sets to define a set of notes on which the group described above will act. Section 5 addresses the *generalized interval system* induced by this action, and Section 6 briefly discusses extensions of this notation to other equal-tempered systems.

E. Chew, A. Childs, and C.-H. Chuan (Eds.): MCM 2009, CCIS 38, pp. 104–114, 2009.

2 Quality Modifiers

In this section and all that follow, mod 7 diatonic class intervals (*dcis*) will be represented by italicized uppercase letters A and B, and mod 12 pitch class intervals (*pcis*) will be represented by italicized lowercase letters a and b. We will shortly define *quality modifiers*, which will be represented by lowercase Greek letters α and β. For interval qualities, lowercase letters d and m will denote diminished and minor intervals, while uppercase letters P, M, and A will denote perfect, major, and augmented intervals.

Intuitively, the adjectives used to describe interval quality—perfect, major, minor, augmented, diminished—convey something about how large a given interval is relative to an "average" interval of the same numerical (generic) size. For example, a m3 is a little smaller than an "average" third, while an A6 is quite a bit larger than an "average" sixth. With the aim of making this intuition precise, we first consider what is meant by an "average" interval.

Since there are seven generic steps to the octave, it seems logical to say that the average interval of a *dci*, call it A, is the fraction $A/7$ of an octave; that is, the average second (*dci* 1) is 1/7 of an octave, the average third (*dci* 2) is 2/7 of an octave, and so on. Equivalently, measured in equal-tempered semitones, the average interval of *dci* A is $12A/7$: the average second is 12/7, the average third is 24/7, and so on. Note that we could have arrived at the same result by averaging the interval sizes within the diatonic scale. For example, in any diatonic scale there are four m3s (of size 3 semitones) and three M3s (4 semitones), and the mathematical average of four 3s and three 4s is again 24/7.

Table 1 tabulates the intervals of all common qualities in the usual 12-note equal-tempered system (12-ETS). The first column gives the common names of the intervals. The next two columns give the *dci* and *pci*, respectively, of the interval—the generic and specific sizes of the interval in the terminology of Clough and Myerson [3, 4]. The fourth column gives the size of the corresponding "average" interval as described above; that is, $12A/7$ where A is the *dci*, or equivalently an interval of size A in a 7-ETS measured in 12 equal-tempered semitones. The fifth column labeled "Difference" tabulates the difference between the *pci* and the average 7-ETS interval, counted positive if the *pci* is larger than average and negative if the *pci* is smaller than average (Column 3 less Column 4). The last column is the *quality modifier* (*qm*) of the interval, which is the difference in Column 5 rounded to the nearest half-integer. Thus, if $[\,\cdot\,]$ is the function that rounds to the nearest integer, then the *qm* of an interval with *dci* A and *pci* a is

$$\alpha = \frac{1}{2}\left[2\left(a - \frac{12A}{7}\right)\right] = a - l(A) \text{ where } l(A) = \frac{1}{2}\left[\frac{24A}{7}\right]. \tag{1}$$

Observe that (1) may be used to determine a from α and A, or to determine $l(A)$ from a and α. Some combinations of a and α are not realized, however, as $l(A)$ takes on a limited number of values. (The values of $l(A)$ for $A = 0, 1, \cdots, 6$ are 0, 1½, 3½, 5, 7, 8½, 10½.)

Table 1. Interval qualities in the familiar 12-ETS

Common Interval Name	*dci*	*pci*	*7-ETS Int.* (mod 12)	*Difference q*	*m*
Quality-Size A		a	$\frac{12A}{7}$	$a-\frac{12A}{7}$	$\alpha = a - l(A)$
d1[1]	0	−1 (11)	0	−1	−1
P1	0	0	0	0	0
A1	0	1	0	+1	+1
d2	1	0	$1\frac{5}{7}$	$-1\frac{5}{7}$	$-1\frac{1}{2}$
m2	1	1	$1\frac{5}{7}$	$-\frac{5}{7}$	$-\frac{1}{2}$
M2	1	2	$1\frac{5}{7}$	$+\frac{2}{7}$	$+\frac{1}{2}$
A2	1	3	$1\frac{5}{7}$	$+1\frac{2}{7}$	$+1\frac{1}{2}$
d3	2	2	$3\frac{3}{7}$	$-1\frac{3}{7}$	$-1\frac{1}{2}$
m3	2	3	$3\frac{3}{7}$	$-\frac{3}{7}$	$-\frac{1}{2}$
M3	2	4	$3\frac{3}{7}$	$+\frac{4}{7}$	$+\frac{1}{2}$
A3	2	5	$3\frac{3}{7}$	$+1\frac{4}{7}$	$+1\frac{1}{2}$
d4	3	4	$5\frac{1}{7}$	$-1\frac{1}{7}$	−1
P4	3	5	$5\frac{1}{7}$	$-\frac{1}{7}$	0
A4	3	6	$5\frac{1}{7}$	$+\frac{6}{7}$	+1
d5	4	6	$6\frac{6}{7}$	$-\frac{6}{7}$	−1
P5	4	7	$6\frac{6}{7}$	$+\frac{1}{7}$	0
A5	4	8	$6\frac{6}{7}$	$+1\frac{1}{7}$	+1
d6	5	7	$8\frac{4}{7}$	$-1\frac{4}{7}$	$-1\frac{1}{2}$
m6	5	8	$8\frac{4}{7}$	$-\frac{4}{7}$	$-\frac{1}{2}$
M6	5	9	$8\frac{4}{7}$	$+\frac{3}{7}$	$+\frac{1}{2}$
A6	5	10	$8\frac{4}{7}$	$+1\frac{3}{7}$	$+1\frac{1}{2}$
d7	6	9	$10\frac{2}{7}$	$-1\frac{2}{7}$	$-1\frac{1}{2}$
m7	6	10	$10\frac{2}{7}$	$-\frac{2}{7}$	$-\frac{1}{2}$
M7	6	11	$10\frac{2}{7}$	$+\frac{5}{7}$	$+\frac{1}{2}$
A7	6	12 (0)	$10\frac{2}{7}$	$+1\frac{5}{7}$	$+1\frac{1}{2}$

[1] In practice d1 can be interpreted as d8. The notation d1 is used here because the *dcis* are reduced mod 7.

When the differences in Column 5 of Table 1 are rounded to the nearest half-integer, the results range from −1½ (for "much smaller than average" intervals) to +1½ (for "much larger than average" intervals). These differences and the half-integers to which they round (*qms*) are reorganized and consolidated in the first two columns of Table 2. The last column of Table 2 reorganizes and consolidates the common names in Column 1 of Table 1 so that they correspond to their *qms* in Column 2 of Table 2. The intervals whose *qms* are 0 are perfect (middle row in Table 2); the intervals whose *qms* are +½ or −½ are major or minor, respectively, and those with larger *qms* (positive or negative) are augmented or diminished; a raised or lowered perfect interval has a *qm* of +1 or −1, respectively, while a raised major or lowered minor interval has a *qm* of +1½ or −1½.

Table 2. Differences, Quality Modifiers, Interval Quality, and Examples

Differences	*qm*	*Quality*	*Common Names*
$-1\frac{5}{7}, -1\frac{4}{7}, -1\frac{3}{7}, -1\frac{2}{7}$	−1½	d (lowered m)	d2, d3, d6, d7
$-1\frac{1}{7}, -1, -\frac{6}{7}$	−1	d (lowered P)	d1, d4, d5
$-\frac{5}{7}, -\frac{4}{7}, -\frac{3}{7}, -\frac{2}{7}$	−½	m	m2, m3, m6, m7
$-\frac{1}{7}, 0, +\frac{1}{7}$	0	P	P1, P4, P5
$+\frac{2}{7}, +\frac{3}{7}, +\frac{4}{7}, +\frac{5}{7}$	+½	M	M2, M3, M6, M7
$+\frac{6}{7}, +1, +1\frac{1}{7}$	+1	A (raised P)	A1, A4, A5
$+1\frac{2}{7}, +1\frac{3}{7}, +1\frac{4}{7}, +1\frac{5}{7}$	+1½	A (raised M)	A2, A3, A6, A7

The *qms* of multiply augmented and diminished intervals can also be determined. For example, the *dci* and *pci* of a doubly augmented sixth (AA6) are $A = 5$ and $a = 11$, respectively. Then

$$\alpha = 11 - \frac{1}{2}\;\frac{24 \cdot 5}{7} = +2½ . \tag{2}$$

So the *qm* of AA6 is 2½. In general, to get the *qm* of a *k*th-augmented (diminished) interval, one adds $k-1$ to (subtracts $k-1$ from) the *qm* of the respective augmented (diminished) interval. Thus, since the *qm* of A6 is 1½, the *qm* of AA6 is 2½.

3 Group Structures

In previous work on musical intervals, Brinkman [5] and Agmon [6] define an interval as an ordered pair in which the first and second coordinates are a *pci* and *dci*, respectively. We say these intervals are in *PD-notation* and denote the group of all such intervals as $I_{\mathrm{PD}} = I_{12} \times I_7$ where I_{12} and I_7 are the groups of *pcis* and *dcis*. This group is a cyclic group of order 84 generated by $[\![7,4]\!]$ (the musical fifth).[2] The binary operator + sums intervals coordinate-wise: the first coordinates are summed mod 12, and the second are summed mod 7. Then the sum of common intervals such as

$$\mathrm{M2}+\mathrm{M3}=\mathrm{A4} \text{ and } \mathrm{M2}+\mathrm{M2}=\mathrm{M3} \tag{3}$$

are represented in PD-notation as

$$[\![2,1]\!]+[\![4,2]\!]=[\![6,3]\!] \text{ and } [\![2,1]\!]+[\![2,1]\!]=[\![4,2]\!]. \tag{4}$$

While it is easy for musicians to determine the quality of the intervals in (4), the notation by itself gives no hint of interval quality. If, however, the *pcis* in the first coordinates of the intervals are replaced by their corresponding *qms* as determined by (1), the qualities of the intervals are immediately obvious in the notation:

$$[\![½,1]\!]\oplus[\![½,2]\!]=[\![1,3]\!] \text{ and } [\![½,1]\!]\oplus[\![½,1]\!]=[\![½,2]\!]. \tag{5}$$

While coordinate-wise addition works for the first sum in (5) ($½+½=1$), it does not work for the second ($½+½\neq ½$). So, it is necessary to construct a binary operator that sometimes adds coordinate-wise and other times does not. Thus $\oplus$ is adopted as the binary operation instead of the usual $+$.

In the process of constructing this binary operation, we first let

$$Q_{12}=\{-5½,-5,-4½,\cdots,5½,6\} \tag{6}$$

be the group of *qm* classes under addition mod 12, and let I_{MD} be the image of the map $\tau: I_{\mathrm{PD}} \to I_{\mathrm{MD}}$ defined by

$$\tau([\![a,A]\!])=[\![\alpha,A]\!] \text{ where } \alpha=a-l(A)\in Q_{12}. \tag{7}$$

We say that the intervals in I_{MD} are in *MD-notation* (MD for modifier-diatonic). From (1), τ is a bijection and maps each interval in PD-notation to the same interval in MD-notation. To discover how intervals are summed in MD-notation so that they mirror the sum of common intervals, we require that τ map the sum of intervals in PD-notation to the sum of the same intervals in MD-notation; that is,

$$\tau([\![a,A]\!]+[\![b,B]\!])=\tau([\![a,A]\!])\oplus\tau([\![b,B]\!]). \tag{8}$$

Finding the binary operator $\oplus$ is somewhat backwards from many problems in group theory texts, which give the binary operations of two groups and ask the reader to find a map (homomorphism) between the groups that preserves the operations. Our task is: given the binary operator of one group and a map that preserves binary operations, find the binary operator of the other group.

Proposition 1. *Let $A,B\in I_7$, and define f as follows:*

$$f(A,B)=l(A)+l(B)-l(A+B). \tag{9}$$

Let $[\![\alpha,A]\!],[\![\beta,B]\!]\in I_{\mathrm{MD}}$. If (8) holds then

---

[2] We use double brackets to denote intervals to avoid confusion with parenthetical ordered pairs which will later be used to represent notes.

$$[\![\alpha, A]\!] \oplus [\![\beta, B]\!] = [\![\alpha + \beta + f(A, B), A + B]\!] \in I_{\mathrm{MD}}. \tag{10}$$

Proof: Let $[\![a, A]\!]$ and $[\![b, B]\!]$ be the preimages of $[\![\alpha, A]\!]$ and $[\![\beta, B]\!]$ under τ, respectively. Then by (1), (7), and (8),

$$\begin{aligned}
[\![\alpha, A]\!] \oplus [\![\beta, B]\!] &= \tau([\![a, A]\!]) \oplus \tau([\![b, B]\!]) \\
&= \tau([\![a, A]\!] + [\![b, B]\!]) \\
&= \tau([\![a + b, A + B]\!]) \\
&= [\![a + b - l(A + B), A + B]\!] \\
&= [\![\alpha + l(A) + \beta + l(B) - l(A + B), A + B]\!] \\
&= [\![\alpha + \beta + f(A, B), A + B]\!] \in I_{\mathrm{MD}}. \blacksquare
\end{aligned}$$

It follows that I_{MD} is a cyclic group of order 84 and $\tau : I_{\mathrm{PD}} \to I_{\mathrm{MD}}$ is an isomorphism. Moreover, for any given $A, B \in I_7$, $f(A, B)$ takes on only one of three values mod 12: ½ , 0, or ½. Thus, f either leaves the sum of the *qms* unchanged ($f(A, B) = 0$), or it *nudges* their sum by the smallest non-zero amount ($f(A, B) = \pm ½$). Hence $f(A, B)$ can be thought of as a measure of the *deviation* from the sum of the *qms*.

Returning to (5), note that $f(1, 2) = 0$ for the first sum (1½ + 3½ – 5 = 0). Thus, the resultant *qm* is simply the sum of the other two. For the second sum, $f(1,1) = –½$ (1½ + 1½ – 3½ = –½); so, the resultant *qm* is ½ less than the sum of the other *qms*.

4 Group Actions

In what follows, pitch classes (*pcs*) will be represented by italicized lowercase letters *m* and *n*. The set of all *pcs* will be denoted U_{12}. Similarly, diatonic classes (*dcs*) will be represented by italicized uppercase letters *M* and *N*, and the set of all *dcs* will be denoted U_7. Now let $U_{\mathrm{PD}} = U_{12} \times U_7$. We say the notes (ordered pairs) in U_{PD} are in *PD-notation*. The action of I_{PD} on U_{PD} is defined by

$$[\![a, A]\!](m, M) = (a + m, A + M) \in U_{\mathrm{PD}} \tag{11}$$

where the first and second coordinates are reduced mod 12 and mod 7 [5, 6].

Consider the following statements:

Statement 1. A P5 above E♭ is B♭.
Statement 2. A M3 above E♭ is G. (12)

In PD-notation, these statements correspond to

$$[\![7,4]\!](3,2) = (10,6) \text{ and } [\![4,2]\!](3,2) = (7,4), \tag{13}$$

which are consistent with (11). Note that neither the interval qualities nor the accidentals attached to the notes in the statements in (12) are conveyed in the notation in (13). While MD-notation solves this problem for intervals, another notation is needed for accidentals attached to notes. Following Hook's [1] work on enharmonic systems, we define *accidental modifiers* (*ams*)—represented with lowercase Greek letters μ and ν—as follows: A note with *k* flats (sharps) attached has an *am* of $\mu = -k$ ($\mu = +k$).

Now let A_{12} be the set of *am* classes mod 12:

$$A_{12} = \{-5, -4, \cdots, -1, 0, 1, \cdots, 5, 6\} \tag{14}$$

Then each note $(M, m) \in U_{\mathrm{PD}}$ is associated with precisely one *am* class in A_{12}. For example, $(7,3) \in U_{\mathrm{PD}}$ is some type of F (*dc* 3) that is 7 half-steps above C (*pc* 7). Therefore the note is F×, implying that the *am* is 2.

To determine the *am* of a note given its *pc* and *dc*, the Clough-Douthett [2] algorithm for maximally even sets is needed:

$$J_{c,d}^{r} = \left\{ J_{c,d}^{r}(k) \right\}_{k=0}^{d-1} \text{ where } J_{c,d}^{r}(k) = \frac{ck + r}{d} \,. \tag{15}$$

The symbol $J_{c,d}^{r}$ is called the *J-representation* of the set, and the subscripts *c* and *d* are the chromatic and diatonic cardinalities, respectively. The superscript *r*, called the *mode index*, is an integer between 0 and $c-1$. For $c = 12$ and 7d = , and for each *r*, $0 \le r \le 11$, $J_{12,7}^{r}$ is a pcset that represents a diatonic scale. For $r = 0$, the pcset represents D♭ major; for $r = 1$ it is A♭, F♯, major, and so on. The set representing the C major scale—the set of all *pcs* associated with no sharps or flats—is obtained when $r = 5$:

$$J_{12,7}^{5} = \left\{ J_{12,7}^{5}(k) \right\}_{k=0}^{6} = \{0, 2, 4, 5, 7, 9, 11\} \,. \tag{16}$$

It follows that the *am* of any $(m, M) \in U_{\mathrm{PD}}$ is

$$\mu = m - J_{12,7}^{5}(M) \,. \tag{17}$$

Then for $(7,3) \in U_{\mathrm{PD}}$ (F×), (17) implies $\mu = 7 - 5 = 2$.

As illustrated in (13), in PD-notation the action of $[\![a, A]\!] \in I_{\mathrm{PD}}$ on the note $(m, M) \in U_{\mathrm{PD}}$ is simply a matter of adding the interval and note coordinate-wise. But in MD-notation, statements 1 and 2 in (12) are expressed as

$$[\![0,4]\!](-1,2) = (-1,6) \text{ and } [\![½, 2]\!](-1,2) = (0,4) \,. \tag{18}$$

Coordinate-wise addition works for the first equation (0 + (–1) = – 1), but not for the second (½ + (–1) ≠ 0). As with interval composition in MD-notation, our task is to find an action that sometimes adds coordinate-wise and other times does not.

Let $\theta : U_{\mathrm{PD}} \to U_{\mathrm{MD}}$ be defined by

$$\theta(m, M) = (\mu, M) \text{ where } \mu = m - J_{12,7}^{5}(M) \in A_{12} \,. \tag{19}$$

Then by (17) $U_{\mathrm{MD}} = \theta(U_{\mathrm{PD}})$, and θ is a bijection that maps each note in PD-notation to the same note in MD-notation. Since (11) defines the appropriate action for common intervals and notes in PD-notation, we need to "translate" (11) into MD-notation:

$$\tau([\![a, A]\!])\,\theta(m, M) = \theta([\![a, A]\!](m, M)) \in U_{\mathrm{MD}} \,. \tag{20}$$

Group actions related in this way are said to be *permutation isomorphic* [7, 8].

Proposition 2. *Let* $A \in I_7$ *and* $M \in U_7$*, and define g as follows:*

$$g(A,M) = l(A) + J^5_{12,7}(M) - J^5_{12,7}(A+M). \tag{21}$$

Let $[\![\alpha, A]\!] \in I_{\text{MD}}$ *and* $(\mu, M) \in U_{\text{MD}}$*. If* (20) *holds, then the action of* I_{MD} *on* U_{MD} *is given by*

$$[\![\alpha, A]\!](\mu, M) = (\alpha + \mu + g(A,M), A+M) \in U_{\text{MD}}. \tag{22}$$

Proof: Let $[\![a, A]\!]$ be the preimage of $[\![\alpha, A]\!]$ under τ, and let (m, M) be the preimage of (μ, M) under θ. It follows from (1), (11), (17), and (20) that

$$\begin{aligned}
[\![\alpha, A]\!](\mu, M) &= \tau[\![a, A]\!]\theta(m, M) \\
&= \theta([\![a, A]\!](m, M)) \\
&= \theta(a+m, A+M) \\
&= (a + m - J^5_{12,7}(A+M), A+M) \\
&= (\alpha + l(A) + \mu + J^5_{12,7}(M) - J^5_{12,7}(A+M), A+M) \\
&= (\alpha + \mu + g(A,M), A+M) \in U_{\text{MD}}. \blacksquare
\end{aligned}$$

Since the action of I_{PD} on U_{PD} is simply transitive, the action of I_{MD} on U_{MD} must also be simply transitive. Also similar to *f*, the function *g* is a measure of deviation from the sum $\alpha + \mu$, but now for each $A \in I_7$ and $M \in U_7$, $g(A,M)$ takes on one of five values mod 12: –1, –½, 0, ½, or 1.

Returning to the first action in (18), $g(4,2) = 0$ (7 + 4 – 11 = 0). It follows that the *am* of the resultant note is the sum of the *qm* of the interval and the *am* of the note on which it acts. On the other hand, $g(2,2) = ½$ in the second action (3½ + 4 – 7 = ½). So the *am* of the resultant note is ½ more than the sum of the *qm* of the interval and the *am* of the note.

5 Generalized Interval Systems

Since in both PD- and MD-notations the actions of the groups of intervals on the sets of notes are simply transitive, there are GISes [9] associated with both notations. The construction of a GIS in PD-notation is straightforward: For $(m, M), (n, N) \in U_{\text{PD}}$, the GIS $(U_{\text{PD}}, I_{\text{PD}}, \text{int}_{\text{PD}})$ in PD-notation is defined by

$$\text{int}_{\text{PD}}((m,M),(n,N)) = [\![n-m, N-M]\!] \in I_{\text{PD}}. \tag{23}$$

As with interval composition and group action, the definition for int_{MD} in the GIS $(U_{\text{MD}}, I_{\text{MD}}, \text{int}_{\text{MD}})$ is a bit more complicated than int_{PD}.

Proposition 3. *Let* $M, N \in U_7$*, and define h as follows:*

$$h(M,N) = J^5_{12,7}(N) - J^5_{12,7}(M) - l(N-M). \tag{24}$$

Let $(\mu,M),(\nu,N)\in U_{\text{MD}}$. *Then the GIS* $(U_{\text{MD}},I_{\text{MD}},\text{int}_{\text{MD}})$ *is defined by*

$$\text{int}_{\text{MD}}\left((\mu,M),(\nu,N)\right)=[\![\nu-\mu+h(M,N),N-M]\!]\in I_{\text{MD}}. \tag{25}$$

Proof: Let (m,M) and (n,N) be the preimages of (μ,M) and (ν,N) under θ, respectively. Then by (7) and (17),

$$\begin{aligned}\text{int}_{\text{MD}}\left((\mu,M),(\nu,N)\right)&=\tau\left([\![n-m,N-M]\!]\right)\\&=[\![n-m-l(N-M),N-M]\!]\\&=[\![\nu+J_{12,7}^{5}(N)-\mu-J_{12,7}^{5}(M)-l(N-M),N-M]\!]\\&=[\![\nu-\mu+h(M,N),N-M]\!]\in I_{\text{MD}}.\blacksquare\end{aligned}$$

In MD-notation, the statement "the interval from E♭ to G is a M3" corresponds to

$$\text{int}_{\text{MD}}\left((-1,2),(0,4)\right)=[\![½,2]\!]. \tag{26}$$

From (24), (25), and the left side of (26), the qm of the resultant interval should be $0-(-1)+h(2,4)=0+1+7-4-3½=½$, which is consistent with the right side of (26). Although tedious, it is straightforward to verify that

$$\text{int}_{\text{MD}}\left((\mu,M),(\nu,N)\right)\oplus\text{int}_{\text{MD}}\left((\nu,N),(\rho,P)\right)=\text{int}_{\text{MD}}\left((\mu,M),(\rho,P)\right) \tag{27}$$

for all $(\mu,M),(\nu,N),(\rho,P)\in U_{\text{MD}}$.

Similar to f and g in (9) and (21), the function h in (24) takes on only a few values mod 12: –½, 0, and ½. But unlike f and g, which measure deviations in the *sums* of modifiers, h measures the deviation in the *difference* of modifiers $\nu-\mu$.

6 Coda

There is a surprise ending for MD-notation; within this notation there is a disguised quartertone system. Recall that there are 84 intervals in I_{MD}, and they all come from the set $Q_{12}\times I_7$. But the cardinality of $Q_{12}\times I_7$ is 168. Thus I_{MD} contains only half the members of the set $Q_{12}\times I_7$. So, what happens if we adopt $\oplus$ as the binary operator for $Q_{12}\times I_7$? In fact, it can be shown that I_{MD} is a subgroup of $I_{\text{MD}}^{*}=(Q_{12}\times I_7,\oplus)$ and that in this parent group, every interval comes in perfect, minor, and major flavors. Moreover, I_{MD}^{*} is a cyclic group of order 168 generated by $[\![0,2]\!]$, which can be interpreted as a P3 and lies halfway between a m3 and a M3. Thus, the length of the P3 is 3½ semitones. That this length is half that of the P5 (7 semitones) is reflected in the following composition in I_{MD}^{*}:

$$[\![0,4]\!]=[\![0,2]\!]\oplus[\![0,2]\!]. \tag{28}$$

To define an action for I_{MD}^{*}, it is necessary to double the size of A_{12}:

$$A_{12}^{*}=\{-5½,-5,-4½,\cdots,5½,6\}. \tag{29}$$

Then the set of quartertone notes is $U^*_{\mathrm{MD}} = A^*_{12} \times U_7$. An *am* of –½ corresponds to a half-flat. For example, $(-½, 2)$ represents a half-flatted E, which lies halfway between E and E♭ (between $(0,2)$ and $(-1,2)$). Similarly, a half-sharped F $(½, 3)$ lies halfway between F and F♯ (between $(0,3)$ and $(1,3)$). With this machinery in place, it is not difficult to construct the GIS $\left(U^*_{\mathrm{MD}}, I^*_{\mathrm{MD}}, \mathrm{int}^*_{\mathrm{MD}}\right)$.

This approach can also be easily generalized to other microtonal universes. Consider Balzano's [10] 20-fold system. The scales in this 20-tet system are nine-note maximally even sets with J-representations $J^r_{20,9}$ where $0 \le r \le 19$. If $a_{\mathrm{B}} \in I_{20}$ and $A_{\mathrm{B}} \in I_9$ are corresponding *pcis* and *dcis* in this system, then the *qm* for this pair is

$$\alpha_{\mathrm{B}} = \frac{1}{2}\ 2\ a_{\mathrm{B}} - \frac{20A_{\mathrm{B}}}{9} \quad = a_{\mathrm{B}} - l_{\mathrm{B}}\left(A_{\mathrm{B}}\right) \text{ where } l_{\mathrm{B}}\left(A_{\mathrm{B}}\right) = \frac{1}{2}\ \frac{40A_{\mathrm{B}}}{9}\ . \tag{30}$$

Then similar to the *dcis* in the 12-tet system, each *dci* in the 20-tet system comes in either major and minor intervals (*ams* ½ and –½) or in a perfect interval (*am* 0), but not both: *dcis* 2, 3, 6, and 7 come in major and minor flavors and *dcis* 0, 1, 4, 5, and 8 come in perfect flavors. In view of the important role the major scale plays in Western music, it would seem reasonable to ask which rotations of Balzano's scales might represent the "major" scales. By observing that the intervals from the root of a major scale in the 12-tet system are either major or perfect, one might speculate that the same is true for Balzano's scales. Then in terms of *pcs* mod 20,

$$\langle 0,2,5,7,9,11,14,16,18 \rangle \tag{31}$$

would be Balzano's "major scale" that begins on *pc* 0. The set in (31) can also be interpreted as the set of natural notes; that is, in MD-notation (31) can be written as the note set

$$\langle (0,0),(0,1),(0,2),(0,3),(0,4),(0,5),(0,6),(0,7),(0,8) \rangle. \tag{32}$$

If *pc* 0 represents the note C in the 20-tet system, (31) can also determine the 20-tet keyboard configuration; (31) is the set of white keys and its complement is the set of black keys.

This approach can also apply to the study of scale systems investigated by Bohlen [11], Mathews et al. [12], Agmon [6], Clough and Douthett [2], Brinkman [5], Zweifel [13], Krantz and Douthett [14], Hook [1], and others.

References

1. Hook, J.: Enharmonic Systems: A Theory of Key Signatures, Enharmonic Equivalence, and Diatonicism. J. Math. Mus. 1, 99–120 (2007)
2. Clough, J., Douthett, J.: Maximally Even Sets. J. Mus. Theory 35, 93–173 (1991)
3. Clough, J., Myerson, G.: Variety and Multiplicity in Diatonic Systems. J. Mus. Theory 29, 249–270 (1985)
4. Clough, J., Myerson, G.: Musical Scales and the Generalized Cycle of Fifths. Am. Math. Monthly 93, 695–701 (1986)

5. Brinkman, A.: A Binomial Representation of Pitch for Computer Processing of Musical Data. Mus. Theory Spectrum 8, 44–57 (1986)
6. Agmon, E.: A Mathematical Model of the Diatonic System. J. Mus. Theory 33, 1–25 (1989)
7. Dixon, J., Mortimer, B.: Permutation Groups. Springer, New York (1996)
8. Peck, R.: Generalized Commuting Groups (unpublished) (2006)
9. Lewin, D.: Generalized Musical Intervals and Transformations. Yale University Press, New Haven (1987)
10. Balzano, G.: The Group-Theoretic Description of 12-Fold and Microtonal Pitch Systems. Computer Mus. J. 4(4), 66–84 (1980)
11. Bohlen, H.: Tonstufen in der Doudezime. Acustica 39, 61–136 (1978)
12. Mathews, M., Pierce, J., Reeves, A., Roberts, L.: Theoretical and Experimental Explorations of the Bohlen-Pierce Scale. J. Acoust. Soc. Am. 84, 1214–1222 (1988)
13. Zweifel, P.: Generalized Diatonic and Pentatonic Scales: A Group-Theoretic Approach. Perspectives New Mus. 34(1), 140–161 (1996)
14. Krantz, R., Douthett, J.: Construction and Interpretation of Equal-Tempered Scales Using Frequency Ratios, Maximally Even Sets, and P-Cycles. J. Acoust. Soc. Am. 107, 2725–2734 (2000)

Determining Feature Relevance in Subject Responses to Musical Stimuli

Morwaread M. Farbood[1] and Bernd Schoner[2]

[1] Music and Audio Research Laboratory (MARL),
35 W. 4th St., New York University, New York, NY 10012, USA
[2] ThingMagic Inc., One Broadway, Cambridge, MA 02142, USA

Abstract. This paper presents a method that determines the relevance of a set of signals (musical features) given listener judgments of music in an experimental setting. Rather than using linear correlation methods, we allow for nonlinear relationships and multi-dimensional feature vectors. We first provide a methodology based on polynomial functions and the least-mean-square error measure. We then extend the methodology to arbitrary nonlinear function approximation techniques and introduce the Kullback-Leibler Distance as an alternative relevance metric. The method is demonstrated first with simple artificial data and then applied to analyze complex experimental data collected to examine the perception of musical tension.

1 Introduction

There are two generic types of responses that can be collected in an experimental setting where subjects are asked to make judgments on musical stimuli. The first is a retrospective response, where the listener only makes a judgment after hearing the musical excerpt; the second is a real-time response where judgments are made while listening. The latter has become increasingly popular among experimental psychologists as an effective means of collecting data. In particular, studies on musical tension have often employed real-time collection methods (Nielsen 1983; Madson and Fredrickson 1993; Krumhansl 1996; Bigand et al. 1996; Bigand & Parncutt 1999; Toiviainen & Krumhansl 2003; Lerdahl & Krumhansl 2007). The validity of this type of data collection is indicated by the high inter- and intra-subject correlation between subject responses and, more importantly, the indication that these responses correspond to identifiable musical structures (Toiviainen & Krumhansl 2003).

In this paper we propose a method to detect and quantify the relevance of individual features in complex musical stimuli where both the musical features describing the stimuli and the subject responses are real-valued. While the method can be used with most types of auditory or visual stimuli and most types of responses,[1] the method discussed here was developed for the purposes of under-

[1] For example, the response signal can be brain activity, as measured by imaging technology (Schoner 2000), a general biological response such as skin conductivity (Picard et al. 2001), or direct subject input by means of a computer interface.

E. Chew, A. Childs, and C.-H. Chuan (Eds.): MCM 2009, CCIS 38, pp. 115–129, 2009.

standing how musical structures affect listener responses to tension. Our analysis is based on the assumption that perceived tension is a function of various salient musical parameters varying over time, such as harmony, pitch height, onset frequency, and loudness (Farbood 2006). It is the objective of this paper to formulate a mathematically sound approach to determine the relative importance of each individual feature to the perception of tension.

In the following sections, we will first provide a methodology based on polynomial functions and the least-mean-square error measure and then extend the methodology to arbitrary nonlinear function approximation techniques. We will first verify our approach with simple artificial data and then apply it to complex data from a study exploring the perception of musical tension.

2 Prior Work

In this paper we rely on prior art from two distinct fields: (A) the statistical evaluation of experimental and continuous data, mostly using variants of linear correlation and regression (Gershenfeld 1999b) and (B) feature selection for high-dimensional pattern recognition and function fitting in machine learning (Mitchell 1997).

(A) is helpful for our task at hand, but its limitation stems from the assumption of linearity. The importance of a feature is determined by the value of the correlation coefficient between a feature vector and a response signal: the closer the correlation value to 1 or to -1, the more important the feature. A variant of this approach—based on the same mathematical correlation—uses the coefficients in a linear regression model to indicate the relevance of a feature.

(B) offers a large amount of literature mostly motivated by high-dimensional, nonlinear machine-learning problems facing large data sets. Computational limitations make it necessary to reduce the dimensionality of the available feature set before applying a classifier algorithm or a function approximation algorithm. The list of common techniques includes Principle Component Analysis (PCA), which projects the feature space on the most relevant (linear) subset of components, and Independent Component Analysis (ICA), which is the nonlinear equivalent of PCA (Gershenfeld 1999b). Both PCA and ICA are designed to transform the feature set for the purpose of estimating the dependent signal, but they do not relate an individual feature to the dependent signal. In fact, most prior work in machine learning is focused on estimating the dependent signal, not the significance of individual features.

Prior art can also be found in the field of information theory. Koller & Sahami (1996) developed a methodology for feature selection in multivariate, supervised classification and pattern recognition. They select a subset of features using a subtractive approach, starting with the full feature set and successively removing features that can be fully replaced by a subset of the other features. Koller & Sahami use the information-theoretic cross-entropy, also known as KL-distance (Kullback & Leibler 1951) in their work.

3 Feature Relevance Measured by Polynomial Least-Mean Square Estimation

In this paper, we estimate the relevance of a particular musical feature x_i by computing the error between the actual subject response signal y and the estimation $\hat{y}$ of the same. We first build a model based on the complete feature set F and derive the least-mean-square error E from $\hat{y}$ and y. We then build models for each of the feature sets F_i, where F_i includes all the features except x_i, and compute the errors E_i based on $\hat{y}_i$ and y. We define the Relevance Ratio $R_i = E/E_i$ and postulate that R_i is a strong indicator of the relevance of x_i for y.

We start by selecting an appropriate model to estimate $\hat{y}$, keeping in mind our goal of overcoming the linearity constraint of common linear techniques. We consider nonlinear function fitting techniques for the underlying estimation framework, and observe that such techniques can be classified into two major categories: linear coefficient models (discussed in this section) and nonlinear models (discussed in the next section). Linear coefficient models and generalized linear models use a sum over arbitrary nonlinear basis functions $f_k(\mathbf{x})$ weighted by linear coefficients a_k,

$$y(\mathbf{x}) = \sum_{k=1}^{K} a_k \; f_k(\mathbf{x}). \tag{1}$$

A prominent example of this architecture is the class of polynomial models, which takes the form

$$f(\mathbf{x}) = a_0 + \sum_{m=1}^{M} a_m \Psi_m(\mathbf{x}), \text{with} \tag{2}$$
$$\Psi_m(\mathbf{x}) = \prod_i x_i^{e_{i,m}}.$$

M denotes the number of basis functions and $e_{i,m}$ depends on the order of polynomial approximation. For example, a two-dimensional quadratic model includes a total of $M = 5$ basis functions: (x_1), (x_2), $({x_1}^2)$, $(x_1 x_2)$ and $({x_2}^2)$. The parameters in this model are typically estimated in a least-mean-square fit over the experimental data set, which is computationally inexpensive for small to medium dimensional feature sets (Gershenfeld 1999b). Using the model we compute $\hat{y} = f(x)$ for all data points $(\mathbf{x}_n, y_n)$, and subsequently derive $E = \sum_N (\hat{y}_n - y_n)^2/N$.

It is a well-known fact that we can cause the error E to shrink to an arbitrarily small value by adding more and more resources to the model—that is, by increasing the number of parameters and basis functions. However, in doing so we are likely to model noise rather than the underlying causal data structure. In order to avoid this problem, we cross-validate our model and introduce a global regularizer that constrains our model to the "right size."

We divide the available data into two data sets. The training data set $(\mathbf{x}, y)_{tr}$ is used to optimize the parameters of the model, whereas the test data set $(\mathbf{x}, y)_{test}$

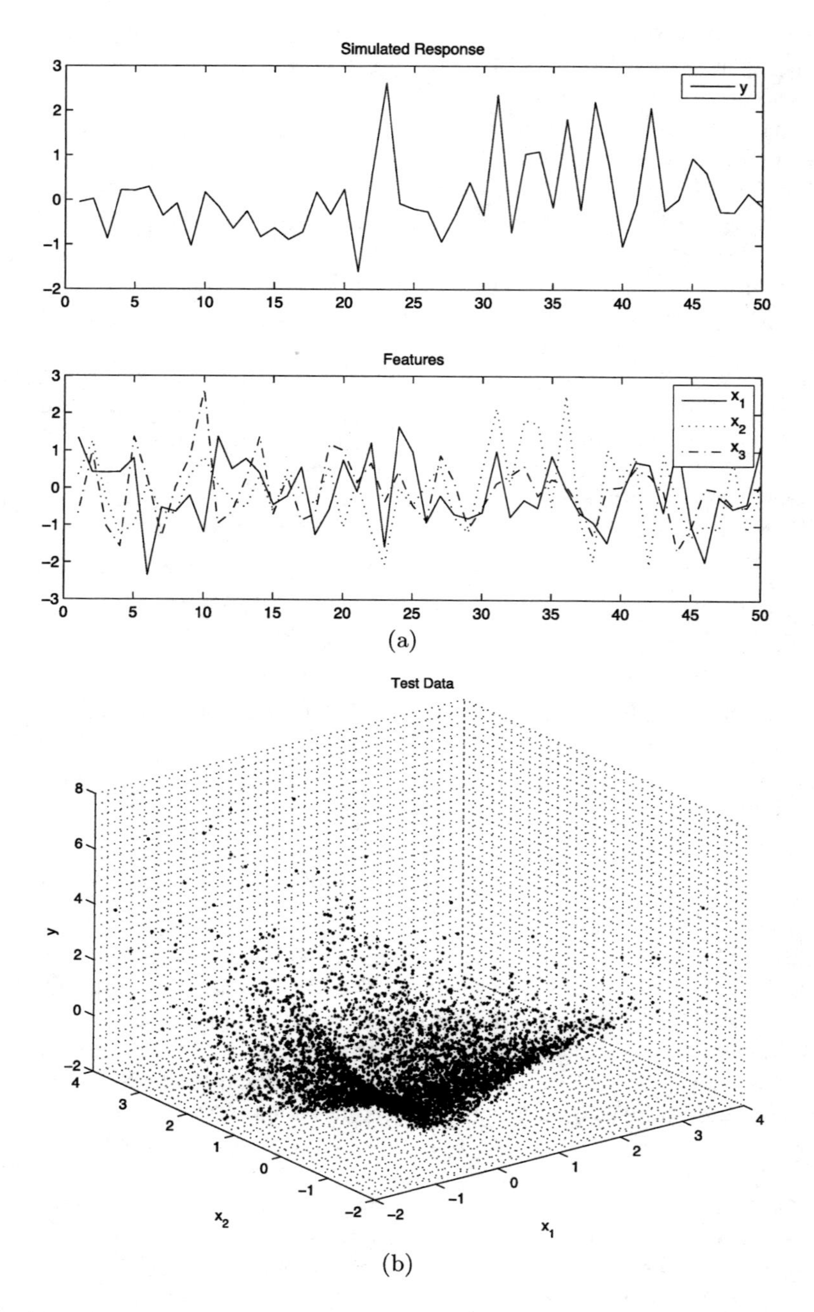

Fig. 1. (a) 1-D plot of features x_1, x_2, x_3, and function y_B and (b) 3-D plot of function y_B (4)

is used to validate the model using E_{test}. As we slowly increase the number of model parameters, we find that the test data estimation error E_{test} decreases initially, but starts to increase as soon as the extra model parameters follow the randomness in the training data. We declare that the model resulting in the smallest estimation error

$$E_m = \sum_{N_{test}} (\hat{y}_{n,m} - y_n)^2 / N_{test} \tag{3}$$

represents the best model architecture for the data set at hand.

Given these considerations, we can now provide a step-by-step algorithm to determine the Relevance Ratio R_i:

1. Divide the available experimental data into the training set $(\mathbf{x}, y)_{tr}$ and $(\mathbf{x}, y)_{test}$. $(\mathbf{x}, y)_{test}$ typically represents $10\% - 30\%$ of the data. If the amount of data set is very limited more sophisticated bootstrapping techniques can be applied (Efron 1983).
2. Build a series of models m based on the complete feature set F, slowly increasing the complexity of the model, i.e. increasing the polynomial order.
3. For each model m compute the error $E_m = \sum_{N_{test}} (\hat{y}_m - y)^2 / N_{test}$. Choose the model architecture m that results in the smallest E_m. Next, build models m_i for all sets $(\mathbf{x}_i, y)$, where the vector $\mathbf{x}_i$ (F_i) includes all features F, except for x_i.
4. Compute $E_i = \sum_{N_{test}} (\hat{y}_i - y)^2 / N_{test}$ for all feature sets F_i and derive the Relevance Ratio $R_i = E_m / E_i$ for all features x_i.

$R_i = 1$ indicates that a feature x_i is irrelevant for the response y. A value of R_i close to 1 indicates little relevance whereas a small value of R_i indicates a high level of relevance. R_i is dimensionless.

Table 1. Application of the polynomial estimator to functions y_A and y_B (4): (a) indicates the error for the different model m based on $\mathbf{x}$; (b) and (c) indicate the resulting Relevance Ratios for features x_1, x_2, and x_3

(a)

Function		Polynomial Order					
		1	2	3	4	5	6
A	Training Set Error	0.8960	0.0398	0.0398	0.0397	0.0395	0.0392
	Test Set Error	0.8938	0.0396	0.0396	0.0398	0.0399	0.0413
B	Training Set Error	0.9989	0.1123	0.1121	0.0740	0.0728	0.0546
	Test Set Error	1.0311	0.1204	0.1210	0.0848	0.0898	0.0924

(b)

Function A	Feature Set		
	F_1	F_2	F_3
Error Training Set	0.8960	0.1452	0.0398
Error Test Set	0.8938	0.1480	0.0396
	x_1	x_2	x_3
Relevance Ratio	0.0443	0.2674	0.9995

(c)

Function B	Feature Set		
	F_1	F_2	F_3
Error Training Set	0.2328	0.8406	0.0745
Error Test Set	0.2478	0.8590	0.0842
	x_1	x_2	x_3
Relevance Ratio	0.3423	0.0988	1.0072

Before we use the method on experimental data, we demonstrate it here on two artificial data sets: A three dimensional set of 5,000 feature data is generated using $x_i = 10 \cdot N(0,1)$, where $N(\mu,\sigma)$ denotes the normal distribution. We define functions A and B as

$$
\begin{aligned}
y_A &= x_1^2 + 5 \cdot x_2 + 0 \cdot x_3 + 30 \cdot N(0,1) \\
y_B &= 100 \cdot log(x_1) + x_2^2 + 0 \cdot x_3 + 20 \cdot N(0,1)
\end{aligned} \tag{4}
$$

Figure 1 shows a one-dimensional and three-dimensional plot of $\mathbf{x}$ and y_B.

Applying our algorithm we obtain the results indicated in Table 1. In the case of function A it can be seen that the polynomial model correctly determines that the data is drawn from a second-order model. For both function A and B, the model correctly assigns a value of $R_3 = 1$ indicating that x_3 was not used to generate y as is indeed the case.

4 Extension to General Nonlinear Estimators and Probabilistic Models

Polynomial models and generalized linear models have many nice properties, including the fact that parameter sets are easily understood. The drawback of these models is that the number of basis terms increases exponentially with the dimensionality of $\mathbf{x}$, making them computationally prohibitive for high-dimensional data sets.

The second category of nonlinear models uses variable coefficients inside the nonlinear basis functions

$$
y(\mathbf{x}) = \sum_{k=1}^{K} f(\mathbf{x}, \mathbf{a}_k). \tag{5}
$$

The most prominent examples of this class of models are artificial neural networks, graphical networks, and Gaussian mixture models (GMM). The models are exponentially more powerful, but training requires an iterative nonlinear search. Here we demonstrate the methodology with GMM's which, as a subclass of Bayesian networks, have the added benefit of being designed on probabilistic principles.

GMM's are derived as the joint probability density $p(\mathbf{x}, y)$ over a set of data $(\mathbf{x}, y)$. $p(\mathbf{x}, y)$ is expanded as a weighted sum of Gaussian basis terms and hence takes on the form

$$
\begin{aligned}
p(y, \mathbf{x}) &= \sum_{m=1}^{M} p(y, \mathbf{x}, c_m) && (6) \\
&= \sum_{m=1}^{M} p(y|\mathbf{x}, c_m) p(\mathbf{x}|c_m) p(c_m) \quad . && (7)
\end{aligned}
$$

Table 2. Application of the GMM estimator to functions y_A and y_B (4): (a) indicates the error for the different model m based on $\mathbf{x}$; (b) and (c) indicate the resulting Relevance Ratios for features x_1, x_2, and x_3

(a)

Function		Number of Clusters									
		2	4	6	8	10	12	14	16	18	20
A	Training Set error	0.414	0.056	0.044	0.043	0.041	0.040	0.041	0.041	0.041	0.040
	Test Set Error	0.413	0.056	0.046	0.044	0.041	0.041	0.042	0.042	0.043	0.041
B	Training Set Error	0.343	0.151	0.095	0.075	0.052	0.044	0.040	0.035	0.033	0.028
	Test Set Error	0.362	0.161	0.106	0.081	0.056	0.050	0.046	0.041	0.038	0.033
		22	24	26	28	30	32	34	36	38	40
A	Training Set error	0.040	0.040	0.039	0.039	0.039	0.039	0.039	0.039	0.039	0.039
	Test Set Error	0.043	0.042	0.041	0.041	0.041	0.041	0.041	0.043	0.042	0.041
B	Training Set Error	0.027	0.029	0.024	0.025	0.024	0.022	0.021	0.024	0.024	0.021
	Test Set Error	0.029	0.035	0.027	0.030	0.028	0.024	0.026	0.029	0.029	0.026

(b)

Function A	Feature Set		
	F_1	F_2	F_3
Error Training Set	0.8950	0.1464	0.0403
Error Test Set	0.8958	0.1515	0.0408
	x_1	x_2	x_3
Relevance Ratio	0.0453	0.2676	0.9928

(c)

Function B	Feature Set		
	F_1	F_2	F_3
Error Training Set	0.2502	0.7630	0.0209
Error Test Set	0.2575	0.8646	0.0233
	x_1	x_2	x_3
Relevance Ratio	0.0912	0.0272	1.0096

We choose

$$p(\mathbf{x}|c_k) = \frac{|\mathbf{P}_k^{-1}|^{1/2}}{(2\pi)^{D/2}} e^{-(\mathbf{x}-\mathbf{m}_k)^T \cdot \mathbf{P}_k^{-1} \cdot (\mathbf{x}-\mathbf{m}_k)/2} \quad , \tag{8}$$

where $\mathbf{P}_k$ is the weighted covariance matrix in the feature space. The output distribution is chosen to be

$$p(\mathbf{y}|\mathbf{x}, c_k) = \frac{|\mathbf{P}_{k,y}^{-1}|^{1/2}}{(2\pi)^{D_y/2}} e^{-(\mathbf{y}-\mathbf{f}(\mathbf{x},\mathbf{a}_k))^T \cdot \mathbf{P}_{k,y}^{-1} \cdot (\mathbf{y}-\mathbf{f}(\mathbf{x},\mathbf{a}_k))/2} \quad , \tag{9}$$

where the mean value of the output Gaussian is replaced by the function $\mathbf{f}(\mathbf{x}, \mathbf{a}_k)$ with unknown parameters $\mathbf{a}_k$.

From this we derive the conditional probability of y given $\mathbf{x}$

$$\begin{aligned} \langle \mathbf{y}|\mathbf{x} \rangle &= \int \mathbf{y}\ p(\mathbf{y}|\mathbf{x})\ d\mathbf{y} \\ &= \frac{\sum_{k=1}^{K} \mathbf{f}(\mathbf{x}, \mathbf{a}_k)\ p(\mathbf{x}|c_k)\ p(c_k)}{\sum_{k=1}^{K} p(\mathbf{x}|c_k)\ p(c_k)} \quad , \end{aligned} \tag{10}$$

which serves as our estimator of $\hat{y}$. The model is trained using the well-known Expectation-Maximization algorithm.

The number of Gaussian basis functions and the complexity of the local models serve as our global regularizers, resulting in the following step-by-step algorithm analogous to the polynomial case discussed before:

1. Divide the data into training set $(\mathbf{x}, y)_{tr}$ and test set $(\mathbf{x}, y)_{test}$.
2. Build a series of models m based on the complete feature set F, slowly increasing the number of Gaussian basis functions.
3. For each model m compute the error $E_m = (\hat{y}_m - y)^2/N_{test}$. Choose the model architecture m that results in the smallest E_m. Build models m_i for all sets $(\mathbf{x}_i, y)$.
4. Compute $E_i = \sum_{N_{test}}(\hat{y}_i - y)^2/N_{test}$ for all feature sets F_i and derive the Relevance Ratio $R_i = E_m/E_i$ for all features x_i.

Applying this new approach to our artificial data sets from before (4), we obtain the results in Table 2.

5 Kullback-Leibler Distance

The linear least-mean-square error metric is without doubt the most commonly used practical error metric, however, other choices can be equally valid. The framework of the Gaussian mixture model allows for the introduction of a probabilistic metric, known as the cross entropy or Kulback-Leibler distance (KL-Distance) (Kullback & Leibler 1951). The KL-Distance measures the divergence between two probability distributions $P(x)$ and $Q(x)$:

$$D_{KL}(P||Q) = \int_x P(x) log \frac{P(x)}{Q(x)} dx \tag{11}$$

where $P(x)$ is typically assumed to be the "true" distribution, and D_{KL} is a measure for how much $Q(x)$ deviates from the true distribution.

For our task at hand we are interested in how much the distribution $p(y|\mathbf{x}_i)$ deviates from $p(y|\mathbf{x}^*)$, where once again $\mathbf{x}_i$ includes all the elements of $\mathbf{x}$ except for x_i. This leads us to the definition

$$D_{KL}(p||p_i) = \int_{\mathbf{x},y} p(\mathbf{x}, y) log \frac{p(y|\mathbf{x})}{p(y|\mathbf{x}_i)} d\mathbf{x}dy \tag{12}$$

and given our definitions above, we obtain

$$\begin{aligned} D_{KL}(p||p_i) &= \int_{\mathbf{x},y} p(\mathbf{x}, y)[log(p(y|\mathbf{x}) - log(p(y|\mathbf{x}_i))]d\mathbf{x}dy \\ &\approx \frac{1}{N} \sum_{n=1}^{N} [log(p(y_n|\mathbf{x}_n) - log(p(y_n|\mathbf{x}_{i,n}))] \quad , \end{aligned} \tag{13}$$

Here we replaced the integral over the density with the sum over the observed data (which itself is assumed to be drawn from the density).

To compute $D_{KL}(p||p_i)$ we need to first estimate $p(y_n|\mathbf{x}_{i,n})$. However, this step consists of estimating the local model parameters only, a relatively minor task. All other parameters needed to numerically evaluate this equation are already part of the model built in the first place.

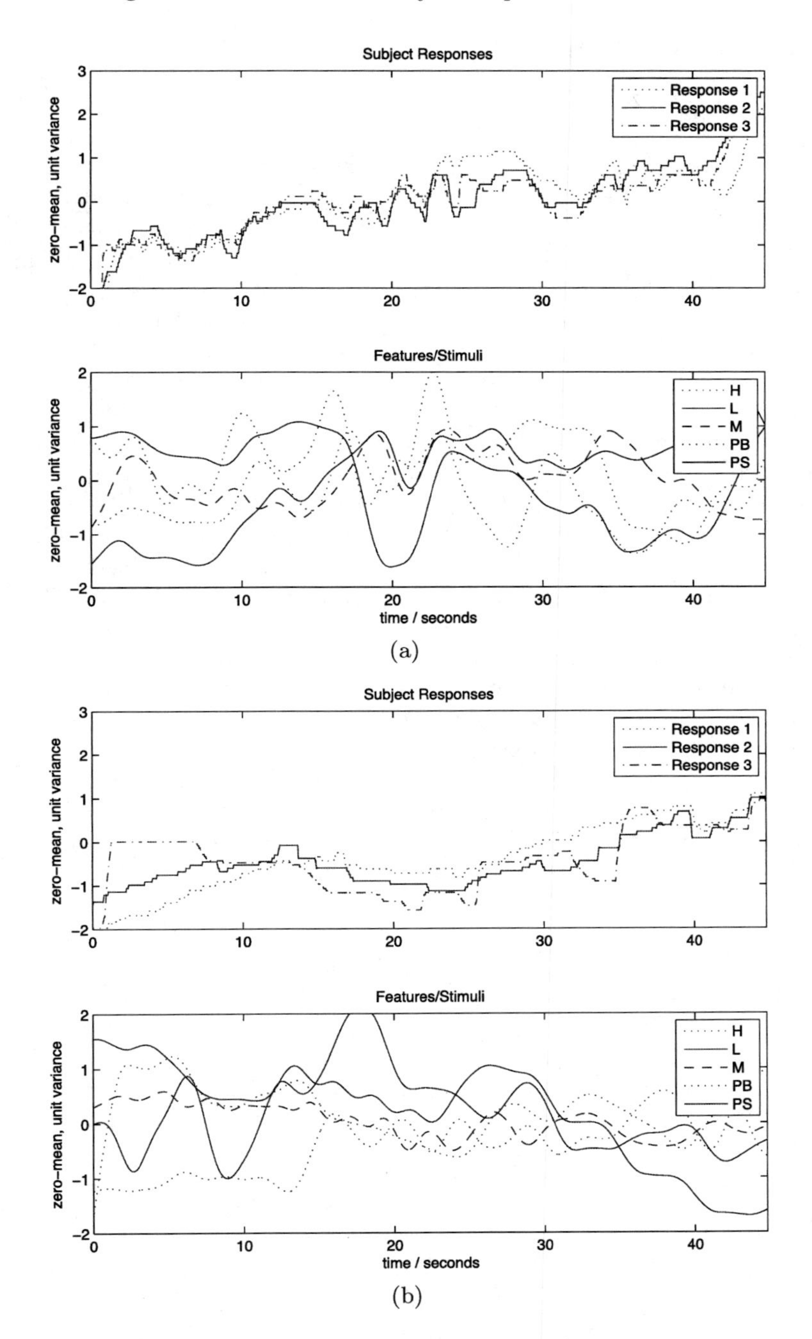

Fig. 2. Features x_i and three subject responses (same subject) for (a) the Brahms excerpt (Fig. 5) and (b) the Bach-Vivaldi excerpt (Fig. 4). H = harmony, L = loudness, M = melodic expectation, PB = pitch height of bass line, PS = pitch height of soprano line.

6 Experimental Results

6.1 Data Set

Data was collected in an experiment that recorded real-time, continuous responses to musical stimuli. Thirty-five subjects, drawn from the faculty and student body at MIT, participated in the experiment. Subjects were asked to move a slider on a computer interface to indicate how they felt tension was changing in the music. They were instructed to raise the slider if they felt a general feeling of musical tension increasing, and to lower it when they felt it lessening. Each musical excerpt was played four times; after each iteration, subjects were asked to rate the confidence level of their response on a scale of 1 to 5. Slider positions were sampled at 50Hz.

Ten musical examples were used as stimuli in the experiment. Six of these examples were short (under 10 seconds) and composed specifically for the study. They featured simple and clear changes in tempo, onset frequency, loudness, harmony, and pitch contour. In addition, there were four excerpts taken from the classical repertoire: Schoenberg Klavierstück, Op. 11 No. 12, Beethoven Symphony No. 1 (Fig. 3), J. S. Bach's organ transcription of Vivaldi's D Major concerto (RV 208) (Fig. 4), and Brahms Piano Concerto No. 2 (Fig. 5). The longer examples were 20 seconds to one minute in length and considerably more complex than any of the short examples.

Musical parameters included in the feature set were harmonic tension, melodic expectation, pitch height of soprano and bass lines, onset frequency, and loudness. Not all features were relevant to all musical examples from the experiment. For the purposes of quantifying harmonic tension and melodic expectation, Lerdahl's (2001)[2] and Margulis's (2005) models were applied respectively.

Fig. 3. Score of Beethoven excerpt

6.2 Results

The key results for all of the complex tonal examples are represented in Table 3. We use both the polynomial models and GMMs and apply our method to various subsets of the feature space. The results are largely robust against variations in

[2] Without the melodic attraction component; this factor is taken into account separately with Margulis's model.

Table 3. Summary of experimental results for the musical tension study. For each experiment we indicate the type of estimation (polynomial or GMM), the global regularizer (polynomial order or number of Gaussians) and the Relevance Ratio of each feature: H = harmony, L = loudness, M = melodic expectation, O = onset frequency, PB = pitch height of bass line, PS = pitch height of soprano line.

Brahms							
Type	POLY	**Relevance Ratio**					
Polynomial order	3	H	L	M	O	PB	PS
Num. Gaussians	N/A	1.0166	1.0099	1.0306	1.0247	1.0251	0.9571
Type	POLY	**Relevance Ratio**					
Polynomial order	3	H	L	M	PB	PS	
Num. Gaussians	N/A	0.8869	0.6133	0.8527	1.0099	0.8366	
Type	POLY	**Relevance Ratio**					
Polynomial order	4	H	L	PB	PS		
Num. Gaussians	N/A	0.8460	0.4795	0.6787	0.6367		
Type	POLY	**Relevance Ratio**					
Polynomial order	4	H	L	M			
Num. Gaussians	N/A	0.8623	0.3228	0.5750			
Type	GMM	**Relevance Ratio**					
Polynomial order	N/A	H	L	M	PB	PS	
Num. Gaussians	16	0.7230	0.2953	0.6583	0.7478	0.9509	
Bach-Vivaldi							
Type	POLY	**Relevance Ratio**					
Polynomial order	2	H	L	M	O	PB	PS
Num. Gaussians	N/A	0.6950	1.1549	0.9703	0.7653	0.8047	0.9472
Type	POLY	**Relevance Ratio**					
Polynomial order	3	H	L	M	PB	PS	
Num. Gaussians	N/A	0.7413	1.1131	0.9780	1.0115	0.9696	
Type	POLY	**Relevance Ratio**					
Polynomial order	3	H	M	PB	PS		
Num. Gaussians	N/A	0.6436	0.8953	0.9265	0.8112		
Type	POLY	**Relevance Ratio**					
Polynomial order	3	H	L	M	PS		
Num. Gaussians	N/A	0.7514	1.0195	0.8625	0.8717		
Type	GMM	**Relevance Ratio**					
Polynomial order	N/A	H	L	M	PS		
Num. Gaussians	N/A	0.6667	1.0439	0.9362	0.8945		
Beethoven							
Type	POLY	**Relevance Ratio**					
Polynomial order	2	H	L	M	O	PB	PS
Num. Gaussians	N/A	1.0575	0.9699	0.9689	0.9644	0.9375	1.0822
Type	POLY	**Relevance Ratio**					
Polynomial order	2	H	L	M	PB	PS	
Num. Gaussians	N/A	1.0607	0.9749	1.0604	1.0580	1.0252	
Type	POLY	**Relevance Ratio**					
Polynomial order	2	H	L	PB	PS		
Num. Gaussians	N/A	0.9502	0.4230	1.0448	1.0289		
Type	GMM	**Relevance Ratio**					
Polynomial order	N/A	H	L	PB	PS		
Num. Gaussians	4	1.2435	0.4087	0.8488	1.1299		

Fig. 4. Score of Bach-Vivaldi excerpt

the model architecture. Relevance is always rewarded with a Relevance Ratio significantly smaller than 1. However, the relative Ratio between features can vary from model to model.

We observe that the model performs best with a modest number of features. The fewer the available feature dimensions, the cleaner the results. We therefore start with a larger feature set and successively remove the least relevant features from the set until the model provides a robust estimate of the feature relevance.

Fig. 5. Score of Brahms excerpt

Mathematically, this phenomenon can be explained by the fact that the features are not statistically independent and that the relevance of one feature may be entirely assumed by an other feature (or a set of features) (Koller & Sahami 1996).

We observe in the case of the Brahms excerpt that loudness is clearly the predominant feature and hence has the smallest Relevance Ratio. In the case of the Bach-Vivaldi excerpt, harmony is primarily responsible for perceived tension. In the Beethoven excerpt, like the Brahms, loudness has the most impact on the response. This makes qualitative sense, as there are no clear changes in the dynamics for the Bach-Vivaldi example, unlike the case for the Brahms and Beethoven, where change in loudness is a salient feature.

The Relevance Ratio confirms that listeners relate salient changes in musical parameters to changes in tension. While there are multiple factors that contribute to how tension is perceived at any given moment, one particular feature may predominate, depending on the context. The Relevance Ratio reveals the overall prominence of each feature in the subject responses throughout the course of a given excerpt. While it could be argued that listeners respond more strongly to certain features (e.g. loudness over onset frequency), it is the *degree of change* in each parameter that corresponds most strongly to tension, regardless of whether the feature is purely musical, as in the case of harmony and melodic contour, or expressive, as in the case of tempo and dynamics.

Summary

We have introduced an new estimator called the Relevance Ratio that is derived from arbitrary nonlinear function approximation techniques and the least-mean-square error metric. To demonstrate the functionality of the Relevance Ratio, it was first applied to a set of artificial test functions where the estimator correctly identified relevant features. In a second step the estimator was applied against a data set of experimental subject responses where we gained valuable insights into the relevance of certain salient features for perceived musical tension. Additionally, we introduced the KL-Distance as an alternative estimator defined in purely probabilistic terms.

References

Bigand, E., Parncutt, R., Lerdahl, F.: Perception of musical tension in short chord sequences: The influence of harmonic function, sensory dissonance, horizontal motion, and musical training. Perception & Psychophysics 58, 125–141 (1996)

Bigand, E., Parncutt, R.: Perceiving music tension in long chord sequences. Psychological Research 62, 237–254 (1999)

Efron, B.: Estimating the Error Rate of a Prediction Rule: Improvements on Cross-Validation. Journal of the American statistical Asociation 78, 316–331 (1983)

Farbood, M.: A Quantitative, Parametric Model of Musical Tension. Ph.D. Thesis, Massachusetts Institute of Technology (2006)

Gershenfeld, N.: The Nature of Mathematical Modeling. Cambridge University Press, New York (1999)

Gershenfeld, N., Schoner, B., Metois, E.: Cluster-Weighted Modeling for Time Series Analysis. Nature 379, 329–332 (1999)

Koller, D., Sahami, M.: Toward Optimal Feature Selection. In: Proceedings of the Thirteenth International Conference on Machine Learning, pp. 284–292 (1996)

Krumhansl, C.L.: A perceptual analysis of Mozart's Piano Sonata K. 282: Segmentation, tension, and musical ideas. Music Perception 13, 401–432

Kullback, S., Leibler, R.A.: On Information and sufficiency. Annals of Mathematical statistics 22, 76–86 (1951)

Lerdahl, F.: Tonal Pitch Space. Oxford University Press, New York (2001)
Lerdahl, F., Krumhansl, C.L.: Modeling Tonal Tension. Music Perception 24, 329–366 (2007)
Madson, C.K., Fredrickson, W.E.: The experience of musical tension: A replication of Nielsen's research using the continuous response digital interface. Journal of Music Therapy 30, 46–63 (1993)
Margulis, E.H.: A Model of Melodic Expectation. Music Perception 22, 663–714 (2005)
Mitchell, T.: Machine Learning. McGraw-Hill, New York (1997)
Nielsen, F.V.: Oplevelse of Musikalsk Spaending. Akademisk Forlag, Copenhagen (1983)
Picard, R.W., Vyzas, E., Healey, J.: Toward Machine Emotional Intelligence: Analysis of Affective Physiological State. IEEE Transactions Pattern Analysis and Machine Intelligence 23(10), 1175–1191 (2001)
Schoner, B.: Probabilistic Characterization an Synthesis of Complex Driven Systems. Ph.D. Thesis, Massachusetts Institute of Technology (2000)
Toiviainen, P., Krumhansl, C.L.: Measuring and modeling real-time responses to music: The dynamics of tonality induction. Perception 32, 741–766 (2003)

Sequential Association Rules in Atonal Music

Aline Honingh, Tillman Weyde, and Darrell Conklin*

Music Informatics research group
Department of Computing
City University London

Abstract. This paper describes a preliminary study on the structure of atonal music. In the same way as sequential association rules of chords can be found in tonal music, sequential association rules of pitch class set categories can be found in atonal music. It has been noted before that certain pitch class sets can be grouped into 6 different categories [10]. In this paper we calculate those categories in a different way and show that virtually all possible pitch class sets can be grouped into these categories. Each piece in a corpus of atonal music was segmented at the bar level and of each segment it was calculated to which category it belongs. The percentages of occurrence of the different categories in the corpus were tabulated, and it turns out that these statistics may be useful for distinguishing tonal from atonal music. Furthermore, sequential association rules were sought within the sequence of categories. The category transition matrix shows how many times it happens that one specific category is followed by another. The statistical significance of each progression can be calculated, and we present the significant progressions as sequential association rules for atonal music.

Keywords: pitch class set categories, atonal music, sequential association rules, similarity measures.

1 Introduction

A typical structure can usually be revealed in tonal music, when it is analyzed harmonically. The chord progressions like the ones shown in Table 1 show some general rules that can often be found in Western tonal music. Atonal music, on the other hand, is not structured around a tonal center like tonal music. Therefore, for atonal music, a progression table like this is impossible. Pitch class set theory can be used to analyze atonal music and more analysis theories have been proposed to analyze atonal music [7]. However, no analogy to chord progression in tonal music has been proposed. In this paper, we will ask ourselves the question whether any kind of progression rules for atonal music can be found which could reveal partly the structure of atonal music.

* Corresponding author: Institute for Logic, Language and Computation University of Amsterdam, The Netherlands. A.K.Honingh@uva.nl

E. Chew, A. Childs, and C.-H. Chuan (Eds.): MCM 2009, CCIS 38, pp. 130–138, 2009.

Table 1. Chord progression in major mode, taken from [9]

Chord	is followed by	sometimes by	less often by
I	IV, V	VI	II, III
II	V	IV, VI	I,III
III	VI	IV	I, II, V
IV	V	I, II	III, VI
V	I	VI, IV	III, II
VI	II, V	III, IV	I
VII	III	I	

1.1 Atonal Music and Pitch Class Set Theory

A distinction is often made between "free" atonal music and twelve tone or serial music. Twelve tone music differs from free atonal music in two important ways: all 12 pitch classes are used and ordered. In this paper, when we speak about atonal music, we mean both free atonal music and serial music.

For the analysis of atonal music, pitch class set theory has been developed. Pitch class set theory has been described in 1973 by Alan Forte [4]. A pitch class is a number between 0 and 11 and is an abstraction of a musical note. All 12 pitch classes represent the semitones from one octave. Collections of pitch class sets (harmonic or melodic) can be analyzed according to pitch class set theory. Forte [4] assumed two types of equivalence (besides the octave equivalence and enharmonic equivalence that belong to pitch classes) related to collections, namely transpositional equivalence and inversional equivalence. Furthermore, the term 'set' covers permutation equivalence and cardinality equivalence. For example, the set $\{0, 4, 7\}$ represents all the chords/melodies that are composed of these three pitch classes (including repetitions). Without these equivalence classes, the number of possible pitch class sets would be huge. But taking these equivalence relations into account, the list of possible pitch class sets are more limited and each set in the list can be characterized by the so-called prime form of the pitch class set (see e.g. [4] for more information). An other way of describing a pitch class set is to characterize it by its intervallic content. The interval-class vector or IcV is an array that expresses the intervallic content of a pitch class set. Since in pitch class set theory an interval is equal to its inverse, an IcV consists of six numbers instead of twelve, with each number representing the number of times an interval class appears in the set. For example, the pitch class set $\{0, 4, 7\}$ has interval class vector [0 0 1 1 1 0] since it consists of 1 'minor third or major sixth', 1 'major third or minor sixth', and 1 'perfect fourth or perfect fifth'. An IcV represents a pitch class set together with all its transformations according to the above mentioned equivalence classes.

Although the list of different pitch class sets according to Forte may be limited, it still consists of 351 sets (that is the list of different prime forms, [4]), and therefore similarity measures are sometimes useful.

2 Pitch Class Set Categories in Atonal Music

Many similarity measures have been developed for pitch class sets, for example Isaacson's IcVSIM [6], Forte's R_n relations [4], Morris' SIM [8], Rahn's MEMB [11], Rogers' cosθ [12] and Scott and Isaacson's Angle [13]. Many of these similarity measures are based on the interval-class vector (IcV), i.e. those measures compare two different IcV's and output a value that characterizes their similarity.

At first sight, these similarity measures seem to be not so much related to each other. They differ in range, intention, way of calculation and more [10]. However, Quinn argues that those similarity measures have actually a lot in common: they tend to group the IcV's in six different categories, each of which can be said to correspond to a cycle of one of the six interval classes [10]. A cycle of interval classes can be thought of in the following way. A cycle of the interval 1 will read: 0,1,2,3,4, ... A cycle of the interval 2 will read: 0,2,4,6,.... A cycle of the interval 3 will read: 0,3,6,9,... , and so on. Using a cluster analysis, Quinn groups the tetrachords and pentachords in six categories according to several different similarity measures. He identifies for each category a prototype. If a certain pitch class set is grouped into a certain category, this pitch class set is similar to the prototype of that category, according to the similarity measure used. The set $\{0, 1, 2, 3, 4\}$ (IcV=[4 3 2 1 0 0]) is the prototype of the Interval Category 1 (IC1) in the pentachord classification, the set $\{0, 2, 4, 6, 8\}$ (IcV=[1 3 1 2 2 1]) the prototype of IC2, and so on. The cycles of IC's that have periodicities that are less than the cardinality of their class (for example, pitch class 4 has a periodicity of 3: {0,4,8}) are extended in the way described by Hanson [5]: the cycle is shifted to pitch class 1 and continued from there. For example, the IC-6 cycle proceeds $\{0, 6, 1, 7, 2, 8...\}$ and the IC-4 cycle proceeds $\{0, 4, 8, 1, 5, 9, 2, ...\}$. Thus for every cardinality, a separate prototype characterizes the category. For example, category IC4 has prototype $\{0, 4\}$ for sets of cardinality 2, prototype $\{0, 4, 8\}$ for set of cardinality 3 and so on. Tables 2 and 3 give an overview of the prototypes of pitch class set categories. Prototypes have been listed for sets from 2 to 10 notes. Pitch class sets with less than 2 notes or more than 10 notes do not make sense. One pitch class set of cardinality 1 exists, $\{0\}$, with interval vector [0 0 0 0 0 0] and it belongs equally to every category. The same is true for cardinality 11: only one prime form pitch class set exists: $\{0, 1, 2, 3, 4, 5, 6, 7, 8, 9, 10\}$ with interval vector [10 10 10 10 10 5] and belongs to every category equally. The pitch class set of cardinality 12 contains all possible pitch classes.

Although the general classification into six categories is clear from the cluster analysis by Quinn [10], some differences can still be found between classifications with respect to different similarity measures. Comparing the clusters that are obtained from the cluster analysis by Quinn on IcVSIM [6] and SATSIM [1], it appears that two sets, $\{0, 1, 2, 5, 7\}$ and $\{0, 1, 3, 6, 8\}$, that are categorized by IcVSIM as IC5 are categorized by SATSIM as IC6 (see [10]). More differences exist in the classifications when a comparison is made with more similarity measures. Aiming to group the pitch class sets uniformly, in this paper a slightly different approach will be used to classify the pitch class sets. We have used the prototypes themselves to classify pitch class sets into the aforementioned six categories by using

Table 2. Prototypes expressed in pitch class sets for the six categories

	prototypes (pc sets)
IC1	$\{0,1\}$, $\{0,1,2\}$, $\{0,1,2,3\}$, etc.
IC2	$\{0,2\}$, $\{0,2,4\}$, $\{0,2,4,6\}$, etc.
IC3	$\{0,3\}$, $\{0,3,6\}$, $\{0,3,6,9\}$, etc.
IC4	$\{0,4\}$, $\{0,4,8\}$, $\{0,1,4,8\}$, etc.
IC5	$\{0,7\}$, $\{0,2,7\}$, $\{0,2,5,7\}$, etc.
IC6	$\{0,6\}$, $\{0,1,6\}$, $\{0,1,6,7\}$, etc.

Table 3. Prototypes expressed in interval class vectors for the corresponding classes of different cardinality

	prototypes (IcV)					
	IC1	IC2	IC3	IC4	IC5	IC6
duochord classes	[1 0 0 0 0 0]	[0 1 0 0 0 0]	[0 0 1 0 0 0]	[0 0 0 1 0 0]	[0 0 0 0 1 0]	[0 0 0 0 0 1]
trichord classes	[2 1 0 0 0 0]	[0 2 0 2 0 0]	[0 0 2 0 0 1]	[0 0 0 3 0 0]	[0 1 0 0 2 0]	[1 0 0 0 1 1]
tetrachord classes	[3 2 1 0 0 0]	[0 3 0 2 0 1]	[0 0 4 0 0 2]	[1 0 1 3 1 0]	[0 2 1 0 3 0]	[2 0 0 0 2 2]
pentachord classes	[4 3 2 1 0 0]	[1 3 1 2 2 1]	[1 1 4 1 1 2]	[2 0 2 4 2 0]	[0 3 2 1 4 0]	[3 1 0 1 3 2]
hexachord classes	[5 4 3 2 1 0]	[0 6 0 6 0 3]	[2 2 5 2 2 2]	[3 0 3 6 3 0]	[1 4 3 2 5 0]	[4 2 0 2 4 3]
heptachord classes	[6 5 4 3 2 1]	[2 6 2 6 2 3]	[3 3 6 3 3 3]	[4 2 4 6 4 1]	[2 5 4 3 6 1]	[5 3 2 3 5 3]
octachord classes	[7 6 5 4 4 2]	[4 7 4 6 4 3]	[4 4 8 4 4 4]	[5 4 5 7 5 2]	[4 6 5 4 7 2]	[6 4 4 4 6 4]
nonachord classes	[8 7 6 6 6 3]	[6 8 6 7 6 3]	[6 6 8 6 6 4]	[6 6 6 9 6 3]	[6 7 6 6 8 3]	[7 6 6 6 7 4]
decachord classes	[9 8 8 8 8 4]	[8 9 8 8 8 4]	[8 8 9 8 8 4]	[8 8 8 9 8 4]	[8 8 8 8 9 4]	[8 8 8 8 8 5]

the chosen similarity measure to calculate to which prototype a pitch class set is closest. When doing this, the categorization of pentachords according to the aforementioned similarity measures IcVSIM and SATSIM are identical, so Quinn's [10] claim about the six categories could be made even stronger. Even more similarity measures could be compared in this respect. We have compared the measures IcVSIM [6], SATSIM [1], ASIM [8] and cosθ [12], and found they all come up with the same classification for the duochords, pentachords, heptachords, octachords, nonachords and decachords, and the classifications for the trichords, tetrachords and hexachords differ at most by 3 pitch class sets. This shows that similarity measures are not too different in this respect, they agree on the classification in the six categories as we find a very high overlap.

We will base our choice of which similarity measure to use, on the ambiguity it produces. It turns out that Rogers' cosθ produces the least ambiguity: when using it to calculate the category of a pitch class set, it outputs virtually always only one category.

3 Sequential Association Rules

Each category can be seen to as having a particular character resulting from the intervals that appear most frequently. Category 1 (see Table 2 for the prototypes) consists of all semitones and is the category of the chromatic scale. Category 2 is the category of the whole-tones or whole-tone scale. Category 3 is the category of the diminished triads or diminished scale. Category 4 is the category of the augmented triads or augmented scale. Category 5 is the category of the diatonic scale. Category 6 is the category of the tritones or D-type all-combinatorial hexachord (see [5]).

As we have shown above, four similarity measures group the pitch class sets into the same categories. Since similarity plays a role in the analysis of music, this might suggest that those categories play a structural role in atonal music. In this paper we will try to discover sequential association rules between those categories in a corpus of atonal music such as to come up with a table of 'category progressions' for atonal music similar to that of Piston [9] for tonal music. A sequential association rule is a progression $a \rightarrow b$, where the probability $p(b|a)$ is higher than chance level, meaning that category b tends to follow category a more often than expected [2].

3.1 The Method

The method has been implemented in Java, using parts of the Musitech Framework [14], and operates on MIDI data. The MIDI file is segmented on the bar level, as a first step to investigate the raw regularities that occur on this level. The pitches from each bar form a pitch class set. From each pitch class set, the interval class vector can be calculated after which the category it belongs to can be calculated. Using Rogers' cosθ as similarity measure we calculate the similarity to all prototypes of the required cardinality. The prototype to which the set is most similar, represents the category to which the set belongs. If the the pitch class set that is constructed from a bar contains less than 2 or more than 10 different pitch classes, the category is not calculated since this does not make any sense, as we explained in Sect. 2. Therefore, if a set (bar) contains more than 10 different pitch classes, the bar is divided into beats and the beats are treated as new pitch class sets. If a set contains less than 2 pitch classes, this set is added to the set that is constructed from the next bar.

First of all, the number of occurrences of all categories are counted, such that we get an overview of the piece in terms of the percentages of occurrence of the different categories. Furthermore, the instances of each progression from one category to another are counted.

A measure for the over-representation of a progression $a \rightarrow b$ is the 'lift'. This measure is taken from [2] and defined as follows:

$$\mathrm{lift}(a \rightarrow b) = \frac{p(b|a)}{p(b)}, \tag{1}$$

where $p(x)$ denotes the probability of category x. The lift can be understood as the number of observed progressions divided by the number of expected progressions due to chance. If the lift is greater than 1, there is a positive correlation, if the lift is smaller than 1, there is a negative correlation.

4 Results

As described in the previous section, the occurrences of each category can be counted. It can be expected that different types of music will show a different occurrence rate for each category. To start with a tonal piece, for example, the distribution of categories of the fourth movement of Beethoven's ninth symphony is shown is Table 4. One can observe that category 5 dominates the whole piece. This turns out to be quite typical for tonal music. In the previous section we have mentioned that each category can be seen as having a specific character and category 5 represents the diatonic scale. Therefore, it is not surprising that a piece of tonal music based on the diatonic scale is dominated by category 5.

Table 4. Distribution of categories of the fourth movement of Beethoven's ninth symphony

category	number of occurrences	percentage of occurrence
1	102	11.40 %
2	69	7.71 %
3	78	8.72 %
4	89	9.94 %
5	552	61.68 %
6	5	0.56 %

For atonal music, we expect something different. We have run the program on atonal music of Schoenberg, Webern, Stravinsky and Boulez. The complete list of music is shown in Table 5. On average, the distribution as shown in Table 6 was found, using this corpus of atonal music. One can see that this distribution is totally different from Table 4 and as such this method might be useful in discrimination tasks. We can see that the music is not dominated anymore by category 5 but a much more equal distribution is present in atonal music.

A transition matrix can be made with our method (Table 7), listing how many times category i is followed by category j. We have calculated the lift matrix as described in the previous section (Table 8) from which one can see which progressions have a positive relation and which have a negative relation.

To answer the question which progressions are meaningful, we have to perform a significance test. We would like to know which progressions have an occurrence rate that is significantly higher or lower than chance level. We use a chi-square test on the data of Table 7 to calculate which progressions cannot be explained by our null hypothesis: the probability of class j following class i does only

Table 5. The atonal music used in the method

composer	piece
Schoenberg	Pierrot Lunaire part 1, 5, 8, 10, 12, 14, 17, 21
Schoenberg	Piece for piano opus 33
Schoenberg	Six little piano pieces opus 19 part 2, 3, 4, 5, 6
Webern	Symphony opus 21 part 1
Webern	String Quartet opus 28
Boulez	Notations part 1
Boulez	Piano sonata no 3, part 2: "Texte"
Boulez	Piano sonata no 3, part 3: "Parenthese"
Stravinsky	in memoriam Dylan Thomas Dirge canons (prelude)

Table 6. Distribution of categories from music of Schoenberg, Webern, Stravinsky and Boulez

category	number of occurrences	percentage of occurrence	standard deviation
1	313	28.25 %	10.56 %
2	117	10.56 %	6.14 %
3	166	14.98 %	7.68 %
4	179	16.16 %	7.97 %
5	138	12.45 %	7.15 %
6	195	17.60 %	6.20 %

depend on the overall number of j's in the music. We calculate the chi-square statistics for every progression separately by making a 2×2 contingency table (with fields $i \rightarrow j$, $i \rightarrow \neg j$, $\neg i \rightarrow j$, $\neg i \rightarrow \neg j$), and calculate the probability from the probability density function of the chi-square distribution with 1 degree of freedom (Table 9). If we take the significance level to be 5%, the progressions that are significantly meaningful are printed in boldface in Table 9.

Now that we can identify the meaningful progressions for our corpus of atonal music, we can make a table for categories analogue to Piston's table for chords. From the lift value in Table 8 can be seen whether a significant progression represents a positive or negative association. These significant rules can be found in Table 9 under the headings "is followed by" (positive association) and "less often by" (negative association). One can see that there is a tendency for categories to follow itself, so that large regions in the music are represented by just one category. This is in accordance with observations by Ericksson [3], who describes 7 categories similar to the ones described above and says that "it is often possible to show that one region [category] dominates an entire section of a piece". Besides these 'repetitions' of categories, one other progression can be identified to present a sequential association rule: the progression from 5 to 4, and four other progressions can be identified to present a negative association, sequential 'avoidance' rules: the progression from category 1 to 2, from 1 to 4, from 1 to 5, and from 5 to 6.

Table 7. The transition matrix

		To					
	category	1	2	3	4	5	6
From	1	109	23	49	36	28	62
	2	27	21	12	15	18	22
	3	49	18	30	21	24	22
	4	44	17	29	39	15	32
	5	33	17	15	29	27	16
	6	47	20	28	34	22	38

Table 8. The lift matrix

		To					
	category	1	2	3	4	5	6
From	1	1.23	0.70	1.04	0.71	0.72	1.13
	2	0.82	1.70	0.68	0.79	1.24	1.07
	3	1.04	1.03	1.21	0.78	1.16	0.75
	4	0.87	0.90	1.08	1.35	0.67	1.02
	5	0.85	1.17	0.73	1.30	1.57	0.66
	6	0.85	0.97	0.96	1.08	0.91	1.11

Table 9. The significance matrix of the results displayed in Table 7

		To					
	category	1	2	3	4	5	6
From	1	**0.001**	**0.020**	>0.5	**0.009**	**0.026**	0.111
	2	0.150	**0.003**	0.097	0.288	0.179	>0.5
	3	>0.5	>0.5	0.135	0.159	0.252	0.079
	4	0.201	>0.5	>0.5	**0.008**	0.059	>0.5
	5	0.163	0.438	0.104	**0.047**	**0.003**	**0.030**
	6	0.167	>0.5	>0.5	0.345	1.222	0.254

Table 10. Category progression in atonal music

Category	is followed by	sometimes by	less often by
1	1	3,6	2,4,5
2	2	1,3,4,5,6	
3		1,2,3,4,5,6	
4	4	1,2,3,5,6	
5	4,5	1,2,3	6
6		1,2,3,4,5,6	

5 Concluding Remarks

Although this work serves as a preliminary study on sequential association rules in atonal music, some interesting things can be said. To sum up the results of this paper, we showed first of all that the 6 different pitch class categories described in [10], can be found in a different way by comparing all pitch class sets to certain prototypes according to a specific similarity measure. Four different similarity measures agree virtually always on the grouping of all possible pitch class sets into these 6 categories. Furthermore, the distribution of notes into these categories appears to be distinguishing between atonal and tonal music and could perhaps be used as a tool for this purpose. Finally, a number of sequential association rules have been found in a corpus of atonal music. A sequential association rule is a progression from category i to j that appears in the music significantly more often than one would expect due to chance. These progression rules may reveal a structure of atonal music that was not known before.

Acknowledgements

We wish to thank our colleague Mathieu Bergeron for presenting this paper at the second international conference of the Society for Mathematics and Computation in Music 2009.

References

1. Buchler, M.: Relative saturation of interval and set classes: A new model for understanding pcset complementation and resemblance. Journal of Music Theory 45(2), 263–343 (2001)
2. Conklin, D.: Melodic analysis with segment classes. Machine Learning 65(2-3), 349–360 (2006)
3. Ericksson, T.: The ic max point structure, mm vectors and regions. Journal of Music Theory 30(1), 95–111 (1986)
4. Forte, A.: The Structure of Atonal Music. Yale University Press, New Haven (1973)
5. Hanson, H.: Harmonic Materials of Modern Music. Appleton-Century-Crofts, New York (1960)
6. Isaacson, E.J.: Similarity of interval-class content between pitch-class sets: the IcVSIM relation. Journal of Music Theory 34, 1–28 (1990)
7. Lerdahl, F.: Atonal prolongational structure. Contemporary Music Review 4(1), 65–87 (1989)
8. Morris, R.: A similarity index for pitch-class sets. Perspectives of New Music 18, 445–460 (1980)
9. Piston, W., DeVoto, M.: Harmony. Victor Gollancz Ltd. (1989) revised and expanded edition
10. Quinn, I.: Listening to similarity relations. Perspectives of New Music 39, 108–158 (2001)
11. Rahn, J.: Relating sets. Perspectives of New Music 18, 483–498 (1980)
12. Rogers, D.W.: A geometric approach to pcset similarity. Perspectives of New Music 37(1), 77–90 (1999)
13. Scott, D., Isaacson, E.J.: The interval angle: A similarity measure for pitch-class sets. Perspectives of New Music 36(2), 107–142 (1998)
14. Weyde, T.: Modelling cognitive and analytic musical structures in the MUSITECH framework. In: UCM 2005 5th Conference Understanding and Creating Music, Caserta, pp. 27–30 (November 2005)

Badness of Serial Fit Revisited

Tuukka Ilomäki

Department of Composition and Music Theory,
Sibelius Academy, Helsinki, Finland
tuukka.ilomaki@iki.fi

Abstract. David Lewin introduced the notion of Badness of Serial Fit, or BSF, to analyze the relation between two twelve-tone rows. It is based on Milton Babbitt's idea of the protocol made of the shared ordered pairs of pitch classes of two rows and aims to evaluate how distinctive the protocol is. While BSF has been mentioned several times in the music theory literature, so far little progress has been made in the analysis of its properties. This paper formalizes BSF in terms of partial orders and links the musical discourse to the pertinent literature in mathematics and computer science. BSF is analyzed in terms of computational complexity and it is shown to be related to the notion of "presortedness" used in the analysis of sorting algorithms. It is proven that the logarithms of the values of BSF define a metric for twelve-tone rows. This new metric is several orders of magnitude finer than any other measure discussed in the literature.

1 Introduction

The relations between twelve-tone rows have been an integral part of the twelve-tone system from the very beginning. The early composers used informal methods to relate rows. However, in the 1940s, Milton Babbitt initiated the process of formalizing the theory of twelve-tone rows and their relations.

Babbitt was interested in ordered pairs throughout his writings. In particular, he introduced the notion of the twelve-tone row as a protocol that defines the order in which the pitch classes appear in it (Babbitt 1962).

A natural way to relate rows is to compare the ordered pairs they comprise. Rothgeb (1967) formalized this idea as the first similarity measure for twelve-tone rows: the more ordered pairs two rows share the more similar they are. David Lewin's (1976) notion of Badness of Serial Fit, or BSF, builds on the ordered pairs of pitch classes as well. However, it does not measure the differences between two rows, but rather picks out their similar features and counts the number of rows that share them. The more similar two rows are, the more common properties they have, and the more distinctive this combination of properties is and, therefore, the fewer rows there are with these properties.

It is rather extraordinary that Lewin introduces a new similarity measure, but does not give a single nontrivial example of calculating it. He measures a row against itself resulting in the value 1 (a row defines a protocol that is satisfied only by itself since no other row contains precisely the same set of ordered pairs; hence, $\mathrm{BSF}(X, X) = 1$ for any twelve-tone row X), and a row against its retrograde

E. Chew, A. Childs, and C.-H. Chuan (Eds.): MCM 2009, CCIS 38, pp. 139–145, 2009.

resulting in the value 479001600 (retrograde-related rows do not share a single ordered dyad and the protocol they define is empty; since any row satisfies an empty protocol, $\mathrm{BSF}(X, RX) = 12! = 479001600$ for any twelve-tone row X). Similarly, later authors, for example Starr (1984) and Morris (2001), have failed to provide nontrivial examples. Ward (1992, 100) enumerates the values of his own variant of BSF for segments of sizes 2 to 6, and remarks that "there are limits to the feasibility of Badness of Serial Fit in the large cardinalities, when potentially hundreds of millions of permutations must be examined."

In general, the more pairs the protocol has, the more refined it is and the fewer rows satisfy it; this relationship is very complex, however, and the size of the protocol is a poor indicator of the BSF value. I will show below that computing the BSF of two arbitrary rows is an intricate task, but with an effective algorithm, it is nowhere close to being as hopeless as Ward implies. In particular, we certainly do not need to examine "hundreds of millions of permutations."

2 Badness of Serial Fit and Partial Orders

Two observations are needed to relate Badness of Serial Fit to partial orders. First, a twelve-tone row as a total order on the set of twelve pitch classes is also a partial order: a special type partial order in which the order of every element is defined, but a partial order nevertheless. Secondly, the intersection of any two partial orders on a given set is a partial order on that set. Consequently, the protocol defined by two rows – the set of shared ordered pairs – is simply the intersection of the partial orders that the rows define. For the sake of brevity, let us consider the "three-tone rows" 012 and 120 (instead of twelve-tone rows). They define partial orders $\{(0,0),(0,1),(0,2),(1,1),(1,2),(2,2)\}$ and $\{(1,1),(1,2),(1,0),(2,2),(2,0),(0,0)\}$, respectively. The intersection of these partial orders is $\{(0,0),(1,1),(1,2),(2,2)\}$ and there are three three-tone rows that satisfy this protocol: 012, 102, and 120. The method for creating the protocol Lewin gives in the appendix of his article is much more cumbersome than simply calculating the intersection of two partial orders.

Badness of Serial Fit has a counterpart in the theory of partial orders: the concept of a linear extension. A linear extension of a partial order is a total order that is a superset of the partial order. The number of linear extensions describes how much the partial order has left undecided. The set of linear extensions is what Starr (1984) labels the total order class: the set of rows that satisfy a protocol. Thus, BSF is the number of linear extensions of the intersection of the protocols that the rows define.

Counting the linear extensions of a given partial order is not a trivial task. In fact, Brightwell and Winkler (1991) have proved that it is a #P-complete problem.[1] Thus, it is difficult to say from looking at a partial order what is

[1] In complexity analysis #P-complete problems are a class of problems in which the number of accepting states of a nondeterministic polynomial time Turing machine are counted. It is believed that there is no polynomial-time algorithm for solving #P-complete problems.

the exact number of its linear extensions. (This translates directly into the fact that it is difficult to say from looking at two twelve-tone rows what the Badness of Serial Fit value is.) However, Pruesse and Ruskey (1997) have developed an efficient algorithm for generating linear extensions; its running time depends on the number of linear extensions to be generated. In technical terms the algorithm is O(N), where N is the number of objects generated. However, as the size of the set increases, the maximum number of linear extensions of a partial order grows exponentially and thus the calculation time required grows exponentially.

The complexity associated with partial orders is reflected also in the number of existing partials orders: the number of possible partial sets on a set grows exponentially with respect to the cardinality of the set. For instance, Erné and Stege (1991A) have calculated that the number of partial orders that can be defined on the set of cardinality twelve is 414864951055853499. Incidentally, since $12! \cdot 12! < 414864951055853499$, this also shows that not all partial orders can be expressed in terms of an intersection of two linear orders.

3 Logarithmic BSF and the Metric

Lewin (1976, 256) suggests using logarithmic values for Badness of Serial Fit.

> For various technical reasons, I suspect that the logarithms of these numbers would provide an even better measure, both intuitively and in light of what seem to me to be some interesting information-theoretic implications. But at the present time, I am nowhere near working out this matter to my own satisfaction.

I do not wish to second-guess the rationale for using the logarithmic values of Badness of Serial Fit that Lewin had in mind. However, at least two reasons can be found: the issues of the metric and distribution.

Definition 1. *Mapping $d\colon X \times X \to \mathbb{R}_+ \cup \{0\}$ defines a metric on set X if it satisfies the four following requirements for all $x, y, z \in X$: (i) $d(x,x) = 0$, (ii) $d(x,y) = d(y,x)$, (iii) $d(x,z) \leq d(x,y) + d(y,z)$, and (iv) $d(x,y) = 0$ implies $x = y$.*

It is easy to show that BSF does not define a metric for two reasons. First, the value of BSF for two identical rows is not zero and, secondly, the triangle inequality does not hold. For example, using integers $0, 1, \ldots, 11$ for pitch classes (A and B standing for the integers 10 and 11, respectively), we have

$$\begin{aligned} &\mathrm{BSF}(0123456789\mathrm{AB}, 0123456789\mathrm{BA}) + \mathrm{BSF}(0123456789\mathrm{BA}, 012345678\mathrm{BA}9) \\ &= 2 + 3 < 6 = \mathrm{BSF}(0123456789\mathrm{AB}, 012345678\mathrm{BA}9). \end{aligned}$$

Using logarithmic values solves both of these problems. Independently of what we select as the base of the logarithm, we obtain $\log 1 = 0$. Secondly, the following theorem by Sidorenko (1992) can be used to prove that triangle inequality holds for the logarithmic values.

Theorem 1 (Sidorenko). *If the incomparability graph of a partial order P can be covered by the incomparability graphs of partial orders $P_1, P_2, \ldots, P_k$, then $e(P) \leq e(P_1)e(P_2)\cdots e(P_k)$.*

Here $e(P)$ denotes the number of linear extensions of partial order P. A key observation here is that the incomparability graph of $X \cap Z$ is covered by the incomparability graphs of $X \cap Y$ and $Y \cap Z$. In order to apply Sidorenko's theorem to the current setting, let us take $k = 2$ and thus obtain the following corollary:

Corollary 1. *If X, Y, and Z are three linear orders on the same set, then the inequality $e(X \cap Z) \leq e(X \cap Y)e(Y \cap Z)$ holds.*

Let us now examine the triangle inequality for Badness of Serial Fit in more detail. We obtain the following inequality from Corollary 1:

$$\mathrm{BSF}(X, Y) \cdot \mathrm{BSF}(Y, Z) \geq \mathrm{BSF}(X, Z).$$

Since the logarithm is a monotonously ascending function and the Badness of Serial Fit values are positive, we can take logarithms on both sides of the inequality, and thereby obtain the following inequality:

$$\log(\mathrm{BSF}(X, Y) \cdot \mathrm{BSF}(Y, Z)) \geq \log(\mathrm{BSF}(X, Z)).$$

By applying the rules of logarithms we then obtain the following inequality:

$$\log(\mathrm{BSF}(X, Y)) + \log(\mathrm{BSF}(Y, Z)) \geq \log(\mathrm{BSF}(X, Z)).$$

Let us define similarity measure Logarithmic Badness of Serial Fit, or LOGBSF, simply as logarithmic values of Badness of Serial Fit. Thus, $\mathrm{LOGBSF}(X, Y) = \log(\mathrm{BSF}(X, Y))$. For the present purposes the base of the logarithm could be any real number greater than 1, but below I will provide some arguments for selecting 2 as the base. Since the LOGBSF values are simply logarithms of the BSF values we can write the above inequality as follows:

$$\mathrm{LOGBSF}(X, Y) + \mathrm{LOGBSF}(Y, Z) \geq \mathrm{LOGBSF}(X, Z).$$

Thus, triangle inequality holds for Logarithmic Badness of Serial Fit.

These inequalities concerning Logarithmic Badness of Serial Fit also give us a better understanding of the Badness of Serial Fit values. Namely, triangle inequality holds for Badness of Serial Fit if the binary operation is not an addition but a multiplication. It also gives us an estimation of how its values behave. Nevertheless, it should be noted that even if BSF can be turned into a metric by using logarithm, metric is not by any means an absolute requisite for a successful measure of similarity. Scattering (Morris 1987) is another example of a similarity measure for twelve-tone rows that does not define a metric (since it is not symmetric).

The second reason for using logarithmic values concerns their distribution. Of course, the values are simply scaled values of the "ordinary" BSF: scaling

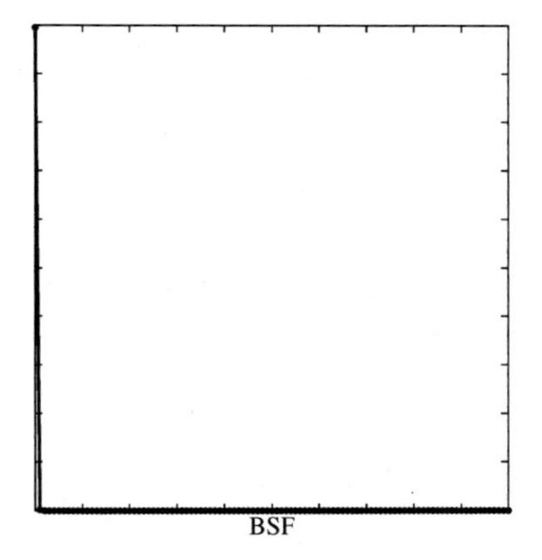

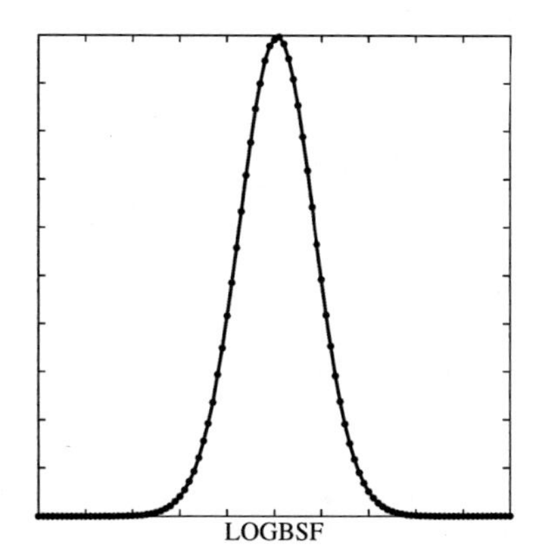

Fig. 1. The distributions of BSF and LOGBSF. The distribution of BSF goes almost along the axes and is therefore difficult to discern in the picture.

the values using logarithms does not, in a sense, give us any new information. However, we obtain a better perspective by using the logarithmic values. As shown in Figure 1, the distribution of values in BSF is extremely skewed, while the distribution of the logarithmic values creates curve resembling the bell curve.

According to BSF the most similar non-identical rows are those in which two adjacent pitch classes have been exchanged. For example, the only difference between rows 0123456789AB and 1023456789AB is the order of the adjacent pitch classes 0 and 1. These two rows are the only ones that satisfy the protocol they define. Hence $\text{BSF}(0123456789\text{AB}, 1023456789\text{AB}) = 2$. If we select 2 as the base we get $\text{LOGBSF}(0123456789\text{AB}, 1023456789\text{AB}) = 1$. Therefore, a minimal difference results in conveniently the value 1.

4 Transformational Similarity and Presortedness of Permutations

So far the discussion has focused on comparing two partial orders. Let us now turn the focus on comparing a linear order to a linear order that could be considered to be normative. We wish to examine how "sorted" a permutation is; in the literature on sorting algorithms this is termed the "presortedness" of a permutation (Mannila 1985). Normativeness depends on the context. Naturally, in the context of twelve-tone rows there is no normative order for the pitch classes (in particular, twelve-tone row 0123456789AB is in no way normative). However, if we consider the transformational relations between twelve-tone rows, we find a natural interpretation for the normative transformation.

At this point, a few notes on the formalization of twelve-tone theory are due. Strictly speaking, rows and row operations are properly formalized as the group of row operations acting on the set of twelve-tone rows. Pitch-class operations are conveniently interpreted as group S_{12} of permutations acting on the set of pitch-class rows and, similarly, order-number operations are interpreted as group S_{12} of permutations acting on the set of order-number rows. However, we can perform calculations as if both rows and row operations were members of the group S_{12} without sacrificing correctness. In a similar vein, it is convenient to note that the order-number row corresponding to pitch-class row as a permutation is its inverse. Using the convention of writing order-number rows in bold, we can write $\mathbf{X} = X^{-1}$. For instance, the order-number row corresponding to pitch-class row 5409728136AB is **2758109463AB** and these two – taken as permutations – are inverse permutations. Nevertheless, it should be kept in mind that rows and row operations are very different kinds of entities.

The concept of *left invariance* is useful in the analysis of similarity measures.[2]

Definition 2. *Metric d on group S_n is left invariant if $d(\pi, \sigma) = d(\tau\pi, \tau\sigma)$ for all $\pi, \sigma, \tau \in S_n$.*

Left invariance guarantees that distances between objects do not depend on how the they are labeled. In Definition 2, permutation τ is applied to permutations π and σ to "relabel" the entities in them.[3] Variables π and σ can be interpreted as pitch-class rows and variable τ as an pitch-class operation. In this context, left invariance means that we are thinking purely in permutational terms and only the ordering relations of the twelve pitch classes matter. Correspondingly, variables π and σ can be interpreted as order-number rows and variable τ as an order-number operation.

All similarity measures that measure the ordering aspect of rows provide left invariance for pitch-class rows. In other words, even if we do not customarily think in such terms, any pitch-class operation, such as a transposition, could be seen as relabeling the pitch classes. Hence, the application of any pitch-class operation to pitch-class rows amounts to a relabeling of the pitch classes, but the order relations between the elements of the rows are not changed.

Assume now that we are measuring the Badness of Serial Fit of rows X and Y. It can be proved that BSF does not depend on how the entities are labeled (the proof would entail showing that a given row satisfies a given protocol if and only if the relabeled row satisfies the relabeled protocol); thus it is left invariant. Let us now relabel the pitch classes in such a way that row Y becomes row id = 0123456789AB. The new rows will now be $Y^{-1}X$ and $Y^{-1}Y$ = 0123456789AB, and the BSF value for the original rows X and Y is identical to that for rows $Y^{-1}X$ and $Y^{-1}Y$. Now, since order-number operations and twelve-tone rows can both be reinterpreted as permutations, we can reinterpret row $Y^{-1}X$ as the order-number operation $\mathbf{YX}^{-1}$ that transforms order-number row $\mathbf{X}$ into order-number row $\mathbf{Y}$ (since pitch-class row X interpreted as a permutation is identical

[2] The concept of left invariance is often known as *right invariance* since right orthography is usually used; see, for example, Mannila (1985) and Estivill-Castro, Mannila and Wood (1993). However, as I use left orthography here I define the concept as left invariance.

[3] In fact, this strategy of relabeling the elements is used by Wong and Ruskey in the implementation of an algorithm devised by Pruesse and Ruskey (1997) to calculate the number of linear extensions of a partial order.

to the order-number row $\mathbf{X}^{-1}$ interpreted as a permutation, applying order-number operation $Y^{-1}X = \mathbf{YX}^{-1}$ to order-number row $\mathbf{X}$ results in $\mathbf{YX}^{-1}\mathbf{X}$, that is the order-number row $\mathbf{Y}$). Therefore Badness of Serial Fit is a measure of the complexity of the order-number operation that maps one row into the other (with respect to the identity transformation that is normative transformation).

Putting the above observations into larger context, BSF measures the presortedness of a permutation. Using logarithmic values makes it a metric; hence, it provides a value describing the complexity of a permutation that is equivalent to an absolute value or norm in other spaces. Furthermore, its resolution is several orders of magnitude finer than any other measure discussed in the literature: the number of distinct values for twelve-tone rows in BSF is 569573 whereas that in other similarity measures is at most a few hundred.

5 Conclusions

The simple idea behind Badness of Serial Fit gives rise to enormous complexity, which makes the measure truly fascinating. Results in mathematics have regularly informed music theory, but only rarely have music-theoretic results lead to new ideas in mathematics. That is the case here as David Lewin's intuition about using logarithmic values in BSF lead to the search and definition of a new metric for permutations.

References

Babbitt, M.: Twelve-Tone Rhythmic Structure and the Electronic Medium. Perspectives of New Music 1(1), 49–79 (1962)

Brightwell, G., Winkler, P.: Counting linear extensions. Order 8(3), 225–242 (1991)

Erné, M., Stege, K.: Counting finite posets and topologies. Order 8(3), 247–265 (1991)

Estivill-Castro, V., Mannila, H., Wood, D.: Right invariant metrics and measures of presortedness. Discrete Applied Mathematics 42, 1–16 (1993)

Lewin, D.: On Partial Ordering. Perspectives of New Music 14(2)/15(1), 252–257 (1976)

Mannila, H.: Measures of Presortedness and Optimal Sorting Algorithms. IEEE Transactions on Computers 34(4), 318–325 (1985)

Morris, R.: Composition with Pitch-Classes: A Theory of Compositional Design. Yale University Press (1987)

Morris, R.: Class Notes for Advanced Atonal Music Theory. Frog Peak Music (2001)

Pruesse, G., Ruskey, F.: Generating Linear Extensions Fast. SIAM Journal on Computing 23(2), 373–386 (1997)

Rothgeb, J.: Some Ordering Relationships in the Twelve-Tone System. Journal of Music Theory 11(2), 176–197 (1967)

Sidorenko, A.: Inequalities for the number of linear extensions. Order 8(4), 331–340 (1997)

Starr, D.: Derivation and Polyphony. Perspectives of New Music 23(1), 180–257 (1984)

Ward, J.: Theories of similarity among ordered pitch class sets. Ph.D. dissertation, The Catholic University of America (1992)

Estimating the Tonalness of Transpositional Type Pitch-Class Sets Using Learned Tonal Key Spaces

Özgür zmirli

Center for Arts and Technology,
Computer Science,
Connecticut College,
270 Mohegan Ave,
New London, CT 06320, USA

Abstract. This paper proposes a method to estimate the tonalness of a pitch-class set using transpositional types. For each set under consideration the method uses the corresponding transpositional type to generate note collections from acoustical instrument sounds and subsequently calculates a learned projection of that set into a low-dimensional space. The structure in this representation is then compared to the structure of a low-dimensional tonal key space learned from audio recordings of labeled tonal music. The term tonalness refers to how strongly the input suggests congruence to pitch use distributions in common-practice tonality. The method is tested on pitch-class sets of cardinality 3 and compared with measures from other work.

Keywords: Tonalness, Tonal Key Spaces, Pitch-class Set, Transpositional Type.

1 Introduction

After hundreds of years of common practice tonality, over the 20th century, composers of atonal music have explored ways to avoid tonal centers in their music. They have purposefully organized the same 12 pitches in ways that perceptually preclude any pitch from assuming a more stable and central role. Serialism and in particular the 12-tone technique was the point at which an even pitch distribution was spelled out by rules. Of course, in the broadest sense of atonality, not all musical works were serial. Some composers even in the late 19th century explored the means of avoiding tonal centers in some of their works. From that time onward, the compositional possibilities now extending from tonal to atonal brought about the questions of how to choose pitch sets to achieve the desired level of atonality. Furthermore, on the analytical side there would be need for a new framework to compose, understand and analyze these works.

Pitch-class set analysis has been devised mainly for use in analyzing atonal music. A pitch-class representation employs principles of octave equivalence and enharmonic equivalence, a property that makes it convenient when dealing with acoustical input. On the whole, the 12 pitch classes form a closed modulo 12 system in which operations

E. Chew, A. Childs, and C.-H. Chuan (Eds.): MCM 2009, CCIS 38, pp. 146–153, 2009.

such as transposition and inversion are defined. Nonetheless, the same formalism can be used to systematically study pitch-class sets in terms of their tonal implications, whether they are taken from a diatonic collection or chosen more arbitrarily.

Given a pitch-class set of any cardinality, a transpositional type (Tn-type) [1] represents a collection of 12 distinct transpositions of the pitches in that set. For example, a Tn-type for a major triad comprises the pitch class sets {0,4,7}, {1,5,8}, {2,6,9} ... {4,8,11}... and can be represented by the generator pattern in prime form - the one that starts at index 0. Each transposition can be shown as $\{0,4,7\}T_n$ for n=0..11.

Temperley [2] proposed a probabilistic framework for measuring tonal implication, tonal ambiguity and tonalness for pitch-class sets. According to Temperley, tonal implication is the key the pitch-class set implies. Ambiguity refers to whether a set implies a single key or more than one key. Tonalness is the degree to which a set is characteristic of common-practice tonality. In this work we draw from these definitions in order to quantify tonalness as a single measure.

Van Egmond and Butler [3] carried out a systematic analysis of pitch-class sets in order to relate them to the common diatonic pitch collections. They listed the connotations of Tn-types of cardinalities between two and six for the major, harmonic minor and ascending minor sets.

Huron [4] has shown that the pitch-class sets that provide the most consonant interval-class collections are the major diatonic scale, the harmonic and melodic minor scales. He further points out that consonant harmonic intervals are found more often in these sets than in other possible sets that can be drawn from the 12 equally tempered pitch chromas.

Brown [5] suggested that there were two approaches to the perception of tonality: structural and functional approaches. The structural approach assumes that a distribution obtained by integration of pitches over time can be used to determine the key. The key with the most similar distribution to the one calculated is selected as the key of the musical fragment. Due to the choice of long integration periods this approach in insensitive to the order of notes. This is in contrast to the functional view which maintains that the sequence and organization of notes play an important role in how people perceive the tonal center.

In this work, we concentrate on a structural approach to explore the tonalness of Tn-sets constructed using real audio. We use a low dimensional representation obtained from accumulated spectral information over long-term windows that span many chords and even phrases in the quantification of tonal ambiguity. For this reason, we do not distinguish between successive and simultaneous occurrences of the pitches in the Tn-sets. Most of the models in the literature that deal with the problem of key finding from audio accumulate spectral information in a similar manner and utilize key profiles as reference points ([6]; see [7] for a survey of key finding methods).

This paper outlines a method for estimation of the tonalness of pitch-class sets constructed using acoustical instrument sounds. Being mindful of the multidimensional and abstract nature of tonality, the term tonalness, in this work, refers to how strongly the input suggests congruence to the pitch use in the common practice of tonal music. Tonal strength can be understood as the opposite of an atonal quality. In other words, tonalness indicates that a listener can identify a tonal center and even follow modulations into other keys. The lack of tonalness or tonal ambiguity, however, would not allow for a clear tonal center to be established and therefore, works of this nature would not have the flexibility of using the full extent of tonal key space. A generalized measure is defined in this work that aims at quantifying this property of pitch class sets.

2 Low Dimensional Tonal Key Space

In this section we describe the construction of a low dimensional tonal key space using audio recordings of tonal music. The recordings comprise of a collection of key-labeled audio in which each of the 12 major and 12 minor keys is represented. The key distribution is shown in Figure 1.

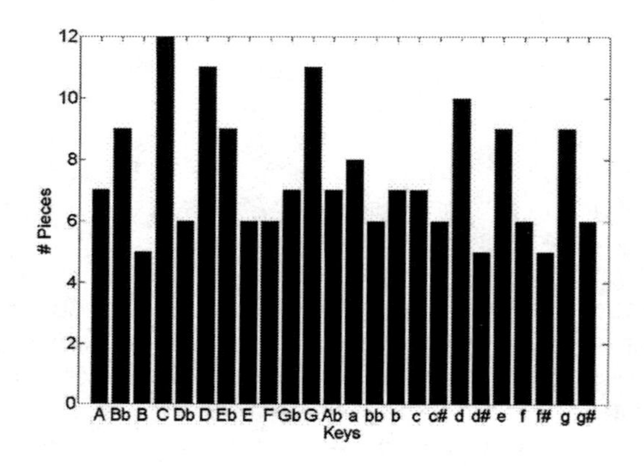

Fig. 1. Key distribution of the recordings used

Chroma based features are very robust representations and have been shown to work well in problems such as key finding, chord recognition, harmonic change detection, audio segmentation and audio alignment (i.e. [8], [9], [10], [11], [7], [12], [13]). Chromagrams try to capture the pitch content of the audio input by mapping pitch-class semitone frequency ranges into their corresponding chromagram bins. They are, however, susceptible to variations in timbre and especially to the spectral distribution of partials. They therefore only approximate a pitch-class distribution of the input music.

To obtain the key space, the audio is first filtered and down-sampled to $f_s = 5512.5$ samples per second. An N=2048 point sliding window Fast Fourier Transform (FFT) with a Hann window and 50% overlap is calculated for the duration of the signal of interest. Next, a 12-element chromagram vector, C, is calculated for each FFT frame with bin j mapping to chromagram bin $cb(j)$:

$$cb(j) = \mathrm{mod}(\, round(12\log_2(jd)), 12) \quad ; \qquad d = \frac{f_s}{f_{A4}N} \tag{1}$$

$$C(i) = \sum_{j;cb(j)=i} X(j) \qquad i = 0..11$$

Where $X(j)$ is the FFT magnitude of bin j and f_{A4} is the reference frequency of A4 : 440 Hz. These chromagrams are then averaged to form a single summary chromagram. It has been shown that further dimensionality reduction of the chromagrams can reveal music theoretical structures such as the circle of fifths and toroidal tonal spaces ([14], [15], [16]). This has also been demonstrated by Burgoyne and Saul [17] using Lerdahl's distances [18].

In this work an n-dimensional representation of a key space is obtained using Principal Component Analysis (PCA). The input matrix $\boldsymbol{Z}$ contains observations, the summary chromagrams of the audio recordings, in its rows. The summary chromagrams are calculated from the first 45 seconds of each piece in a database of 180 tonal pieces recorded from the Naxos site (www.naxos.com). This is done without regard to any modulations that might be taking place during that time. A summary chromagram consists of the averages of the individual chromagram calculations (Eq. 1). While the key distribution is not completely flat, care was taken for each key to be sufficiently represented. The data is standardized by subtracting the mean and dividing by the standard deviation for each variable. Next, the eigenvalues and eigenvectors of the covariance matrix $\boldsymbol{R}$ of the input matrix $\boldsymbol{Z}$ are calculated. The eigenvectors are then rearranged in descending order of the eigenvalues and scaled to have unit length. The mapping matrix $\boldsymbol{A}$ is constructed from the eigenvectors of the n largest eigenvalues. Finally, the input is transformed to the new axes by the transformation $\boldsymbol{ZA}$. The entity n is the number of dimensions to be kept in the transformation. While eliminating dimensions, it is important to monitor the total variance explained by the dimensions to be kept so as to understand how much of the original variation is explained in the retained data.

3 Mapping Tn-Type Sets to the Tonal Key Space

The mapping matrix obtained from the audio database in the previous section can be used to map data points that were not part of the original data during the PCA calculation. The audio for a Tn-type is generated by mixing musical instrument sounds for the notes in this set over two octaves ranging from A2 to Ab4. Piano tones were used in the experiments reported here although other instruments were also tested and not found to change the results significantly. For a given pitch-class set, $\boldsymbol{P}$, in prime form, the note mixtures are realized with 12 different starting positions according to $\boldsymbol{P}\mathrm{T}_n$, n=0..11. A chromagram is obtained for each of these chromatic positions to form another input matrix $\boldsymbol{Y}$. Finally, the projection is found using the transformation $\boldsymbol{YA}$. The idea is that this projection will show congruence to the circular structure of the tonal key space if the tonalness of the Tn-type is significant.

Figure 2 shows the data for the learned tonal key space and examples of a few trivial pitch sets. Only the first two dimensions are shown because they are easy to understand visually although they only account for 52% of the variance. This means that these plots do not completely represent the data but are visually informative in the first two dimensions. Two of the examples are the commonly used tonal pitch sets for the diatonic major and harmonic minor. The last one is the chromatic pitch set with cardinality 12. We observe that the learned tonal space in the first two dimensions is roughly circular. Furthermore, we also observe that, in general, pitch collections that have unambiguous tonal implications tend to form circular patterns

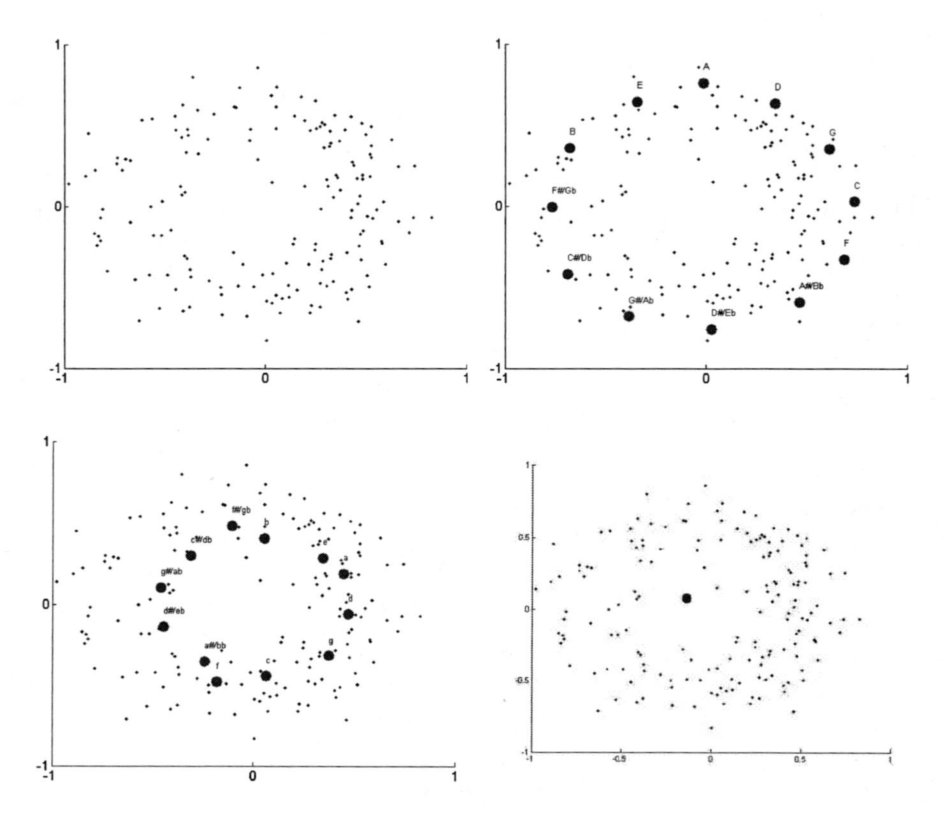

Fig. 2. Visualization in the first two dimensions. The transformed data points from the tonal pieces (top left). The data points for the diatonic major set of cardinality 7 projected onto the original space (top right). The projection for the harmonic minor set (bottom left). The projection for the chromatic set of cardinality 12 (bottom right).

with radii comparable to the structure of the learned space. On the other hand, pitch sets that have weak tonal implications, such as the chromatic set, tend to form small clusters (in the first two dimensions) possibly due to the variance being spread to other dimensions - their inter-distance patterns not being similar to the learned ones.

4 Evaluation

The method is evaluated by comparing the results to a generalized version of root ambiguity measure proposed by Parncutt [19]. We interpret root ambiguity of a pitch-class set to inversely reflect the degree of tonalness. In a Tn-type, if the root ambiguity is low then the implication of key will be strong for every transposition. This will result in a pattern resembling a diatonic set in the tonal key space. On the other hand if the ambiguity is high then it will not manifest such a pattern. Parncutt originally proposed this measure for sequential tones but we employ it without

Table 1. Pitch-class sets of cardinality 3 used in this study. Sets are given by Forte name and prime form without inversional equivalence

Number	Forte Name	Prime Form	Number	Forte Name	Prime Form
1	3-1	012	11	3-7	025
2	3-2	013	12	3-7B	035
3	3-2B	023	13	3-8	026
4	3-3	014	14	3-8B	046
5	3-3B	034	15	3-9	027
6	3-4	015	16	3-10	036
7	3-4B	045	17	3-11	037
8	3-5	016	18	3-11B	047
9	3-5B	056	19	3-12	048
10	3-6	024			

distinguishing between sequential or simultaneous due to our structural approach to the problem. Tonal profiles are calculated from Krumhansl and Kessler's [20] key profiles. The profiles represent probabilities that each chromatic scale degree will be perceived as tonic. A profile for a particular Tn-type is calculated as follows: The range of the key profiles is mapped to the interval [0,1]. Then a weight, $w(k)$ (k=0..23), for each of the 24 profiles is calculated by summing the profile value of the pitches present in the Tn-type. The tonal profile is given by the weighted sum of all major and minor key profiles, where the weights are $w(k)$. The root ambiguity measure is calculated by dividing the sum of 12 elements of the tonal vector by the maximum element and then taking the square root. We employ the same method to calculate the tonal profiles using other profiles found in the literature, namely, Temperley's profile [21] and Aarden's profile [22]. These profiles have been suggested for improving the Krumhansl and Kessler profiles and Aarden's profile is based on different assumptions. We include them for comparison reasons.

Table 1 shows the Tn-types of cardinality 3 used for the evaluation. The first column shows the index and the second column shows Forte's naming for the Tn-types. The last column lists the pitch classes present in the corresponding Tn-types arranged to start from pitch class 0. Figure 3 shows the ambiguity for the proposed model and various key profiles. The Tn-type number in this figure corresponds to the index in the first column in Table 1.

From the observation that the data is projected in roughly circular patterns onto the first two dimensions with the radii correlating with tonalness, the ambiguity for the proposed model is calculated by simply summing the variances in n dimensions, mapping the maximum and minimum to a [0,1] range and subtracting the total from 1. All other profiles were also mapped to the same range for comparison. The number of dimensions from 2 to 7 were tested. While a 2-dimensional solution gave satisfactory results, n=4, resulted in the highest correlation for the Krumhansl and Kessler profile based ambiguity measure. The 4 dimensions were able to explain 76% of the variance of the data. The correlation coefficients between the model's results and the ambiguity calculated from other profiles are Krumhansl & Kessler 0.83, Temperley: 0.79, Aarden: 0.84 and diatonic: 0.78 ($p < 0.0001$ in all of them). It can be seen that all measures are in agreement with the least ambiguous Tn-type $\{027\}T_n$ and mostly in agreement with

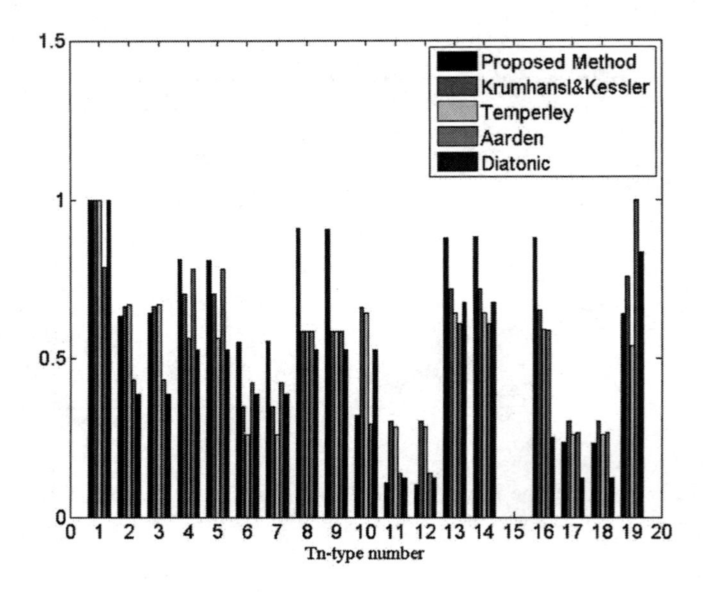

Fig. 3. Ambiguity measures for all Tn-types in Table 1. The proposed method is compared to ambiguity measures using 4 other profiles.

next low-ambiguity sets: {025}T_n, {035}T_n, {037}T_n and {047}T_n. Also all except one measure are in agreement with the most ambiguous Tn-type {012}T_n.

4 Conclusions

We have outlined a method to estimate the tonalness of transpositional type pitch-class sets realized with real audio. A low-dimensional space is used to gain intuition into the topological nature of the key space and the transformed data. The method is tested on pitch-class sets of cardinality 3 and compared with measures from other work. The results are reported as ambiguity measures which indicate the inverse of tonalness. That is, a Tn-type with a high ambiguity measure is less likely to have strong tonalness in reference to common-practice tonality. The model's output correlates well with ambiguity measures derived from other key profiles and flat diatonic profiles. Future work will involve experiments with higher cardinality and also recorded music chosen from different tonal and atonal styles.

References

[1] Rahn, J.: Basic Atonal Theory. Longman, New York (1980)

[2] Temperley, D.: The Tonal Properties of Pitch-Class Sets: Tonal Implication, Tonal Ambiguity, and Tonalness. In: Hewlett, W.B., Selfridge-Field, E. (eds.) Computing in Musicology. Tonal Theory for the Digital Age, vol. 15, pp. 24–38 (2008)

[3] Van Egmond, R., Butler, D.: Diatonic Connotations of Pitch-class Sets. Music Perception 15, 1–29 (1997)

[4] Huron, D.: Interval-class Content in Equally Tempered Pitch-class Sets: Common Scales Exhibit Optimum Tonal Consonance. Music Perception 11, 289–305 (1994)

[5] Brown, H.: The Interplay of Set Content and Temporal Context in a Functional Theory of Tonality Perception. Music Perception 5(3), 219–250 (1988)

[6] zmirli, Ö.: Audio Key Finding Using Low-Dimensional Spaces. In: Proceedings of the International Conference on Music Information Retrieval, Victoria, Canada (2006)

[7] Gómez, E.: Tonal Description of Music Audio Signals, Ph.D. Dissertation, Pompeu Fabra University, Barcelona (2006)

[8] Bartsch, M.A., Wakefield, G.H.: To Catch a Chorus: Using Chroma-based Representations for Audio. In: Proceedings of the IEEE Workshop on Applications of Signal Processing to Audio and Acoustics, New Paltz, NY (2001)

[9] Fujishima, T.: Realtime Chord Recognition of Musical Sound: A System Using Common Lisp Music. In: Proceedings of the International Computer Music Conference, Beijing, China, pp. 464–467 (1999)

[10] Pauws, S.: Musical Key Extraction from Audio. In: Proceedings of the Fifth International Conference on Music Information Retrieval, Barcelona, Spain (2004)

[11] Harte, C., Sandler, M., Gasser, M.: Detecting Harmonic Change in Musical Audio. In: Proceedings of AMCMM 2006, Santa Barbara, California, USA (2006)

[12] Sheh, A., Ellis, D.P.W.: Chord Segmentation and Recognition using EM-Trained Hidden Markov Models. In: Proceedings of the International Conference on Music Information Retrieval, Baltimore, Maryland, USA (2003)

[13] Hu, N., Dannenberg, R.B., Tzanetakis, G.: Polyphonic Audio Matching and Alignment for Music Retrieval. In: Proceedings of the IEEE Workshop on Applications of Signal Processing to Audio and Acoustics, New Paltz, NY, USA (2003)

[14] zmirli, Ö.: Cyclic Distance Patterns Among Spectra of Diatonic Sets: The Case of Instrument Sounds with Major and Minor Scales. In: Hewlett, W.B., Selfridge-Field, E. (eds.) Computing in Musicology. Tonal Theory for the Digital Age, vol. 15, pp. 11–23 (2008)

[15] Purwins, H., Graepel, T., Blankertz, B., Obermayer, K.: Correspondence Analysis for Visualizing Interplay of Pitch Class, Key, and Composer. In: Luis-Puebla, E., Mazzola, G., Noll, T. (eds.) Perspectives in Mathematical Music Theory (2003)

[16] Purwins, H., Blankertz, B., Obermayer, K.: Pitch Class Profiles and Inter-Key Relations. In: Hewlett, W.B., Selfridge-Field, E. (eds.) Computing in Musicology. Tonal Theory for the Digital Age, vol. 15, pp. 73–98 (2008)

[17] Burgoyne, J.A., Saul, L.K.: Visualization of Low Dimensional Structure in Tonal Pitch Space. In: Proceedings of the International Computer Music Conference (ICMC 2005), Barcelona, Spain, pp. 243–246 (2005)

[18] Lerdahl, F.: Tonal Pitch Space. Oxford University Press, New York (2001)

[19] Parncutt, R.: Revision of Terhardt's Psychoacoustical Model of the Root(s) of a Musical Chord. Music Perception 6, 65–94 (1988)

[20] Krumhansl, C.L., Kessler, E.J.: Tracing the Dynamic Changes in Perceived Tonal Organization in a Spatial Representation of Musical Keys. Psychological Review 89, 334–368 (1982)

[21] Temperley, D.: The Cognition of Basic Musical Structures. MIT Press, Cambridge (2001)

[22] Aarden, B.J.: Dynamic Melodic Expectancy. Ph.D. Thesis, Ohio State University, Music (2003)

Musical Experiences with Block Designs

Franck Jedrzejewski[1], Moreno Andreatta[2], and Tom Johnson[3]

[1] CEA-INSTN, F-91191 Gif-sur-Yvette Cedex, France
[2] IRCAM-CNRS, 1 Place Stravinski, F-75004 Paris, France
[3] Editions 75, 75 rue de la Roquette, F-75011, Paris

Abstract. Since the pioneer works of composer Tom Johnson, many questions arise about block designs. The aim of this paper is to propose some new graphical representations suitable for composers and analysts, and to study the relationship between pcsets and small t-designs. After a short introduction on the combinatorial aspects of t-designs, we emphasize the musical perspectives open by these mathematical objects.

1 t-Designs: A Brief Survey

A *t-design* t-(v, k, λ) is a pair $D = (X, \mathcal{B})$ where X is a set of v elements (i.e. a v-set) and $\mathcal{B}$ a set of k-subsets of X called blocks such that every t-subset of X is contained in exactly λ blocks. D is simple if it has no repeated block.

The 2-design is called a *Balanced Incomplete Block Design* (BIBD) or simply a *Block Design* and denoted (v, k, λ). If the index $\lambda = 1$, t-designs are called *Steiner Systems.* For $k = 3$, t-$(v, 3, 1)$ are *Triple Systems* (TS), 2-$(v, 3, 1)$ are *Steiner Triple Systems* (STS) and 2-$(v, 4, 1)$ are *Steiner Quadruple System* (SQS). A *symmetric design* is a BIBD (v, k, λ) such that the number of blocks is equal to the cardinality of the set $(b = v)$. There are no known examples of non trivial t-designs with $t \geq 6$ and $\lambda = 1$. But it is known that 5-$(24, 8, 1)$ is a Steiner System. Two t-designs $(X_1, \mathcal{B}_1)$ and $(X_2, \mathcal{B}_2)$ are said to be *isomorphic* if there is a bijection $\varphi : X_1 \rightarrow X_2$ such that $\varphi(\mathcal{B}_1) = \mathcal{B}_2$. One of the simplest block design is *Fano plane.* It is a 2-$(7, 3, 1)$ design whose blocks are written (vertically) by this matrix:

$$\begin{pmatrix} 0\,0\,0\,1\,1\,2\,3 \\ 1\,2\,4\,2\,5\,3\,4 \\ 3\,6\,5\,4\,6\,5\,6 \end{pmatrix}$$

If the set X of a $(X, \mathcal{B})$ design is identified with musical objects such as pitch classes, modes, rhythms, etc. the combinatorial structure of blocks can be used to create a path through this musical material, linking blocks by their common objects. Composer Tom Johnson has explored these properties in *Block Design for piano* built on the 4-(12, 6,10) design defined by 30 base blocks and one automorphism of the permutation group over 12 elements, namely, in cyclic notation, $\sigma = (0\ 1\ 2\ 3\ 4\ 5\ 6\ 7\ 8\ 9\ 10)(11)$. In *Kirkman's ladies,* he uses a *large* (15, 3, 1) design with 13×35 blocks. In *Vermont Rhythms*, he uses 42×11

E. Chew, A. Childs, and C.-H. Chuan (Eds.): MCM 2009, CCIS 38, pp. 154–165, 2009.

rhythms based on the (11,6,3) design, a system worked out by Jeffery Dinitz and his student Susan Janiszewski. Another example is the mapping of Messiaen's modes on the set X: Mode 2 with (6,3,2) 10 blocks, Mode 3 with (9,3,1) 12 blocks, Mode 4 with (8,4,3) 14 blocks, Mode 5 with (8,4,6) 28 blocks Mode 6 with (8,3,6) 56 blocks, Mode 7 with (10,4,2) 15 blocks.

As we have previously remarked, a t-design has only four parameters t-(v, k, λ). From these quantities, we can easily derive some combinatorial properties. For example, the number of blocks that contain any i-set is given by

$$b_i = \lambda \binom{v-i}{t-i} / \binom{k-i}{t-i}, \quad i = 0, 1, ..., t \tag{1}$$

where $\binom{a}{b} = a!/b!(a-b)!$ indicates the binomial coefficient. In particular, the number of blocks of a t-design is

$$b = \lambda \frac{v!}{(v-t)!} \frac{(k-t)!}{k!} \tag{2}$$

And by setting

$$r = \lambda \frac{(v-1)!}{(v-t)!} \frac{(k-t)!}{(k-1)!} \tag{3}$$

we get the following relation

$$bk = vr \tag{4}$$

As we have seen, two t-designs are isomorphic if there is a bijection between there blocks, and this reduces the research of representative. From a set theoretical perspective, the knowledge of a t-design $D = (X, \mathcal{B})$ leads to the knowledge of its complement $D^c = (X, X \backslash \mathcal{B})$ where $X \backslash \mathcal{B}$ is the set of blocks

$$X \backslash \mathcal{B} = \{B^c, \quad B \in \mathcal{B}\}$$

The complement of $t - (v, k, \lambda)$ design is the $t - (v, v-k, \mu)$ design with

$$\mu = \lambda \binom{v-t}{k} / \binom{v-t}{k-t} = \lambda \frac{(v-k)!}{(v-t-k)!} \frac{(k-t)!}{k!} \tag{5}$$

Remark that D and D^c have the same number of blocks, and for $t = 2$, the block design D with b blocks

$$b = \frac{v(v-1)\lambda}{k(k-1)}, \quad r = \lambda \frac{(v-1)}{(k-1)}, \quad bk = vr \tag{6}$$

has a complement D^c with b blocks and $(v, v-k, b-2r+\lambda)$. For example, the complement of the Fano Plane $(7, 3, 1)$ is $(7, 4, 2)$ with blocks $\{0, 1, 2\}^c = \{3, 4, 5, 6\}$, etc.

An *automorphism* of a design D is a permutation of the point set that preserves the blocks. The group of all automorphims of D will be indicated by $Aut(D)$.

For example, the $D = (7, 3, 1)$ design has an automorphism group $Aut(D)$ equal to the goup $L_3(2)$ of 168 elements with presentation

$$L_3(2) = \langle a, b \mid a^2 = b^3 = (ab)^7 = [a, b]^4 = 1 \rangle \tag{7}$$

where a and b are the permutations (in cyclic notation) of seven elements

$$a = (0\ 3\ 4\ 1\ 2\ 5\ 6), \quad b = (1\ 2\ 0\ 3\ 5\ 6\ 4) \tag{8}$$

We end this section by a characterisation of Steiner Systems. The proof of these theorems can be found in [2] and [7].

Theorem 1 (Wilson). *Let p^m be a prime power. If 3-$(v+1, p^m+1, 1)$ and 3-$(w+1, p^m+1, 1)$ are Steiner Systems then 3-$(vw+1, p^m+1, 1)$ is a Steiner System.*

Theorem 2 (Kirkman, 1847). *A Steiner Triple System of order v exists if and only if $v \equiv 1, 3$ (mod 6), (i.e. $v = 6n+1$ or $v = 6n+3$, i.e. for 7, 9, 13, 15, etc.)*

Examples of Steiner Systems: Let $q = p^m$ be a prime power

- $2 - (q^n, q, 1)$, $n \geq 2$
- $3 - (q^n + 1, q + 1, 1)$, $n \geq 2$
- $2 - (q^n + ... + q + 1, q + 1, 1)$, $n \geq 2$
- $2 - (q^3, q + 1, 1)$,
- $2 - (2^{r+s} + 2^r - 2^s, 2^r, 1)$, $2 \leq r < s$ (Denniston systems)

2 Drawing t-Designs

Until now t-designs have rarely been used for musical purposes. Moreover, there exists no canonical way to draw a t-design. Usually, musical transformations are not considered in the mathematical litterature of t-designs. We will restrict to the most common musical transformations, namely

1. Transpositions:
$$T_n(x) = x + n \pmod{v} \tag{9}$$
2. Inversions
$$I_n(x) = -x + n \pmod{v} \tag{10}$$
3. Affine transformations
$$M_{m,n}(x) = mx + n \pmod{v} \tag{11}$$

In *Kirkman's Ladies*, a strong unity of the score is obtained by considering *parallel classes,* i.e. sets of blocks that partition the point set. A design (v, k, λ) is *resolvable* if its blocks can be partitioned into parallel classes. For example, the (9,3,1) design is resolvable, as shown on table 1.

Table 1. The (9,3,1) design

0,1,2	0,3,6	0,4,8	0,5,7
3,4,5	1,4,7	1,5,6	1,3,8
6,7,8	2,5,8	2,3,7	2,4,6

The Kirkman problem (see also [8]) has been stated in 1850 by Thomas P. Kirkman: *Fifteen young ladies in a school walk out abreast for seven days in succession : it is required to arrange them daily, so that no two walk twice abreast.* Since that time, we define a *Kirkman Triple System* (KTS) as a resolvable Steiner Triple System, also called the *social golfer problem* in computer science. The following theorem limits the cardinality of the point set.

Theorem 3. *A* Kirkman Triple System *of order* v *exists if and only if* $v \equiv 3$ *(mod 6)*

For $v = 15$, it has been shown that there are eighty Steiner Triple Systems (15,3,1). One solution is given in table 2:

Table 2. The Kirkman Triple system $(15, 3, 1)$

Monday	0,1,2	3,9,11	4,7,13	5,8,14	6,10,12
Tuesday	0,3,4	1,8,10	2,10,14	5,7,11	6,9,13
Wednesday	0,5,6	1,7,9	2,11,13	3,12,14	4,8,10
Thursday	1,3,5	0,10,13	2,7,12	4,9,14	6,8,11
Friday	1,4,6	0,11,14	2,8,9	3,7,10	5,12,13
Saturday	2,3,6	0,7,8	1,13,14	4,11,12	5,9,10
Sunday	2,4,5	0,9,12	1,10,11	3,8,13	6,7,14

The musical question is: how to draw this solution showing each parallel class and considering musical transformations between them? Reinhard Laue [9] studied some visualizations of Steiner Systems which make resolvability obvious, and Tom Johnson [6] gave some drawings considering sub-networks in t-designs. For a simplier design such as (6,3,2), which is the best representation? Is it a graph where the set of vertices is the point set, or a graph where vertices are blocks ? (fig. 1).

In figure 1, the opposite borders are supposed to be glued together in the sense of the arrows, in such a way that if you leave the bottom through the line [3, 4] of the triangle $\{2, 3, 4\}$, you enter by the top through the same line in the triangle $\{1, 3, 4\}$. Each triangle has three neighbours. A compositional problem would be to find *Hamiltonian paths* (i.e. paths that visit each vertex exactly once) or *Hamiltonian circuits* (i.e. cycles that visit each vertex exactly once and return to the starting vertex), when vertices are blocks of a t-design. In figure 1, it corresponds to the second graph (on the right) or to the dual graph of the first graph (on the left).

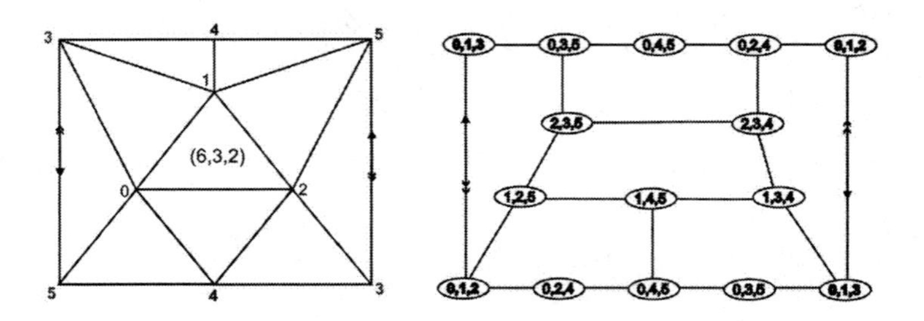

Fig. 1. Two dual representations of the (6,3,2) design

Another point of view is to consider affine transformations between blocks. In the previous example (6,3,2), it leads to a graph with two connected components (fig. 2).

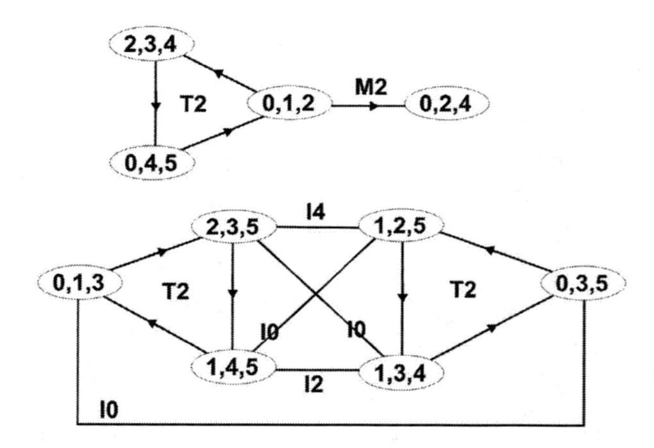

Fig. 2. The affine transformations in the (6,3,2) design

The graph shows all affine transformations modulo 6. The translation T_2 (adding 2 modulo 6) has an inverse T_4 and each inverse transformation I_n acts both ways. This kind of graph is suitable for neo-riemannian analysis. In the next section, we will see another type of graph, where the musical transformations are permutations on base blocks.

3 Cyclic Representations

In some cases, blocks can be constructed from generators under the action of a group. This is the case when $q = p^\alpha$ is a prime power, and the action is the translation T_1 of the cyclic group. The Steiner Triple Systems 2-$(q^2 + q +$

$1, q+1, 1)$ are examples of projective geometries and denoted by $PG(2, q)$. More generally, *symmetric designs* are projective geometries with parameters $PG(m-1, q)$ corresponding to block designs

$$2-\left(\frac{q^m-1}{q-1}, \frac{q^{m-1}-1}{q-1}, \frac{q^{m-1}-1}{q-1}\right) \quad (12)$$

The following table (table 3) shows the first designs. Observe that there is no Steiner Triple System for $PG(2, 6)$, since 6 is not a prime power.

Table 3. Generators of $PG(2, p)$

$(7,3,1)$	$PG(2,2)$	$(0,1,3)$
$(13,4,1)$	$PG(2,3)$	$(0,1,3,9)$
$(21,5,1)$	$PG(2,4)$	$(0,1,4,14,16)$
$(31,6,1)$	$PG(2,5)$	$(0,1,3,8,12,18)$
$(57,8,1)$	$PG(2,7)$	$(0,1,3,13,32,36,43,52)$
$(73,9,1)$	$PG(2,8)$	$(0,1,3,7,15,31,36,54,63)$
$(91,10,1)$	$PG(2,9)$	$(0,1,3,9,27,49,56,61,77,81)$

The parameters of designs are given in the first column, the second column gives the prime power p^α written $PG(2, p^\alpha)$ and the last column gives a generator. The action of the translation T_1 in $\mathbb{Z}_{p^2+p+1}$ yields to the set of blocks. Namely for (7,3,1), the blocks are $B = \{0, 1, 3\}$, $T_1(B) = \{1, 2, 4\}$, $T_1^2(B)$, etc. Can this construction be generalized for $(7, 3, n)$ design with $n > 1$? Unfortunately not. Look at the first values of n. For $n = 1$, the design $(7, 3, 1)$ is generated by $B = \{0, 1, 3\}$ and the translation $T_1(x) = x + 1 \mod 7$, which is also the permutation in cyclic notation $\sigma = (0\ 1\ 2\ 3\ 4\ 5\ 6)$. This design is represented by a heptagone with outer triangles, corresponding to the blocks. For $n = 2$, the design (7,3,2) is not generated by one block and a translation. However, it is generated by two blocks and two actions: the block $B_1 = \{0, 1, 2\}$ and the permutation $\sigma = (0\ 1\ 5\ 3\ 4\ 2\ 6)$ and the block $B_2 = \{0, 1, 3\}$ and the permutation $\sigma^2 = (0\ 5\ 4\ 6\ 1\ 3\ 2)$ which is the square of the previous permutation. The drawing of (7,3,2) is a triangulation of two concentric heptagones, the vertices of each heptagone are labelled by the cyclic notation of the previous permutations. For $n = 3$, the design (7,3,3) is generated by the action of $\sigma = (0\ 1\ 3\ 5\ 2\ 6\ 4)$ on the blocks $B_1 = \{0, 1, 3\}$ and $B_2 = \{0, 1, 2\}$, and the action of $\sigma^4 = (0\ 2\ 1\ 6\ 3\ 4\ 5)$ on the block $B_3 = \{0, 2, 3\}$. The drawing (fig. 3) shows the design $(7, 3, 1)$.

Another question is to determine the generators of a t-design. We sumarize now some results: Netto Theorem and Singer Difference Sets.

Theorem 4 (Netto, 1893). *Let p prime, $n \geq 1$, $p^n \equiv 1$ (mod 6). Let $\mathbb{F}_{p^n}$ be a finite field on X of size $p^n = 6t + 1$ with 0 as its zero element and α a primitive root of unity. The sets*

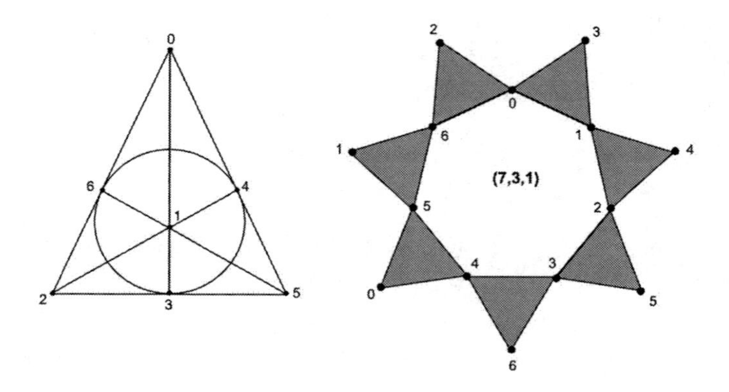

Fig. 3. The (7,3,1) design

$$B_i = \{\alpha^i, \alpha^{i+2t}, \alpha^{i+4t}\} \mod p^n \tag{13}$$

for $i = 1, 2, ..., t-1$ *are generators* ($T_j(B) = j + B \bmod p^n$) *of the set blocks of an* $STS(p^n)$ *on* X.

The proof of this theorem is given in [3]. As an example of how the theorem works, consider the (7,3,1) design. As $p = 7$, $n = 1$, $t = 1$, and $\alpha = 3$ is a generator of $\mathbb{F}_7$, then the set $\{1, \alpha^2, \alpha^4\} \bmod 7 = \{1, 2, 4\} \simeq \{0, 1, 3\}$ up to transposition, is a cyclic generator of $\mathcal{B}$.

Singer Difference Sets are introduced in [4]. Let p be a prime, and m a non-negative integer. Let $f(x)$ be a primitive polynomial of degree m in $\mathbb{F}_p$.

$$f(x) = x^m + a_1 x^{m-1} + ... + a_{m-1} x + a_m \tag{14}$$

Consider the recurrence relation

$$\begin{aligned} u_0 &= 1, \; u_1 = \cdots = u_{m-1} = 0 \\ u_n &= -(a_1 u_{n-1} + \cdots + a_{m-1} u_1 + a_m) \end{aligned} \tag{15}$$

Theorem 5. *The* Singer Difference Set

$$B = \left\{ i, \; 0 \le i < \frac{p^m - 1}{p - 1}, \; u_i = 0 \right\} \tag{16}$$

is a generator of the set of blocks.

Example. For $p = 7$, $m = 3$, $f(x) = x^3 + 3x + 2$ is a primitive polynomial of $\mathbb{F}_7$. The sequence defined by the relations $u_0 = 1, \; u_1 = 0, \; u_2 = 0,$

$$u_n = -3u_{n-2} - 2u_{n-3} = 4u_{n-2} + 5u_{n-3} \mod 7 \tag{17}$$

leads to the first values $u_3 = 5, \; u_4 = 0, \; u_5 = 6, \; u_6 = 4$. The index i such that $u_i = 0$ determines the base block of the cycle $B = \{1, 2, 4\}$.

4 Pcsets and Designs

We would like now to study the relationship between Forte' pcsets and t-designs. Precisely, we would like to investigate the question: is there a t-design t-(v, k, λ) with $v \leq 12$ such that k-blocks include all k-pcsets in Forte classification ? Such a design is called a *Forte design*. If the point set is identified with pitch classes ($v \leq 12$), each block can be considered a chord. If all chords are described by the design, the design is a Forte design. A computer program analyzing the 2-designs given by the *Encyclopedia of t-designs* shows that the 2-designs do not lead to a Forte design. In table 4, the first column gives the parameters of the design, the second column the number of blocks b in the design, the third column gives the complement of the design. In the fourth column is the Forte name of at least a missing k-chord. Stars indicate autocomplementation, and n a positive integer. To have a complete pcset of k-chords, at least two sets of blocks are required. For example, the $(9, 3, 1)$-design has under the action of $\sigma_1 = (2\ 6)(3\ 8)(4\ 7)(0)(1)(5)$

$$\mathcal{B}_1 = \begin{pmatrix} 0\,0\,0\,0\,1\,1\,1\,4\,2\,3\,5\,2 \\ 1\,2\,3\,4\,3\,4\,2\,6\,6\,5\,7\,3 \\ 6\,8\,7\,5\,8\,7\,5\,8\,7\,6\,8\,4 \end{pmatrix} \tag{18}$$

all Forte's trichords except 3-7 and 3-12. And under the action of $\sigma_2 = (2\ 7\ 8\ 6\ 5\ 4\ 3)(0)(1)$

$$\mathcal{B}_2 = \begin{pmatrix} 0\,0\,0\,0\,1\,1\,1\,2\,3\,4\,2\,5 \\ 1\,2\,3\,4\,2\,3\,4\,7\,5\,6\,3\,6 \\ 7\,5\,6\,8\,6\,8\,5\,8\,7\,7\,4\,8 \end{pmatrix} \tag{19}$$

it contains all Forte's trichords except 3-8 and 3-11. That way, using two sets of blocks of the same design, a composer can use the whole set of trichords.

To conclude this section, we would like to mention the link of t-design with Mathieu Groups. First, as it has been underlined in [5] Olivier Messiaen's *Ile de feu 2* use two permutations in cyclic notation

Table 4. Missing at least a Forte chord

(v,k,λ)	b	$(v,k,\lambda)^c$	*Missing*
$(6,3,2n)$	$10n$	$(6,3,2n)^*$	3-5
$(7,3,n)$	$7n$	$(7,4,2n)$	3-1
$(8,4,3n)$	$14n$	$(8,4,3n)$	4-5
$(9,3,n)$	$12n$	$(9,6,5n)$	3-2
$(9,4,3n)$	$18n$	$(9,5,5n)$	4-3
$(10,4,2n)$	$15n$	$(10,6,5n)$	4-2
$(10,5,4n)$	$18n$	$(10,5,4n)^*$	5-1
$(11,5,2n)$	$11n$	$(11,6,3n)$	5-2
$(12,3,2n)$	$44n$	$(12,9,24n)$	3-2
$(12,4,3n)$	$33n$	$(12,8,14n)$	4-1
$(12,6,5n)$	$22n$	$(12,6,5n)^*$	6-1

$$a = (1\ 7\ 10\ 2\ 6\ 4\ 5\ 9\ 11\ 12)(3\ 8)$$
$$b = (1\ 6\ 9\ 2\ 7\ 3\ 5\ 4\ 8\ 10\ 11)(1\ 2) \tag{20}$$

which generate Mathieu's group M_{12} of order 95040. In the same way, *Les yeux dans les roues* (O. Messiaen, *Livre d'orgue VI*) is built on six permutations (a permutation and five *actions*): $\sigma_0 = (1\ 11\ 6\ 2\ 9\ 4\ 8\ 10\ 3\ 5)$ and for $j = 1, ..., 5$, $\sigma_j = A_j\sigma_0$, the actions are defined by: *Extremes au centre*: $A_1 = (2\ 12\ 7\ 4\ 11\ 6\ 10\ 8\ 9\ 5\ 3)$, *Centre aux extrêmes*: $A_2 = (1\ 6\ 9\ 2\ 7\ 3\ 5\ 4\ 8\ 10\ 11)$; *Rétrograde*: $A_3 = (1\ 12)(2\ 11)(3\ 10)(4\ 9)(5\ 8)(6\ 7)$, *Extrêmes au centre, rétrograde:* $A_4 = A_1A_3$ and *Centre aux extrêmes, rétrograde*: $A_5 = A_2A_3$. If we set $a = A_2^{-1}A_1$ and $b = A_2^3A_1A_2^2A_1$ these permutation generate the Mathieu group M_{12} of presentation

$$M_{12} = \langle a, b \mid a^2 = b^3 = (ab)^{11} = [a, b]^6 = (ababab^{-1})^6 = 1\rangle \tag{21}$$

Table 5 gives the links between Mathieu Groups and t-designs.

Table 5. Mathieu groups and t-designs

Groups	*Order*	*t-design*	*# blocks*
M_{11}	7 920	4-(11,5,1)	66
M_{12}	95 040	5-(12,6,1)	132
M_{22}	443 520	3-(22,6,1)	77
M_{23}	10 200 960	4-(23,7,1)	253
M_{24}	244 823 040	5-(24,8,1)	759

The two first Mathieu groups are built with eleven or twelve points. Neither M_{11}, nor M_{12} are Forte designs. In M_{11} 11 pcsets are missing (5-1, 5-3, 5-5, etc.), and in M_{12} 12 pcsets are missing (6-1, 6-4, 6-7, etc.). In M_{12} if we take only three notes in each block, we get neither 3-12, nor 3-1.

5 A Compositional Application

To show a specific compositional application for all this, and also to summarize the combinations and graphs that come together in block designs, we offer a brief analysis of the third movement of Johnson's *Twelve for Piano* (2008) (the score is reproduced in Annexe). This one-page piece uses the precise four-note chords produced by one of the over 17 million possible solutions of the (12,4,3) design, where 12 elements (notes) are partitioned into 33 subsets (chords) of four elements (notes), such that each pair of notes appears in exactly three of the chords.

To write this music, the composer needed to map the system, so as to see how the 33 chords related to one another, and to do this he drew a graph

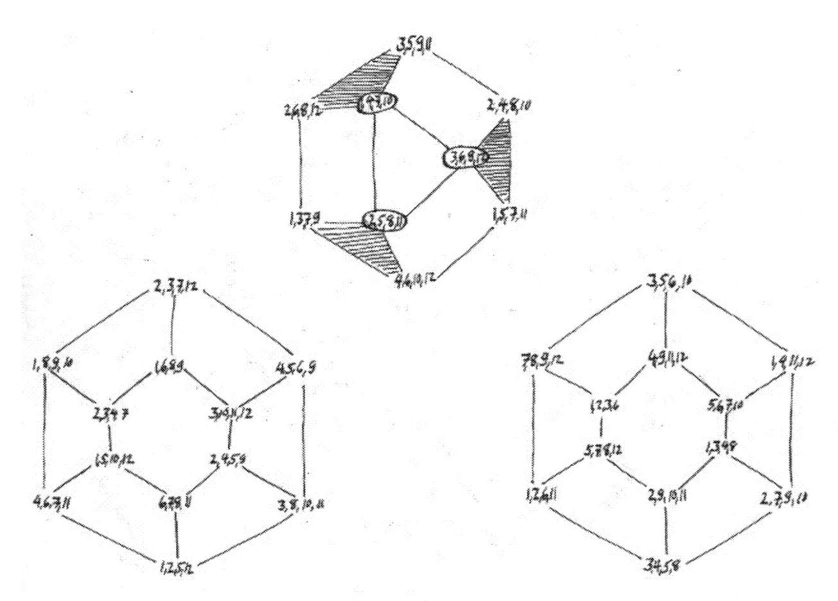

Fig. 4. Graph for *Twelve for piano*

by connecting chords when they had no notes in common. The graph would be different for each of the 17 million solutions, but In this case it takes the shape of the three hexagonal formations shown here. The three shaded triangles represent three parallel classes, three cases where chords with no notes in common come together as a collection of all 12 notes. These nine chords form the central section of the piece, beginning with (1,3,7,9) (4,6,10,12) (2,5,8,11). The remaining 24 chords, those in the other two hexagons, form the opening and closing sections.

The first four phrases of the piece, the first 12 chords, come from the hexagon at the lower right, beginning with the inner ring: (5,7,8,12) (1,2,3,6) (4,9,11,12) and (5,6,7,10) (1,3,4,8) (2,9,10,11) followed by the outer ring: (3,4,5,8) (1,2,6,11) (7,8,9,12) and (3,5,6,10) (1,4,11,12) (2,7,9,10). The final four phrases follow the hexagon at the lower left in this same manner. We have not shown the numbers on the accompanying score, but this is rather easy to decipher, since 1 is the lowest note of the scale and 12 is the highest.

Simply following the connections in this way produces a number of remarkable symmetries, symmetries that are difficult to imagine in a rigorous 12-tone music, and surely impossible in any non-rigorous music.

- Consider first of all the cadences. The first and second phrases both end on D-F-sharp, and the third and fourth phrases both end on B-flat-D. The final four phrases in the piece rhyme in this same way.
- The notes marked "a" appear twice in the same phrase. These same notes are omitted either in the phrase just before or in the phrase just after. Each of the 12 notes appears exactly 11 times in the piece.
- The intervals marked "b" occur at the same place in two subsequent phrases.

• The "c" interval of the first phrase appears again in the third phrase, and there is a similar pair of "c" intervals in the last section of the piece.

• In the middle section, containing three complete sets of 12 notes, one finds three "d" intervals, three "e" intervals and three "f" intervals, though it is difficult to explain why they fall as they do. But then, it is difficult to explain all these other symmetries as well. The music produced by this block design simply does not behave like music we already know.

References

1. Andreatta, M.: Méthodes algébriques dans la musique et musicologie du XXe siècle, Thèse, Ircam/Ehess (2003)
2. Blanchard, J.L.: A construction for Steiner 3-designs. J. Combin. Theory A 71, 60–67 (1995)
3. Colbourn, C., Rosa, A.: Triple Systems. Clarendon Press, Oxford (1999)
4. Colbourn, C., Dinitz, J.: Handbook of Combinatorial Designs, 2nd edn. Chapman & Hall, Boca Raton (2007)
5. Jedrzejewski, F.: Mathematical Theory of Music. Ircam/Delatour, Paris (2006)
6. Johnson, T.: Networks, Harmonies found by Tom Johnson. Editions 75, Paris (2006), www.editions75.com
7. Kirkman, T.P.: On a problem in combinations. Cambridge and Dublin Math. J. 2, 191–204 (1847)
8. Kirkman, T.P.: On the puzzle of the fifteen young ladies. London Philos. Mag. and J. Sci. 23, 198–204 (1862)
9. Laue, R.: Resolvable t-designs. Des. Codes Crypt. 32, 277–301 (2004)

III

A Generalisation of Diatonicism and the Discrete Fourier Transform as a Mean for Classifying and Characterising Musical Scales

Julien Junod[1], Pierre Audétat[2], Carlos Agon[1], and Moreno Andreatta[1]

[1] Equipe des Représentations Musicales, IRCAM, Paris, France
[2] Jazz Department, University of Applied Sciences Western Switzerland

Abstract. Two approaches for characterising scales are presented and compared in this paper. The first one was proposed three years ago by the musician and composer Pierre Audétat, who developed a numerical and graphical representation of the 66 heptatonic scales and their 462 modes, a new cartography called the *Diatonic Bell*. It allows sorting and classifying the scales according to their similarity to the diatonic scale.

The second approach uses the Discrete Fourier Transform (DFT) to investigate the geometry of scales in the chromatic circle. The study of its coefficients brings to light some scales, not necessarily the diatonic one, showing remarkable configurations. However, it does not lead to an evident classification, or linear ordering of scales.

1 Introduction

Over centuries, western musicians have extensively used half a dozen of heptatonic scales, but combinatorics teach us that they represent only a tenth of the totally available musical material. Many catalogues exist, but they often reduce to numerical tables, that may not be easy to handle for composers.

The musician and composer Pierre Audétat [2] developed a numerical and graphical representation of all 66 heptatonic scales and their 462 associated modes. Such a cartography, called the *Diatonic Bell*, opens a field of experiment equally relevant for composition and analysis, and presents interesting developments for teaching.

The first part of this paper deals with the classification and ordering of scales obtained with the diatonic bell, presenting a mathematical formulation of Audétat's original empirical work. The second part investigates scales in the chromatic circle using the Discrete Fourier Transform (DFT) in order to exhibit certain scales with remarkable properties.

David Lewin proposed this tool in 1958 for analysing intervallic relationships. The idea was pursued by Ian Quinn [7] for classifying chords and by Emmanuel Amiot [1] for redefining Clough and Douthett's maximal evenness [4]. Inspired by this work, we will see how DFT coefficients reflect the geometric configuration of a scale in the chromatic circle, and how they can be used to characterise scales.

E. Chew, A. Childs, and C.-H. Chuan (Eds.): MCM 2009, CCIS 38, pp. 166–179, 2009.

The two methods differ structurally, the former being tonal, the latter atonal. We will discuss in the conclusion some points of convergence between these two approaches.

2 The Diatonic Bell

Modes play a key role in Jazz. The first *diatonic bell* was produced by hand in an effort to investigate the 66 heptatonic scales and their 462 modes. The reader interested in how this system displays a network of musical relations offering new opportunities in composition and may facilitate the modal approach of improvisation, is invited to consult the online documentation.[1] We will focus on the step by step procedure, along with the mathematical formulas necessary for a complete construction.

The general idea is to consider every scale as an alteration of a reference, *natural* scale. We will call it *diatonic*, but it may be another maximally even scale. Scales are ordered according to their increasing degree of alteration, from the maximally even to its maximally compact counterpart.

Two different musical spaces are successively used in the process. We first enumerate scales in the finite chromatic circle before moving to the infinite *diatonic spiral* — a generalisation of the spiral of fifths to microtonal contexts — for the graphical representation. This is to avoid the ambiguity induced by enharmony: alterations of the diatonic scale such as $G\sharp = [5+1]_{12} = [6]_{12} = [7-1]_{12} = A\flat$ are not distinguishable in the chromatic circle, whereas they represent two different points on the diatonic spiral.

Two conditions need to be fulfilled before we can compare scales:

1. They need to be centred, or aligned on the symmetry axis of the diatonic scale.
2. We have to make sure that their representation on the diatonic spiral is as compact as possible.

2.1 Input Parameters

Only two parameters are necessary. The size c of the chromatic universe of pitch classes, and the size d of the scale. The procedure works under certain conditions:

1. (a) d must be odd. This is to avoid a hole at the origin (symmetry axis) in the diatonic spiral.
   (b) d must be prime. It guarantees the existence and the unicity of a centred scale in each transposition class, and we avoid scales with internal symmetries (e.g. Messiaen's modes with limited transposition), as a by-product.
2. The parameters must be coprime (i.e. $< c, d >= 1$). This guarantees existence and unicity of a reference scale.

[1] http://www.cloche-diatonique.ch/

2.2 Find All Scales

The chromatic gamut is modelled by the chromatic circle $\mathcal{C}_c = \mathbb{Z}/c\mathbb{Z}$, or cyclic group. The pitch classes are modular integers $[x]_c := x + c \cdot \mathbb{Z}$. A scale S is an unordered subset of $\mathcal{C}_c$. We define the set of all d-notes scales of $\mathcal{C}_c$ as

$$\mathcal{S}_c^d := \{S \subseteq \mathcal{C}_c | d = Card(S)\}. \tag{1}$$

This set has cardinality $\binom{c}{d}$ and contains all possible transpositions of a same scale, a c times redundant information. The cyclic group $\mathbb{Z}_c$ of transpositions acts on $\mathcal{S}_c^d$, and the quotient space will be indicated by $\mathcal{S}_c^d/\mathbb{Z}_c$. A transposition class contains all scales equivalent by translation $T_{[l]_c}$ where $[l]_c \in \mathbb{Z}_c$:

$$S' \sim_{\mathbb{Z}_c} S :\Leftrightarrow \exists [l]_c \in \mathbb{Z}_c : S' = T_{[l]_c}(S) \quad S, S' \in \mathcal{S}_c^d. \tag{2}$$

The most economical way to enumerate all transposition classes is to generate all intervallic structures that uniquely define each class. This can be done by searching for all integer partitions of c into d parts, see [6].

2.3 Find All Centred Scales

In each transposition class $[S]_{\mathbb{Z}_c} \in \mathcal{S}_c^d/\mathbb{Z}_c$, find the unique scale $S^\star$ centred around $[0]_c$: Its chromatic coordinates (pitch classes) sum to zero. The fact that d is coprime with c guarantees the existence and unicity of such a centred scale for each transposition class.

$$\mathcal{S}^{\star}{}_c^d := \{S \in \mathcal{S}_c^d | \sum_{[x]_c \in S} [x]_c = [0]_c\} \tag{3}$$

2.4 Find the Reference Scale

We search for the *maximally even* scale $S_0^\star$ [4]. Here again, the condition $< d, c >= 1$ guarantees existence and unicity of such a scale [1]. It will also be *generated* in the sense of [3], and the most compact in the diatonic spiral. We set it to be the reference scale in our representation. It can be found using the discrete Fourier transform $F\{S\}$ of a scale $S \in \mathcal{S}_c^d$

$$\begin{aligned} F\{S\} : \mathcal{C}_c &\longrightarrow \mathbb{C} \\ [k]_c &\longmapsto \sum_{[x]_c \in \mathcal{C}_c} \mathbb{1}_S([x]_c) \cdot e^{-i\frac{2\pi}{c} \cdot x \cdot k} \end{aligned} \tag{4}$$

where $\mathbb{1}_S$ is the indicator (or characteristic) function of the subset S. The maximally even scale will maximise the module of the d-th coefficient.

$$S_0^\star := argmax_{S^\star \in \mathcal{S}^{\star}{}_c^d} |F\{S^\star\}([d]_c)| \tag{5}$$

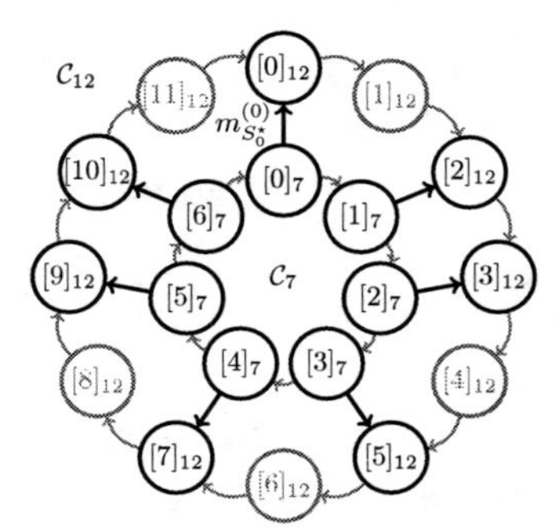

Fig. 1. The diatonic scale's dorian mode is the reference centred mode $m^{([0]_d)}_{S^\star_0}$ in the usual context ($c = 12$ and $d = 7$). It begins with a D ($[0]_{12}$).

2.5 Find the Reference Mode

Order makes the difference between scales and modes. While a scale is defined as an unordered subset, the cyclic ordering of its steps is essential to distinguish between its d modes. We define a mode m_S of a scale $S \in \mathcal{S}^d_c$ to be a function $m_S : \mathcal{C}_d \longrightarrow \mathcal{C}_c$ whose image is exactly the subset S

$$Im(m_S) = S \tag{6}$$

and which preserves the cyclic sequence of the element of the circles (consider them as cyclic oriented graphs G).

$$\begin{aligned} V(G) &= \{[0]_c, \ldots, [c-1]_c\} \\ ([x]_c, [x']_c) \in A(G) &\Leftrightarrow [x']_c = [x+1]_c. \end{aligned} \tag{7}$$

Since d was chosen to be prime, all d modes of a scale S are distinct (no limited transposition modes). A cyclic permutation $\pi = ([0]_d[1]_d \ldots [d-1]_d)$ of the diatonic circle $\mathcal{C}_d$ connects them altogether.

$$m^{([k]_d)}_S := m^{([0]_d)}_S \circ \pi^k \quad \forall k \in \mathbb{Z} \tag{8}$$

We choose the centred mode $m^{([0]_d)}_{S^\star_0}$, to be the one starting at $[0]_c$:

$$m^{([0]_d)}_{S^\star_0}([0]_d) = [0]_c. \tag{9}$$

Fig. 1 shows the example of the diatonic scale.

2.6 Find All Centred Modes

For every scale $S^\star \in \mathcal{S}^{\star d}_c$, find the centred mode $m^{([0]_d)}_{S^\star}$ that implies the minimum amount of alterations of the reference centred mode $m^{([0]_d)}_{S^\star_0}$.

$$m^{(0)}_{S^\star} = argmin_{m_{S^\star}} \sum_{[k]_d \in \mathcal{C}_d} d_{\mathcal{C}_c}(m_S([k]_d), m^{(0)}_{S_0}([k]_d)) \tag{10}$$

where $d_{\mathcal{C}_c}$ is the circle distance:

$$\begin{aligned} d_{\mathcal{C}_c} : \mathcal{C}_c \times \mathcal{C}_c \to \mathbb{N} \\ ([x]_c, [x']_c) \longmapsto argmin_{n \in [x]_c, n' \in [x']_c} |n - n'|_{\mathbb{Z}} \end{aligned} \tag{11}$$

2.7 Construct All Representations

Once we have a centred mode $m^{(0)}_{S^\star}$, we can associate the chromatic coordinate $[x]_c = m^{(0)}_{S^\star}([k]_d)$ of each step $[k]_d$ with its original pitch classes $[x_0]_c$ in the reference mode $m^{(0)}_{S_0^\star}$ and compute the chromatic alteration $[a]_c$ necessary to obtain the new pitch classes

$$[x]_c = [x_0 + a]_c \tag{12}$$

a processes graphically depicted in Fig. 2.

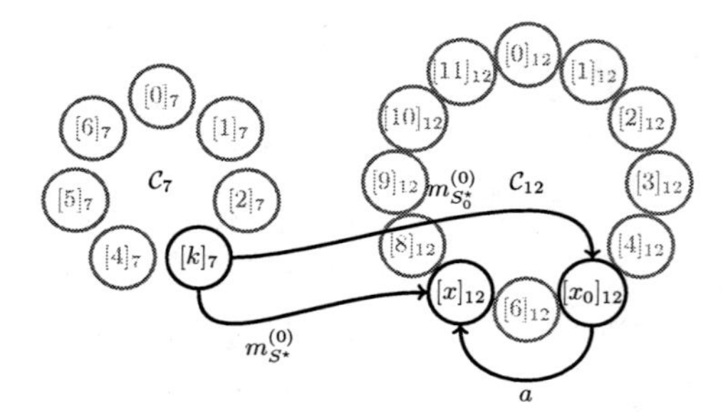

Fig. 2. Chromatic alteration. Two sharps ($[a]_{12} = [+2]_{12}$) alter a G ($[x_0]_{12} = [5]_{12}$).

Before changing our representation space for the diatonic spiral, modelled by the integers $\mathbb{Z}$, an unfolding operation of the chromatic circle is needed. We already defined a distance on $\mathcal{C}_c$; we still need to know the direction from one chromatic coordinate $[x]_c$ to another $[x']_c$.

$$\begin{aligned} sgn_{\mathcal{C}_c} : \mathcal{C}_c \times \mathcal{C}_c \to \{-1, +1\} \\ ([x]_c, [x']_c) \longmapsto \begin{cases} +1 & [x' - x]_c \in \{[0]_c, \ldots, [\lfloor \frac{c-1}{2} \rfloor]_c\} \\ -1 & \text{otherwise} \end{cases} \end{aligned} \tag{13}$$

Both functions combine into the unfolding operation $u_{\mathcal{C}_c}$.

$$\begin{aligned} u_{\mathcal{C}_c} : \mathcal{C}_c \longrightarrow \mathbb{Z} \\ [x]_c \longmapsto sgn_{\mathbb{Z}_c}([x]_c) \cdot d_{\mathbb{Z}_c}([x]_c) \end{aligned} \tag{14}$$

It is now possible to compute the original diatonic coordinate ξ_0 and the diatonic alteration α on the diatonic spiral for every step $[k]_d \in \mathcal{C}_d$ of a mode.

$$\begin{aligned} \alpha := d \cdot u_{\mathcal{C}_c}([a]_c) \\ \xi_0 := u_{\mathcal{C}_c}([d]_c^{-1} \cdot [x_0]_c) \end{aligned} \tag{15}$$

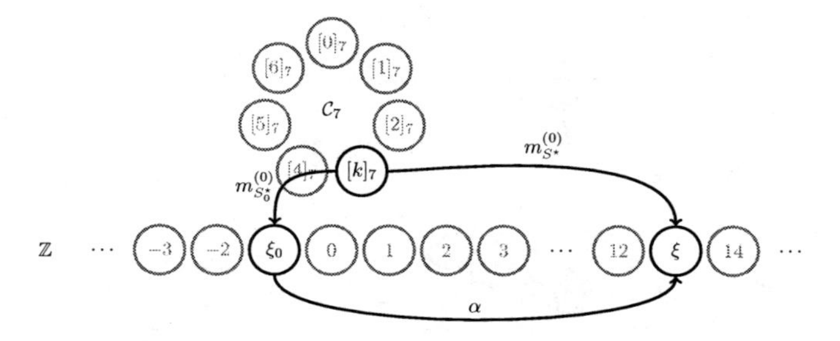

Fig. 3. The diatonic alteration corresponding to Fig. 2. The diatonic spiral is modelled by the discrete line of integers. $G\sharp\sharp$ is indexed by $13 = -1 + 7 \cdot 2$.

The same relation as in (12) holds for the diatonic space. The final diatonic coordinate is given by

$$\xi := \xi_0 + \alpha, \tag{16}$$

a process depicted in Fig. 3. Note that it is impossible for two different pairs (ξ_0, α) to correspond to a same ξ. On the diatonic spiral we have $G\sharp = -1 + 7 = +6 \neq -6 = +1 - 7 = A\flat$. This is due to the fact that $\mathbb{Z} = \{-3, \ldots, +3\} \oplus 7\mathbb{Z}$.

2.8 Order All Scales

The distribution of a centred mode's diatonic coordinates can be used to define a linear ordering on the set of centred scales, from the most compact (the diatonic) to the most widely spread (the chromatic). This order is preserved by inversion, and in case a scale is not symmetric, we need to distinguish between two members of an antisymmetric pair. Thus, each transposition class $[S]_{\mathbb{Z}_c}$ receives two indices: The first one designates the rank of the dihedral class $[S^\star]_{D_c}$ (equivalence through transposition and/or inversion) in the compactness order, whereas the second one tells if it is a palindrome (0), or which member of an antisymmetric pair (-1 and $+1$) it is.

We want to express a scale's compactness around the symmetry axis $0_\mathbb{Z}$. So we compare diatonic coordinates from the edge to the centre. The permutation $o : \mathcal{C}_d \to \mathcal{C}_d$ orders them by decreasing absolute value.

$$\left|\xi(o([0]_d))\right| \geq \left|\xi(o([1]_d))\right| \geq \ldots \left|\xi(o([d-1]_d))\right| \tag{17}$$

We define an ordering of scales by comparing these ordered vectors:

$$S > S' :\Leftrightarrow \exists k \in \mathbb{N} : \rho([k]_d) > \rho'([k]_d) \text{ and } \rho([\tilde{k}]_d) = \rho([\tilde{k}]_d), \forall \tilde{k} < k \tag{18}$$

where $\rho = \xi \circ o$. In case of an antisymmetric pair, the scale containing the greatest positive coordinate is given index $+1$, and the scale with the greatest negative coordinate index -1. Fig. 4 shows an example of his construction.

This procedure was first applied to the heptatonic scales, the result can be seen in Fig. 5.

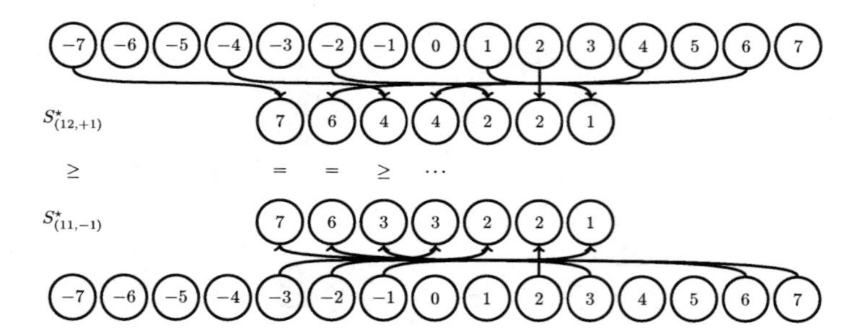

Fig. 4. Two successive classes, 11 and 12, of antisymmetric pairs −1 and +1 are being compared by testing for the spread of their diatonic distribution

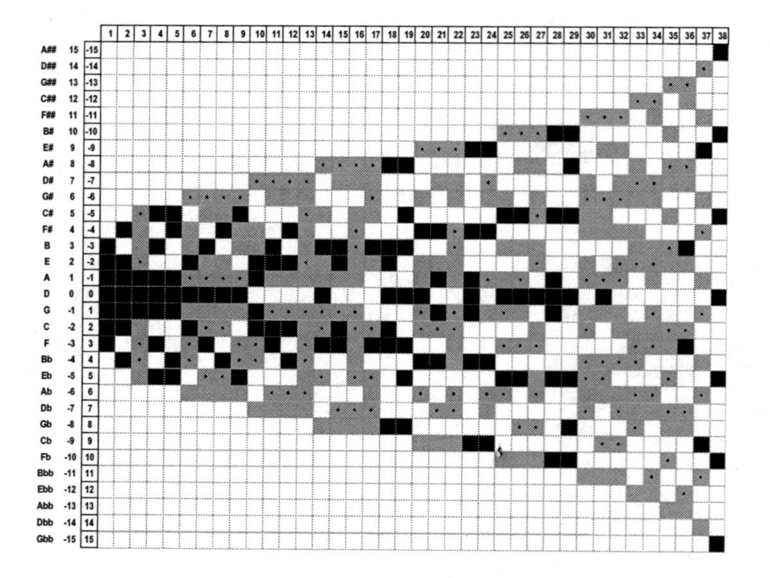

Fig. 5. ©2006 Pierre Audétat. His original diatonic bell for heptatonic scales, as proposed in [2]. Each cell represents a note and the mode corresponding to it. Each column contains a dihedral class, consisting either of a single symmetric scale or a pair of inverse scales. Alterations increase from the diatonic scale on the left to the maximally altered chromatic scale on the right. Each row represents a diatonic coordinate. The origin of the vertical axis is D, units are in steps of fifths. Black cells are symmetric notes, gray cells anti-symmetric notes, the bullet distinguishes the negative scale from the positive.

2.9 Modal Transposition

The online catalogue offers many musical examples of a same melody transformed into each of the 462 heptatonic modes. There are two possibilities to transform pitches in order to preserve their role from the diatonic to the target scale, depending on the presence or absence of notes foreign to the diatonic scale (black keys). In the first case, only the scale (along with its complement) can be mapped. Information about a possible mode gets lost. In the second case, it is possible to play with modes, and even to transpose a melody from one mode to the other within a same scale.

In the diatonic scale, we can identify every pitch class $[x_0]_c$ with a specific step $[k]_d$ of a given mode $m_{S_0^\star}^{([n_0]_c)}$ of the reference scale $S_0^\star$, and then map it to the same step of a given mode $m_{S^\star}^{([l]_c)}$ in the target scale $S^\star$.

$$\begin{array}{ccccccccc} \text{orig. pitch} & & \text{orig. pc} & & \text{mode's step} & & \text{new pc} & & \text{new pitch} \\ \mathbb{N} & \longrightarrow & \mathcal{C}_c & \xrightarrow{(m_{S_0^\star}^{([n_0]_c)})^{-1}} & \mathcal{C}_d & \xrightarrow{m_{S^\star}^{([l]_c)}} & \mathcal{C}_c & \longrightarrow & \mathbb{N} \\ x_0 & \longmapsto & [x_0]_c & \longmapsto & [k]_d & \longmapsto & [x]_c & \longmapsto & x \end{array}$$

Note that some freedom is left for converting the pitch classes back into integer pitches in the last step. The octave equivalence can be used to alter the melody as least as possible.

3 The DFT Analysis of Scales

The Discrete Fourier Transform (4) is a measure of periodicity. Traditionally, its modulus has been used in greater extend than its phase, because of its is greater ability to pinpoint some quantities invariant under transposition and inversion. Characteristics of scales or chords in music theory, energy in signal processing.

On the other hand, the phase may often be as complex and difficult to interpret as the original data. Making again an analogy with signal processing, phases are not perceptually relevant for stationary sounds, but are critical when it comes to transients. In our case, it depends on the particular transposition of a scale. This arbitrariness disappears when we use the centred representatives of each transposition class used in the diatonic bell. Hence, the phase provides information about the symmetric character of a scale.

In order to interpret the DFT coefficients, we first identify the chromatic circle with the unit circle in the complex plane, see Fig. 6.

$$\begin{aligned} \mathcal{C}_c &\longrightarrow \mathbb{C} \\ [k]_c &\longmapsto e^{i\frac{2\pi}{c}k} \end{aligned} \tag{19}$$

Computing the $[k]_c$-th DFT coefficient reduces to the vector addition of d unit vectors pointing to the (possibly multi-) set $[k]_c \cdot S^\star$, as shown by [1].

Is the index k coprime with c, the sum (4) will be computed on a shuffled regular c-polygon. Otherwise, it is computed on a polygon having fewer vertices, possibly populated with more than one pitch class. Such situations are described in [7]. They are called *balances*, because the DFT coefficients then point to a lack of equilibrium in the pitch class distribution.

If we display all pitch classes that accumulate in a given angle, we get stars with $\frac{c}{k}$ branches, as in Fig. 7. Pitch classes occupying symmetric positions at

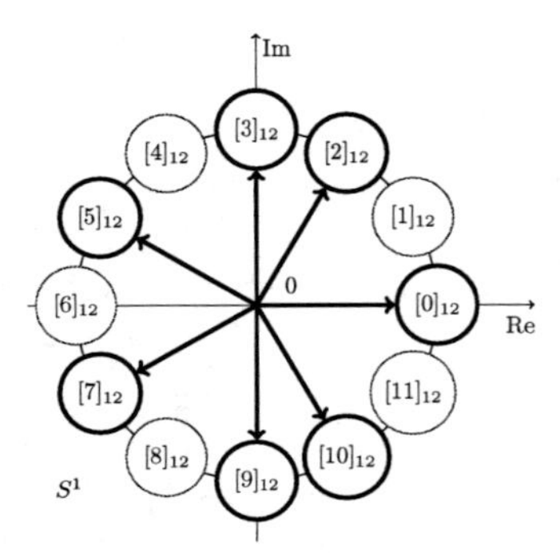

Fig. 6. The embedding of the diatonic scale $S^{\star}_{1,0}$ in the unit circle S^1 of the complex plane $\mathbb{C}$. A unit vector $e^{-i\frac{2\pi}{c}\cdot x}$ points to each chromatic coordinate $[x]_c$.

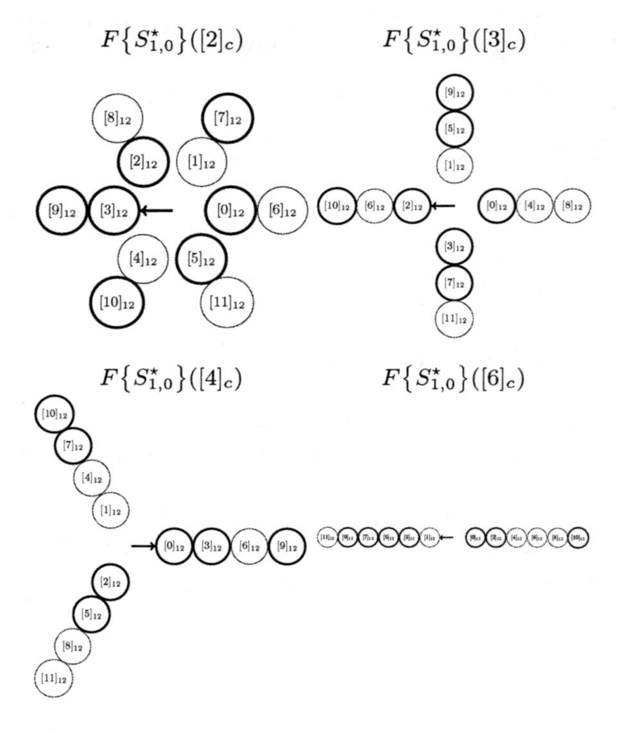

Fig. 7. The four DFT balances of the diatonic scale $S^{\star}_{1,0}$. The arrow represents the $[k]_{12}$-th coefficient, in this case a unit vector always pointing to a single unbalanced pitch class.

diameters, or regular triangles cancel each other out. The vector resulting from their sum points at the origin and yields a null DFT coefficient. Only pitch classes in “excess”, that are not balanced by some other ones, contribute to the coefficient.

The choice of coprimes c and d has a direct consequence on the balance of the diatonic scale $S^{\star}_{1,0}$: there will always be at least one unbalanced pitch for a coefficient not coprime with c. The scale size d was also chosen to be odd, so that it is impossible to cancel all pitch classes out with opposite pairs.

$$F\{S\}([k]_c) \neq 0 \quad \forall [k]_c \in \mathcal{C}_c : k|c \tag{20}$$

But a triple cancellation is possible in the hexagonal $[2]_{12}$-th balance. This is achieved by the melodic minor, $S^{\star}_{2,0}$, as well as $S^{\star}_{12,\pm 1}$ and $S^{\star}_{29,0}$.

The coefficients of the DFT show a particularly nice behaviour for two operations common in music. Both preserve the dihedral class numbering of the diatonic bell.

1. Inversion. It is connected to the scales symmetry. The real part of the DFT coefficients is an even function, the imaginary part an odd one. In case of a palindromic (symmetric) scale, it hence must disappear. Corresponding phases of asymmetric pairs will have opposite signs.
2. Complementation. Moving from a pentatonic $S^{\star}$ to a heptatonic scale $S^{\star c}$ preserves DFT modules, and inverses phases of non null even coefficients This follows from the linearity of the DFT,

$$d \cdot \delta_{k,0} = F\{\mathcal{C}_c\}([k]_c) = F\{S\}([k]_c) + F\{S^c\}([k]_c) \quad \forall [k]_c \in \mathcal{C}_c \tag{21}$$

and the additional rotation of π radians necessary to centre the complement:

$$S^{\star c \star} = -S^{\star c}. \tag{22}$$

Also notice that the indicator function of a scale is a real function, so its DFT is symmetric: there are only $\lfloor \frac{c}{2} \rfloor + 1$ independent coefficients.

Having restated those general principles, we now turn to the interpretation of particular phases and modules. We keep coefficient $F\{S^{\star c}\}([0]_c)$ aside. It always points towards the positive real direction (null phase), and its length measures the (already given) scale's cardinality.

3.1 Phases

Coefficient $F\{S^{\star}\}([\frac{c}{2}]_c)$ tells if there is an excess of even or odd pitch classes. In the first case, the phase will be null, in the second case, the coefficient points to the negative region of the real axis, and the phases is $\pm\pi$.

For all other coefficients, the phase indicates the direction of the resulting unbalanced excess. Fig. 7, shows how class B ($[9]_{12}$) is unbalanced for the second coefficient: it is the famous tritonus B-F that populates twice one corner of the hexagon. The coefficient will thus point to -1 and the phase be equal to $\pm\pi$.

Since coefficients of palindromic scales are real, their phases will be either 0 (positive) or $\pm\pi$ (negative). For asymmetric scales, the phases will be opposites.

3.2 Modules

We will measure three different aspects of the geometric configuration of scales with help of the modules a DFT. They all have to deal with the idea of uniform distribution of pitch classes across the chromatic circle. The integers d and c are coprime, which prevents us from finding an absolutely regular d-polygon, where the three criteria would be confounded.

Symmetry. The first coefficient of the DFT becomes the sum of unit vectors pointing to each of the pitch classes.

$$\begin{aligned} \sigma : \mathcal{S}_c^d &\longrightarrow \mathbb{R} \\ S &\longmapsto \big|F\{S\}([1]_c)\big| \end{aligned} \tag{23}$$

A lower index indicates a higher degree of symmetry, the perfect case being achieved when the sum is null (all vectors cancel out). In $c = 12$, only the double harmonic scale ($S^{\star}_{5,0}$) shows a perfect balance (Fig. 8). The chromatic scale ($S^{\star}_{38}$) being compactly grouped on one side of the circle shows the worst results.

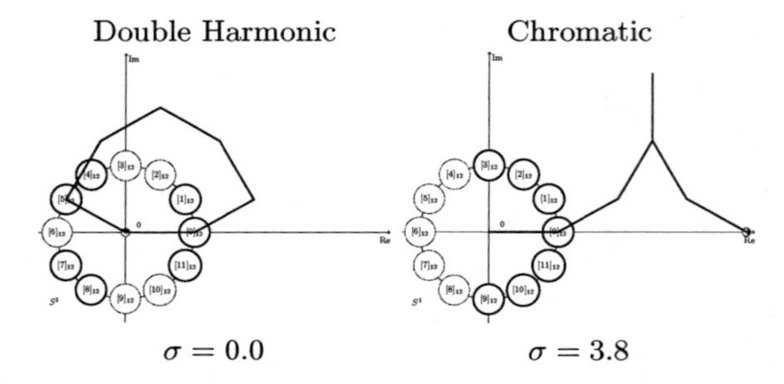

Fig. 8. Vector addition and symmetry index σ. The perfectly symmetrical double harmonic scale is built with an augmented triad $[0]_{12}, [4]_{12}, [8]_{12}$ that forms a regular triangle and two triton pairs $[1]_{12}, [7]_{12}$ and $[5]_{12}, [11]_{12}$. In the least symmetrical chromatic scale, only the triton $[3]_{12}, [9]_{12}$ is neutralised, leaving five unbalanced pitch classes.

3.3 Periodicity

It is well known that the DFT measures periodicity. The higher the modulus of the $[k]_c$-th coefficient, where $k|c$, the more $\frac{c}{k}$-periodic is the pitch class distribution. We define an index measuring the periodicity of a scale with:

$$\begin{aligned} \pi : \mathcal{S}_c^d &\longrightarrow \mathbb{R} \\ S &\longmapsto \max_{k|c} \big|F\{S\}([k]_c)\big| . \end{aligned} \tag{24}$$

A higher index shows higher periodicity. It is maximal for the unitonic scale ($S^{\star}_{4,0}$), see Fig. 9.

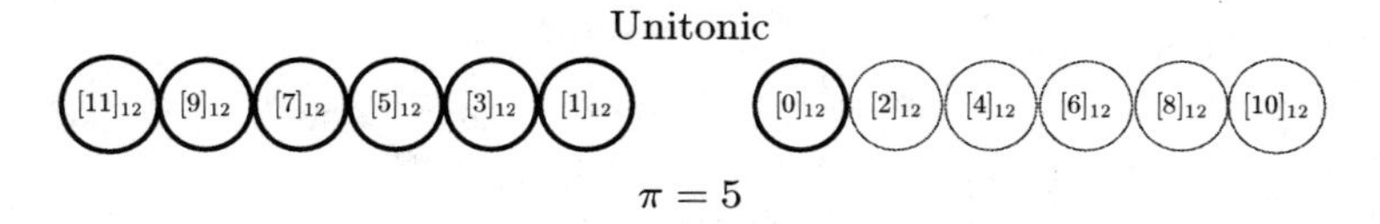

Fig. 9. Periodicity π and the $[6]_{12}$-th balance. The unitonic scale contains all odd pitch classes, that form one of the two whole tone scales, whose periodicity is $\frac{12}{6} = 2$. This achieves an excess of 5 odd pitch-classes, the maximum reachable in $c = 12$. The diatonic scale, whose maximal evenness ensures no excess greater than 1 on any coefficient, obtains the worst score, see Fig. 7.

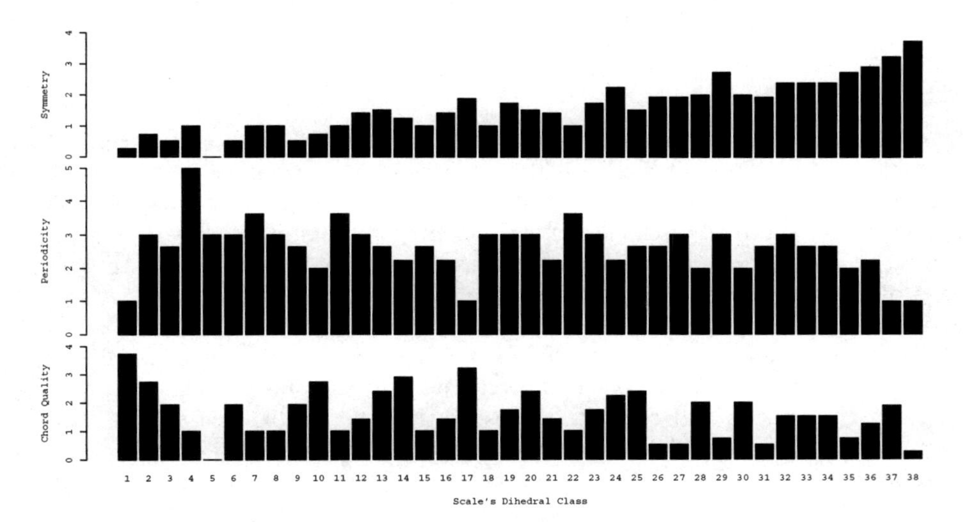

Fig. 10. Comparison of the three module based DFT indices σ, π and ε versus the diatonic bell's linear ordering of dihedral classes. Numbering of heptatonic scales goes from 1 on the left for the diatonic scale $S^{\star}_{1,0}$, towards the chromatic scale $S^{\star}_{38,0}$ on the right.

3.4 Chord Quality

As mentioned in Sec. 2.4, the $[d]_c$-th modulus called *chord quality* by Quinn [7] serves also for a new definition of maximall evenness.

$$\begin{aligned} \varepsilon : \mathcal{S}^d_c &\longrightarrow \mathbb{R} \\ S &\longmapsto \left|F\{S\}([d]_c)\right| \end{aligned} \tag{25}$$

It is related to the symmetry index σ through an affine permutation of the coefficients.

Despite some correlation appearing between the symmetry index σ and the diatonic bell's ordering, the three indices do not lead to a progressive classification from a diatonic to chromatic character, see Fig. 10.

4 Conclusion

The diatonic bell and the DFT differ in their structure. Whereas the underlying space of the former is infinite, the usual definition of a DFT requires finiteness. Nevertheless, both are constructed on an analogous principle: the *balance*. The idea of a physical balance lies behind the process of centring scales in the diatonic bell, and this image also helps for interpreting Fourier coefficients.

By lifting up scales from the chromatic circle to the spiral of fifths, the diatonic bell adds a tonal structure to the atonal combinatorics of musical set theory. Although the DFT is defined on the chromatic circle and, in this sense, is purely atonal, it shares some elements with the diatonic bell, namely the relevance of symmetry and the ability of pinpoint the diatonic flavour of some scales.

4.1 Symmetry

The role pitch class D plays as a symmetry axis in both the chromatic and diatonic worlds is clearly shown. This remarkable fact was already noticed by the french music theorist and composer Camille Durutte in his treatise of 1855 [5], where he described pitch classes with 31 integers, ranging from -15 to $+15$, centred around $D = 0$, and ordered by fifths. The diatonic bell's horizontal axis thus already appeared in the first historic attempt to formalise pitch classes algebraically.

The symmetry axis is also essential for the DFT, since it lies on the real axis of the complex plane. Inversion then corresponds to complex conjugation.

Using the centred and compact representatives of the diatonic bell has two advantages. The comparison between transpositional classes makes sense and interpreting phases of the DFT coefficients becomes more accessible: it eliminates a great amount of uninformative components that would have been induced by an arbitrary rotation. The most striking fact is that the coefficients of palindromic scales are purely real.

4.2 Measuring the Diatonic Character of a Scale

The diatonic bell displays scales as a deformation of the diatonic scale and arranges them according to the their degree of compactness on the spiral of fifths, ranging over all dihedral classes from the diatonic to the chromatic. Our initial intention was to use this linear ordering to define a measure of a scale's diatonic or chromatic character. The former being the maximally even scale, the latter the minimally even one, we expected to observe the same trend with DFT coefficients measuring regularity in the geometric configurations. As shown in Fig. 10, DFT-based indices did not confirm the bell's ordering. The chromatic or diatonic character of a scale does not reduce to a one-dimensional question, at least not this way.

On the other hand, almost all measures succeed in isolating the diatonic and chromatic scales as poles. What happens in between is less clear, but both approaches converge in distinguishing a group of particular scales, formed by the six first scales located on the left side of the diatonic bell. They correspond exactly to those used in the western tradition: diatonic ($S^\star_{1,0}$), melodic minor ($S^\star_{2,0}$), harmonic major ($S^\star_{3,-1}$) and minor ($S^\star_{3,+1}$), unitonic ($S^\star_{4,0}$) and double harmonic ($S^\star_{5,0}$).

One reason may be that they have to be the most compact, so that the tonic pitch class D, is surrounded with its dominant A and subdominant G, a feature essential for tonal music. Note that only three other scales show a similar behaviour: $S^\star_{22,\pm1}$ and $S^\star_{28,0}$. Optimums of the geometrical measurements defined with help of DFT modules in Sec. 3 systematically exhibit scales from this same *harmonic* block.

- Diatonic is the most even: $\varepsilon(S^\star_{1,0}) = 3.73$.
- Minor melodic is also one of the three balanced scales with regard to the triton periodicity: $F\{S^\star_{2,0}\}([2]_{12}) = 0$.
- Unitonic is the most periodic: $\pi(S^\star_{4,0}) = 5.00$.
- Double harmonic is the most symmetric: $\sigma(S^\star_{5,0}) = 0.00$.

In that case, the diatonic bell's requirement for compactness seems to agree with those of he DFT for regularity. This follows from the property of the diatonic scale to be generated by a succession of fifths, and that this sequence is not degraded too much for the first scales. Convergence between musical practice and mathematical interest as personified by the diatonic scale seems to extend also to the neighbour scales.

We are currently working on the implementation of these two approaches (the diatonic bell and the DFT) within OpenMusic visual programming language, as a package included in the MathTools environment. These new tools will allow the user to automatically generate diatonic bells and musical transpositions for the heptatonic and pentatonic scales. Divisions of the octave other than $c = 12$ will also be handled, as long as the requirements on the parameters are fulfilled (Sec. 2.1). One simply should take care about the exponential growth of the diatonic bell in microtonal context.

References

1. Amiot, E.: David lewin and maximally even sets. Journal of Mathematics and Music 1(3), 157–172 (2007)
2. Audétat, P.: La cloche diatonique. Jazz Deptartment, University of Applied Sciences of Western Switzerland (2006)
3. Carey, N., Clampitt, D.: Aspects of well-formed scales. Music Theory Spectrum 11(2), 187–206 (1989)
4. Clough, J., Douthett, J.: Maximally even sets. Journal of Music Theory 35, 93–173 (1991)
5. Durutte, C.: Esthétique musicale. Technie, ou lois générales du système harmonique. Mallet-Bachelier, Paris (1855)
6. Knuth, D.E.: Generating All Combinations and Partitions. The Art of Computer Programming, vol. 4(fascicle 3). Addison-Wesley, Reading (2005)
7. Quinn, I.: A Unified Theory of Chord Quality in Equal Temperaments. PhD thesis, Eastman School of Music, University of Rochester (2004)

The Geometry of Melodic, Harmonic, and Metrical Hierarchy

Jason Yust

Abstract. Music is hierarchically structured in numerous ways, and all of these forms of organization share essential mathematical features. A geometrical construct called the Stasheff polytope or associahedron summarizes these similarities. The Stasheff polytope has a robust mathematical literature behind it demonstrating its wealth of mathematical structure. By recognizing hierarchies that arise in music, we can see how this rich structure is realized in multiple aspects of musical organization. In this paper I define hierarchic forms of melodic, harmonic, and metrical organization in music, drawing on some concepts from Schenkerian analysis, and show how each of them exhibits the geometry of the Stasheff polytope. Because the same mathematical construct is realized in multiple musical parameters, the Stasheff polytope not only describes relationships between hierarchies on a single parameter, but also defines patterns of agreement and conflict between simultaneous hierarchies on different parameters. I give musical examples of conflict between melodic and rhythmic organization, and show how melodic and harmonic organization combine in melody and counterpoint.

1 General Characteristics of Musical Hierarchy

Hierarchies operate in a number of different modalities in tonal music, and forms of musical hierarchy generally share some essential properties. This fact opens the door to a rich application of a mathematical conception of hierarchy to music.

Hierarchical accounts in the basic modalities melody, harmony, and metrical rhythm play crucial roles in understanding tonal music. All of these types of hierarchy share principal underlying features: (1) they involve objects ordered in time, (2) they are better represented by hierarchies on intervals between successive objects rather the objects themselves, and (3) they are accurately captured by strictly binary bracketings. Together these features imply a robust mathematical description captured by the geometry of the Stasheff polytope or associahedron.

The graph-theoretical object of a *rooted tree* represents the concept of hierarchy in its most general sense. Property (1) for musical hierarchies implies that they correspond more specifically to *plane trees*, trees in which the children of each node are ordered from left to right (where for our purposes left corresponds to prior in time, etc.). Plane trees are equivalent to ways of bracketing series of objects (the objects corresponding to the leaves of the tree).

As for property (2), consider the passing-tone figure in Figure 1. Formal accounts of Schenkerian theory often represent the hierarchy of this type of musical figure with a plane tree that takes musical notes as its objects, as in Figure 2 (Cohn and Dempster

E. Chew, A. Childs, and C.-H. Chuan (Eds.): MCM 2009, CCIS 38, pp. 180–192, 2009.

Fig. 1. The passing tone of second species counterpoint

Fig. 2–3. Plane trees with musical notes or motions between notes as objects

1998, Marsden 2005, Lerdahl and Jackendoff 1983, Rahn 1979, Smoliar 1980). This type of hierarchy involves an arbitrary assignment of either of the consonant tones as the parent of the dissonant one, obfuscating the symmetry of the musical figure. Larson 1997 criticizes this aspect of the theory of Lerdahl and Jackendoff 1983, and Lerdahl's response (1997) is dismissive and inadequate. Proctor and Higgins 1988 and Yust 2006 have pointed out that such "reductionist" theories inaccurately represent Schenker's own analytical practice. In more recent computational applications that take this formal model as the standard (Marsden 2005 and 2007), the requirement for multiply-embedded arbitrary decisions results in a combinatorial explosion within the acceptable analysis of short musical phrases. Yust 2006 proposes a solution to the the interpretive awkwardness of hierarchies with notes as objects: bracketing intervals between successive notes, rather than the notes themselves, as shown in Figure 3.

Meter presents a precisely analogous situation: it is not a bracketing on timepoints but on timespans (intervals between timepoints). In the rhythm of Figure 1, a whole-note duration divides into two half-note durations; the weak-beat timepoint does not group with either the preceding or following strong beat by virtue of the meter. Unlike models of tonal hierarchy proposed by music theorists, models of metrical hierarchy typically take intervals between events (durations) as their objects rather than the events themselves (timepoints). Presumably the difference treatment of parameters comes from musical notation, which reifies durations and pitches rather than timepoints and intervals. As a result, studies that discuss relationships and conflict between meter and tonal structure (Komar 1971; Lerdahl and Jackendoff 1983; Rahn 1979; Schachter 1976, 1980, 1987; Yeston 1976) present models in which time and pitch are treated in fundamentally different ways. Construing tonal hierarchy in terms of intervals rather than notes corrects this situation, so that rhythmic and tonal patterns can be compared directly in terms of hierarchic structure.

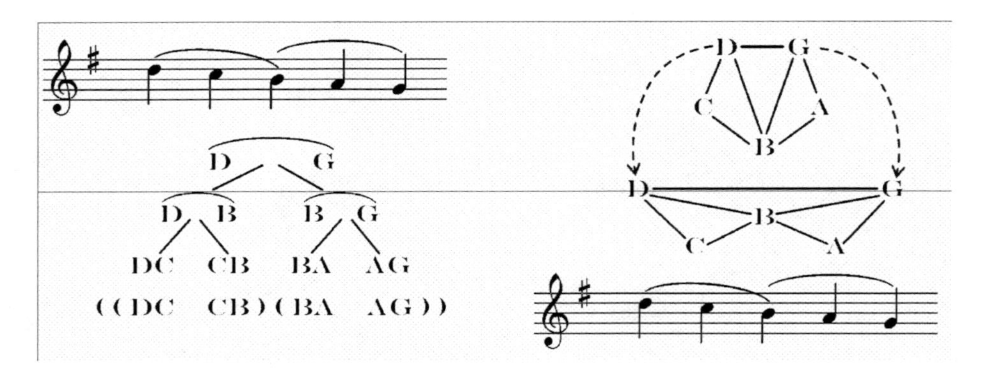

Fig. 4. The bijection between binary plane trees and triangulations of a polygon as representations of musical hierarchy

Property (3) provides a useful heuristic for present purposes. Although non-binary relationships can exist in melodic, harmonic, and rhythmic hierarchies, binary structures are the most important and demonstrate the essential properties of hierarchy in these modalities. Therefore I will focus here on the binary plane tree, in which all nodes that are not leaves (called *internal nodes*) have exactly two children, a left child and a right child. The application of the Stasheff polytope is more general however, describing the relationships between all possible plane trees. (See the end of section 3 below).

Though the objects the musical hierarchies explored here are intervals rather than more concrete musical objects, networks visually convey their musical implications more clearly when the nodes correspond to concrete objects. The bijective equivalence of binary plane trees and *triangulations of polygons* makes it possible to create a network on musical objects that gives a hierarchy on directed intervals between them. Figure 4 illustrates this bijection—which replaces the nodes of the tree with edges—and shows how to redraw the polygon to show the left-to-right ordering on the vertices clearly. Note that internal edges in the triangulation (which correspond to internal nodes in the tree) can be described as upward- or downward-slanting.

2 Harmonic, Melodic, and Metrical Forms of Hierarchy

2.1 Harmonic Hierarchy

Schenker (1987) had the fruitful insight of taking the passing-tone figure of second species counterpoint as a paradigm for hierarchical relationships in music. However, the passing-tone figure can only serve as a paradigm of tonal hierarchy if it is somehow generalized, because literal passing figures alone do not embed one another recursively. Schenker (1979, 1997) addressed this by harmonizing the dissonant passing note, as in the *Ursatz* of Figure 5, and *horizontalizing* the resulting vertical consonance so that he could then fill it with new passing motion, as shown in Figure 6. The process implies a tonal hierarchy, also shown in Figure 6, that generalizes the passing-tone figure as a *symmetrical binary division of step-class intervals*. That is,

Fig. 5. Schenker's *Ursatz* (From *Der Freie Satz*)

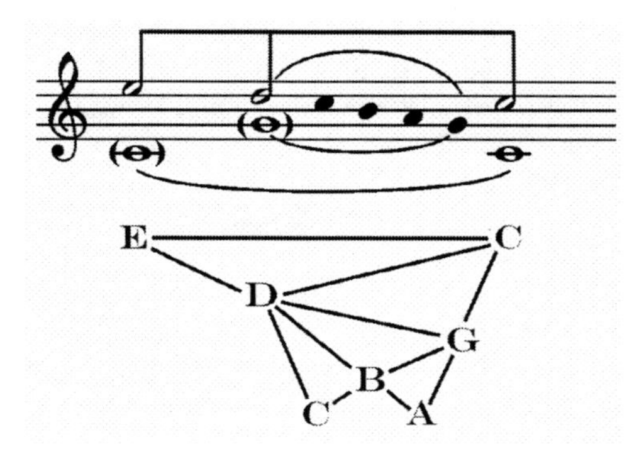

Fig. 6. Schenker's *Ursatz* with a horizontalized dominant

the passing tone divides the step-class interval of a third into two equal step-class intervals (seconds) and similarly, the second divides into two fifths, the harmonization of the passing tone, the fifth divides into two thirds which complete the triad that supports the passing tone, and new passing motion then fills the horizontalized thirds. (This process of division is an order-3 automorphism of the group of step-class intervals, a cyclic group of order seven; see Yust 2007)

I will call this type of hierarchy *harmonic structure*. Because its building blocks include triads and series of fifths, it is useful for describing harmonic patterns in tonal music, particularly sequential harmonic patterns. Its main shortcoming, from the perspective of music analysis, is also one of its assets: it cuts across compositional voices and registers, collapsing tonal content into a single hierarchy that can only show one melodic progression at a time. Therefore, a description that recognizes counterpoint and the continuity of voices requires a different form of hierarchy on tones, a purely melodic type of hierarchy, to complement this one.

2.2 Melodic Hierarchy

A melodic hierarchy is one on tones strictly ordered in time and belonging to one voice. One particularly useful melodic hierarchy emerges from the principles, first, that the line it represents should be conceived as constantly *in motion* at every level (so that the line can include no repeated tones or neighboring motions); and second,

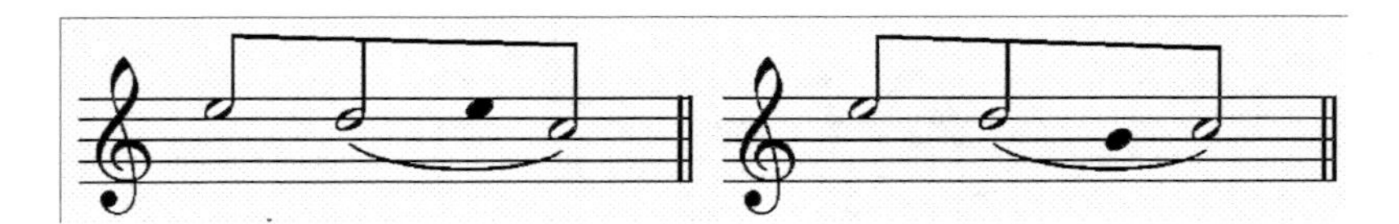

Fig. 7. Incomplete neighbor figures for melodic elaboration

that the hierarchy reflect the process—characteristic of Schenkerian analysis—of locating stepwise connections and stepwise passing motions within a musical line at various levels.

Again, the passing-tone figure is a model, and the problem of its inability to embed itself reappears. Expanding seconds with incomplete neighbors gives a solution that prioritizes stepwise motion. Figure 7 shows two such possibilities for the descending second. We can equate these different incomplete neighbor figures under a system of equivalence classes consisting of notes separated by diatonic fourths. These equivalence classes identify the position of tones within a system of conjunct tetrachords, so we can call them *tetrachordal positions*. For melodies of restricted range, a sequence of three distinct tetrachordal positions may represent a literal stepwise passing motion, any incomplete neighbor figure involving a leap of a third, or arpeggiations of a close-position triad. Unlike harmonic hierarchies, melodic hierarchies built out of tetrachordal passing motions do not recognize octave equivalence, and so are most useful in confined registral ranges.

2.3 Metric Hierarchy

Another useful form of hierarchy is the *metric* hierarchy of an unsyncopated rhythm. For present purposes, we will consider only strictly binary metrical schemes. (See the end of the next section regarding mixed binary-triple schemes). An exemplary binary metrical scheme is 4/4 meter, possibly extending hypermetrically to regular four- and eight-bar phrases and so forth. An *unsyncopated rhythm* within such a scheme is a succession of timepoints (1) that begins and ends on the metrically strongest timepoints within its span, and (2) where no unarticulated timepoint is stronger than the articulated one immediately preceding it. Like the assumption of strict binary organization, the no-syncopation assumption sacrifices musical generality for mathematical simplicity and can be corrected after the basic mathematical framework is established.

3 Musical Realizations of the Stasheff Polytope

These three forms of hierarchy enable a musical exploration of the Stasheff polytope. The vertices of the Stasheff polytope correspond to the possible binary hierarchies, and its geometrical properties represent much of the structure implicit in the concept of hierarchy. The work of mathematician Jean-Louis Loday and others demonstrates the great depth of this structure. (Stasheff 1963; Loday 2004, 2005, 2007; Loday and Ronco 2002) This convex polytope can be constructed in any dimension. In

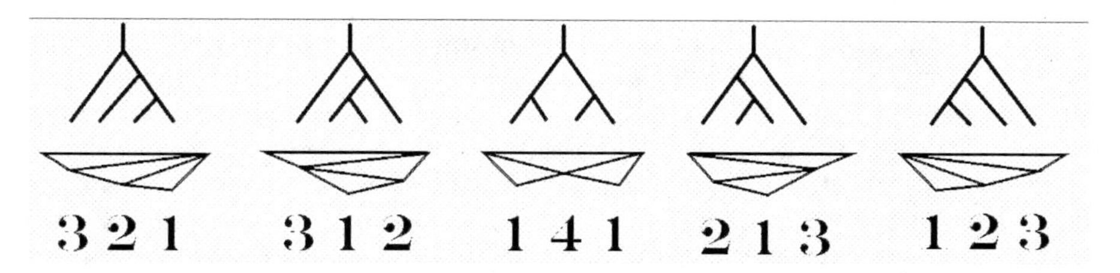

Fig. 8. The possible hierarchies on five events as trees and triangulations

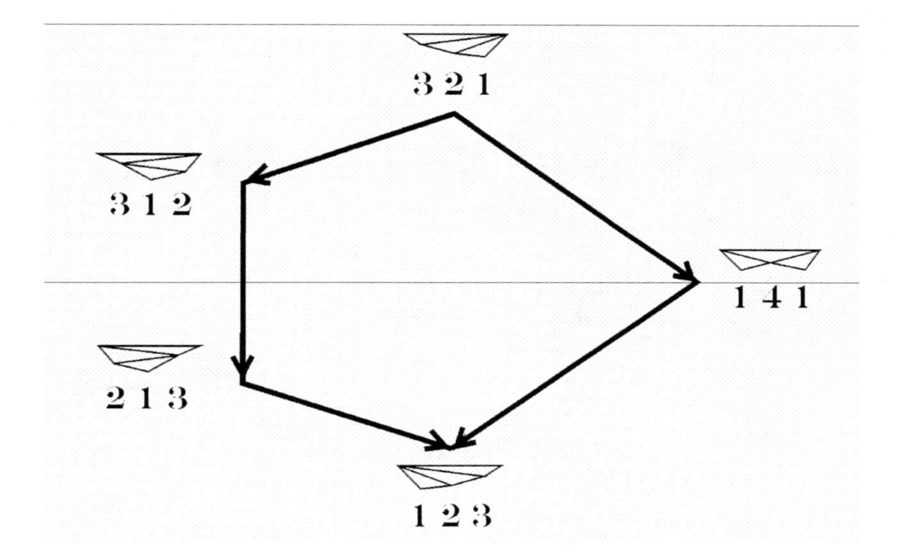

Fig. 9. The two-dimensional Stasheff polytope

n-dimensional space, the vertices of the polytope correspond to binary trees with $n + 2$ leaves, or triangulations of an $(n + 3)$-gon.

For example, let us construct the two-dimensional Stasheff polytope, whose points correspond to the binary plane trees with four leaves or the triangulations of a pentagon. Figure 8 shows the possible hierarchies on series of five musical objects, both as trees and as triangulations. We define a point in three-dimensional space for each triangulation where the coordinates of the point correspond to the vertices of the triangulation from left to right (which in turn correspond to musical events in the sequence) excluding the first and last (uppermost) vertices. The value for each coordinate is given by taking the highest edges to the left and right of that point, counting the number of boundary edges below them, and multiplying the two numbers. Figure 8 shows the resulting coordinates for triangulations of the pentagon.

The Stasheff polytope is the convex hull of these points. The dimension of the polytope is one less than the number of coordinates because the points are constrained to lie on the plane $x_i = n(n + 1)/2$ (in dimension n). In musical examples, the numbers define relative weights (e.g., metrical accent) for the corresponding objects. The edges in the geometry represent elementary "flip" operations that move exactly one internal edge in the triangualtion. A *leftward* flip replaces an upward-slanted edge with a downward-slanted one; a *rightward* flip does the opposite. Figure 9 gives the 2-dimentsional polytope, with arrows pointing in the direction of leftward flips; as Loday (2007) observes, they establish an important partial ordering on the hierarchies.

It is possible to define operations that relate rhythmic and melodic sequences whose hierarchies share an edge in the polytope (i.e., are related by a flip). Figure 10 presents a series of melodies that traverse the two-dimensional polytope starting at (1 4 1) and moving counterclockwise around the perimeter until arriving back at (1 4 1) with simultaneous rhythmic and tonal operations. Figure 11 shows the Stasheff polytope in three dimensions. Figure 12 gives an (arbitrary) sample traversal that circles the three-dimensional polytope through a series of flips in coordinated rhythmic-melodic structures.

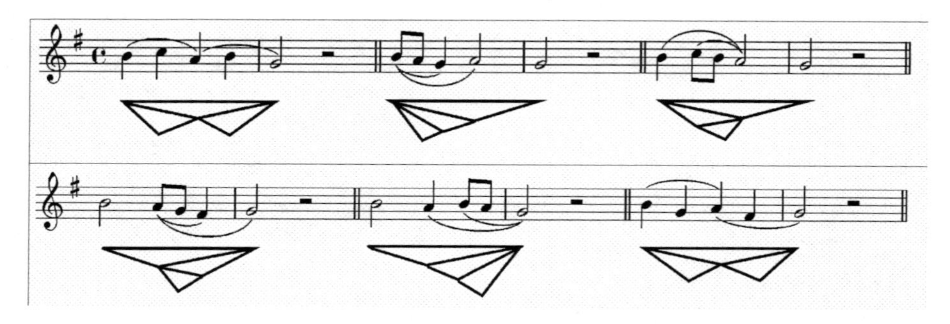

Fig. 10. A melodic traversal of the two-dimensional Stasheff polytope

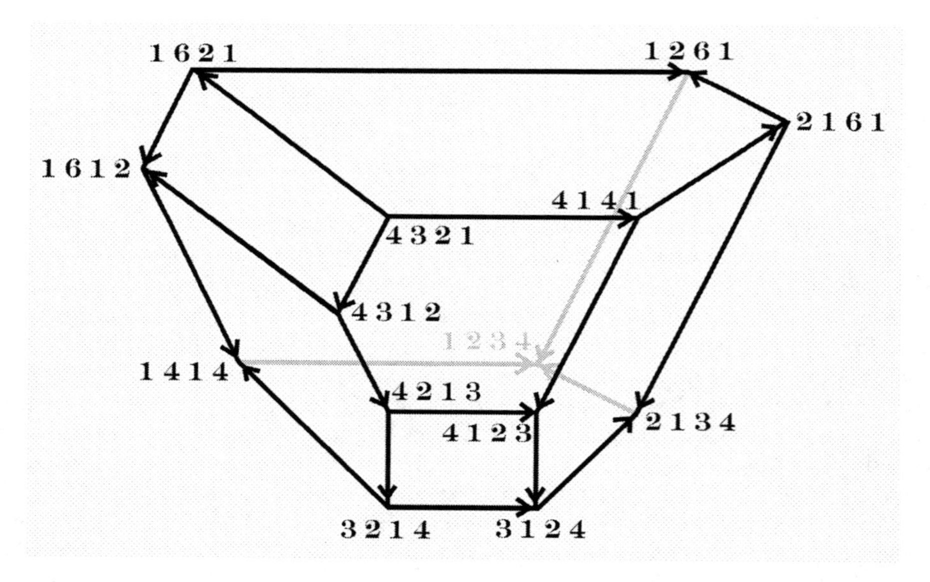

Fig. 11. The three-dimensional Stasheff polytope

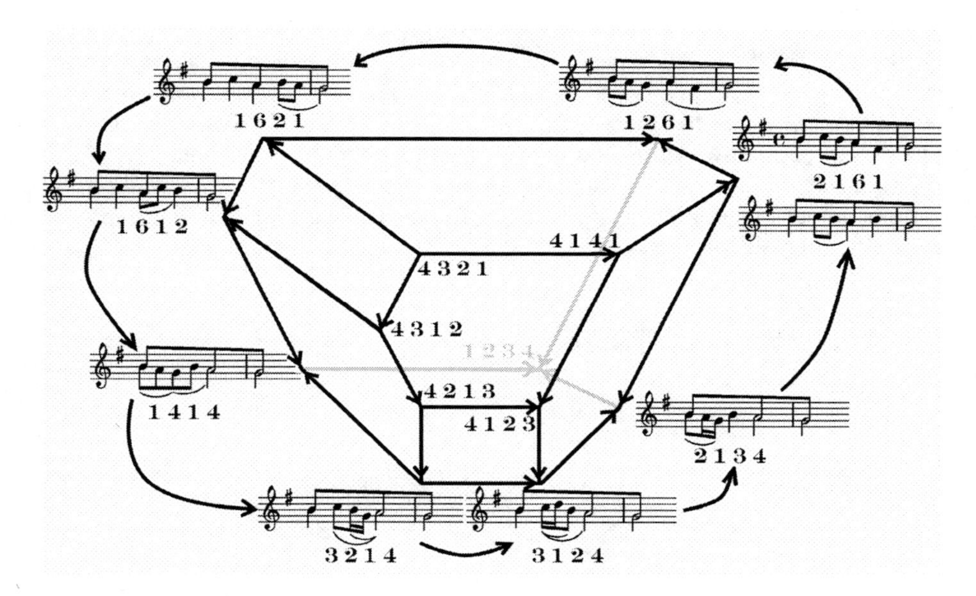

Fig. 12. A melodic traversal of the three-dimensional Stasheff polytope

The Stasheff polytope structures the relationships not only between binary hiearchies, but between any hierarchy representable by a plane tree. All types of structure described here admit of mixed binary-ternary hierarchies, which are necessary for dealing with, e.g., triple and compound meter. The cells of the polytope correspond to the plane trees with $n + 1$ leaves, with higher-dimensional cells corresponding to trees with fewer internal nodes. The binary trees, with a maximum of internal nodes, correspond to the 0-cells (points) of the polytope. Mixed binary/ternary trees correspond to higher dimensional cells (edges, faces, etc.) depending on the number of ternary branchings they include.

4 Relating Hierarchies on Different Musical Parameters

4.1 Conflict between Melodic and Metric Structures

The generalization of the concept of hierarchy to different musical parameters means that not only can one geometrically relate different hierarchies on a given parameter, but also hierarchies existing simultaneously on different parameters. Figure 13 shows the rhythmic operations of Figure 10 on a sequence with an unchanging melodic structure, resulting in various kinds of conflict between simultaneous rhythmic and melodic structure.

Schenker used such rhythmic operations on occasion to add emphasis to his tonal-hierarchical analyses of musical passages. Figure 14 reproduces a portion of his analysis of the theme from the C minor fugue of Bach's *Well-Tempered Clavier*. (It excludes the upper part of the compound melody and normalizes syncopations). The background of the theme is a descending line G–F–E♭. Schenker gradually adds new elements in coordinated rhythmic and melodic structure, and then transforms rhythmic structure to misalign the two hierarchies. Thus, rhythm helps to clarify the structure of the melody at each introduction of a new element, while progressive transformations bring it gradually in line with the actual rhythm of the fugue theme. The end result is a combination of hierarchies far apart in the five-dimensional Stasheff polytope—a heavily left-weighted tonal structure against a right-weighted rhythm.

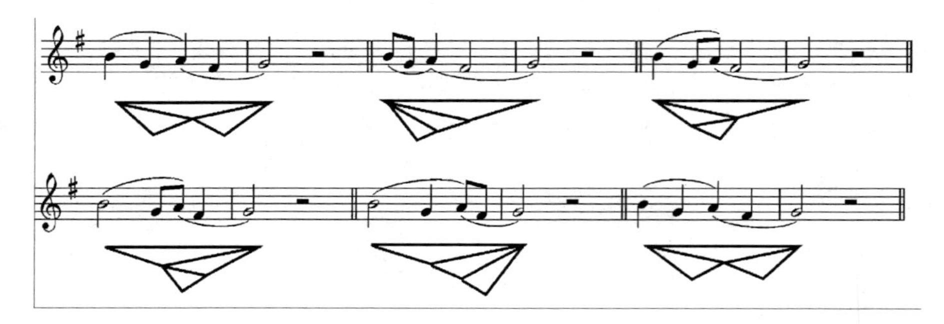

Fig. 13. Rhythmic transformations against a constant melodic structure

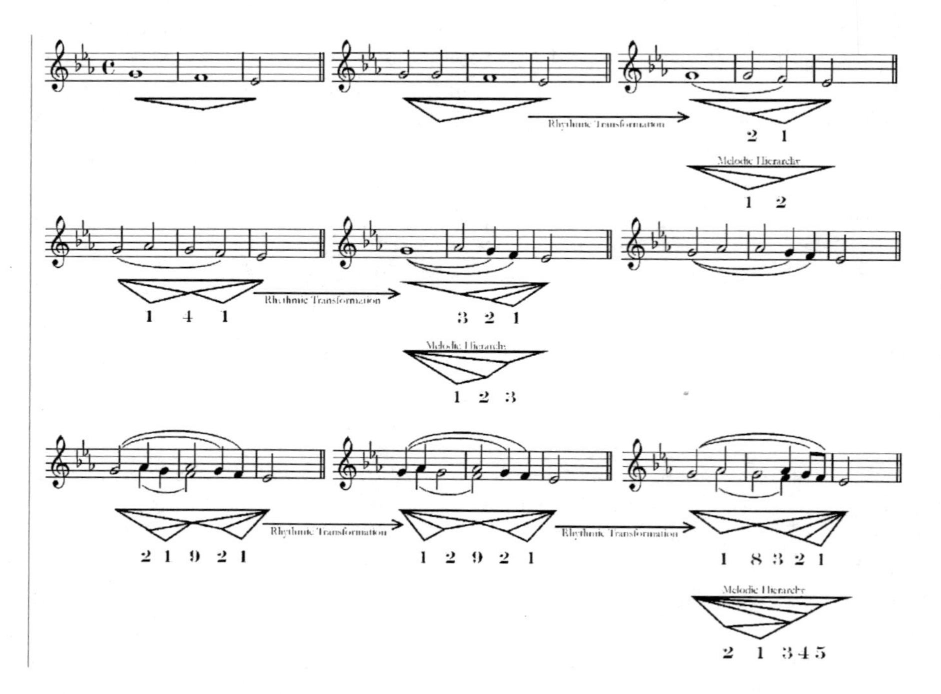

Fig. 14. Rhythmic transformations in Schenker's analysis of the C minor fugue subject

4.2 Melodic Structures in Counterpoint

Geometrical relationships between hierarchies can also describe aspects of tonal counterpoint. Counterpointing melodies with melodic and rhythmic structures in perfect coordination produce successions of consonant thirds and sixths, as in the first counterpoint of Figure 15. Transforming the melodic and rhythmic structures in one voice against the other voice creates oblique dissonances. Adding conflict between melodic and rhythmic structures *within* one voice to the conflict between voices produces accented dissonances. For example, Figure 16 transforms the rhythm of the upper voice in the fourth counterpoint of Figure 15.

Fig. 15. Transformations of melodic/rhythmic hierarchy in counterpoint with a constant melody

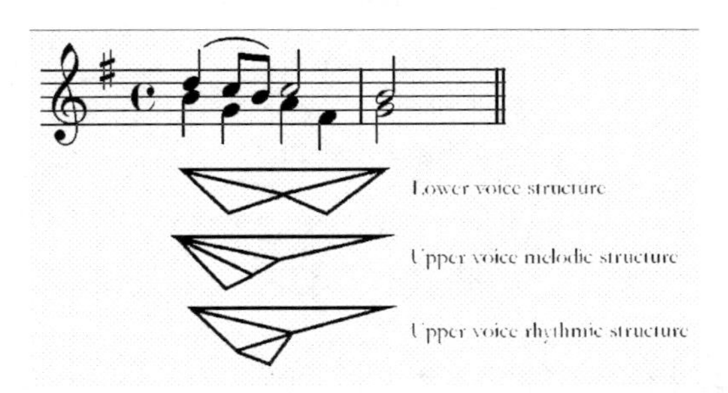

Fig. 16. Accented dissonance as a relationship of three hierarchies

4.3 Relationships between Melodic and Harmonic Structure

Melodic structures also relate to harmonic structures through the properties of their hierarchies. Melodic and harmonic structures, however, relate through the contraction of their hierarchies, and therefore invoke the relationships between Stasheff polytopes in different dimensions. The fifths-sequence figure distinguishes harmonic hierarchies from melodic ones, so the contraction of thirds from below the fifths sequences of a harmonic hierarchy creates an associated melodic hierarchy. Figure 17 shows this relationship; notice that the contraction here preserves the stepwise passing motions of the harmonic structure and introduces some new stepwise relationships, such as the step from leading tone to tonic. The contracted edges appear as vertical intervals in the musical notation of Figure 17.

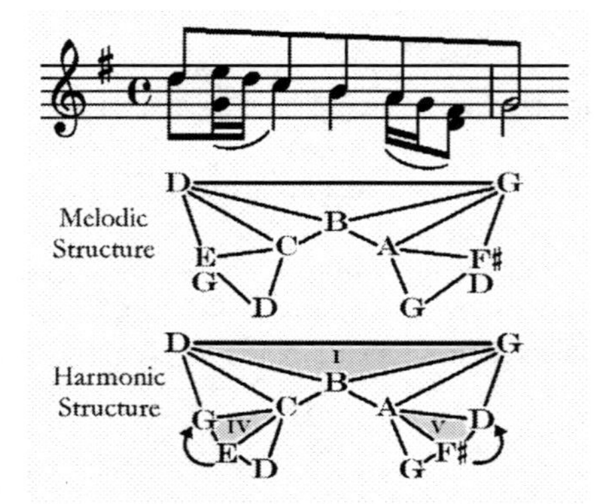

Fig. 17. Melodic structure as a contraction of harmonic structure

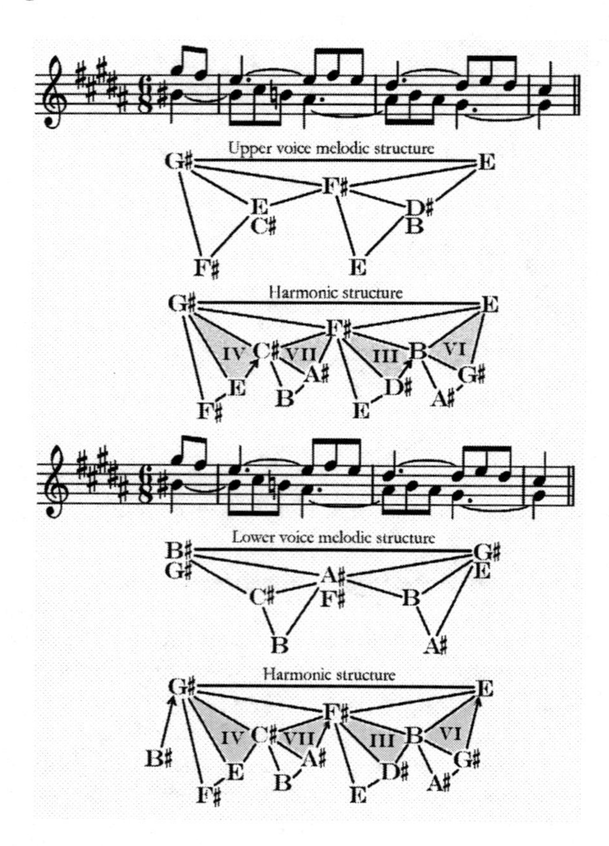

Fig. 18. Two melodies in counterpoint (from the G♯ minor prelude of Bach's *WTC I*) as differing contractions of a harmonic structure

In general, there might be no such contractions of a harmonic structure that do not annihilate parts of it. In such an instance, a compete representation of the harmonic structure requires multiple melodic structures in counterpoint, each a contraction of a common harmonic structure. Figure 18 shows a sequence from the G♯ minor prelude of the *WTC I* (mm. 19–21). The imitative counterpoint of the two upper voices consists of two different contractions of the underlying harmonic structure, and the two voices together express the complete structure of the harmonic sequence.

Each contraction in Figure 18 requires two vertex deletions. Geometrically, a vertex deletion defines an n-simplex in the n-dimensional Stasheff polytope (consisting of all the hierarchies reducing to a common structure with the deletion of one foreground vertex). The intersections of these n-simplexes reflect the corresponding simplicial structure on the $(n-1)$-polytope. Two successive vertex deletions therefore define an $(n-1)$-simplex of intersecting n-simplexes. Two melodic structures in counterpoint define a precise harmonic structure when the uncontracted melodic structures correspond to two (simplexes of simplexes of . . .) simplexes in some Stasheff polytope that intersect in precisely one point. For instance, the first half of the harmonic structure in Figure 18 (from G♯ to F♯) corresponds to a point on the 4-dimensional Stasheff polytope. The deletions of this structure for each melodic voice, (G♯-F♯-E-C♯-F♯ and G♯-C♯-B-A♯-F♯) each correspond to 3-simplexes (tetrahedrons) of intersecting 4-simplexes in the 4-dimensional polytope. The two sets of sixteen points encompassed by each of these intersects in one point, which corresponds to the overall harmonic structure of the passage.

5 Conclusion

The geometry of hierarchy is fascinating by virtue of its depth of mathematical structure alone. Indeed, we have only scratched the surface of that mathematical structure here. Given the myriad ways that music is hierarchically organized this geometry illuminates it from many angles, showing the nature of relationships between different ways of hierarchically organizing rhythms or tonal patterns, conflict between the simultaneous rhythmic and tonal organization of a melody or multiple melodies in counterpoint, and the relationships between different types of hierarchical organization that apply to tonal patterns. All of these relationships not only demonstrate the importance of hierarchy in tonal music, but also give the lie to the idea, implicit in Schenkerian analysis, that musical hierarchy is a unitary phenomenon. Indeed, a single hierarchy is too simple an object to represent the complex interactions that bring life to music in the tonal idom.

References

Cohn, R., Dempster, D.: Hierarchical unity, plural unities: Towards a reconciliation. In: Bergeson, K., Bohlman, P.V. (eds.) Disciplining Music: Musicology and its Canons. University of Chicago Press, Chicago (1992)

Komar, A.: Theory of Suspensions: A Study of Metrical and Pitch Relations in Tonal Music. Princeton University Press, Princeton (1971)

Lee, C.W.: The associahedron and triangulations of the n-gon. European Journal of Combinatorics 10/6, 551–560 (1989)
Larson, S.: The problem of prolongation in 'tonal' music: Terminology, perception, and expressive meaning. Journal of Music Theory 41(1), 101–136 (1997)
Lerdahl, F.: Issues in prolongational theory. Journal of Music Theory 41(1), 141–155 (1997)
Lerdahl, F., Jackendoff, R.: A Generative Theory of Tonal Music. MIT Press, Cambridge (1983)
Loday, J.-L.: Realization of the Stasheff polytope. Archiv der Mathematik 83, 267–278 (2004)
Loday, J.-L.: Inversion of integral series enumerating binary plane trees. Séminaire Lotharingien de Combinatoire 53 Article B53d, 1–16 (2005)
Loday, J.-L.: Parking functions and triangulation of the associahedron. Contemporary Mathematics 431, 327–340 (2007)
Loday, J.-L., Ronco, M.O.: Order structure on the algebra of permutations and of planar binary trees. Journal of Algebraic Combinatorics 15, 253–270 (2002)
Marsden, A.: Generative structural representation of tonal music. Journal of New Music Research 34(4), 409–428 (2005)
Marsden, A.: Empirical study of Schenkerian analysis by computer. In: The joint meeting of the Society for Music Theory and the American Musicological Society, Nashville (2008)
Proctor, G., Lee Riggins, H.: Levels and the reordering of chapters in Schenker's Free Composition. Music Theory Spectrum 10, 102–126 (1988)
Rahn, J.: Logic, set theory, music theory. College Music Symposium 19(1), 114–127 (1979)
Schachter, C.: Rhythm and Linear Analysis: A Preliminary Study. In: Mitchell, W., Salzer, F. (eds.) Music Forum 4 (1976); reprinted in Straus, J. (ed.): Unfoldings: Essays in Schenkerian Theory and Analysis, pp. 17–53. Oxford University Press, New York (1999)
Schachter, C.: Rhythm and Linear Analysis: Durational Reduction. In: Mitchell, W., Salzer, F. (eds.) Music Forum 5 (1980); reprinted in Straus, J. (ed.): Unfoldings: Essays in Schenkerian Theory and Analysis, pp. 54–79. Oxford University Press, New York (1999)
Schachter, C.: Rhythm and Linear Analysis: Aspects of Meter. In: Mitchell, W., Salzer, F. (eds.) Music Forum 6 (1987); reprinted in Straus, J. (ed.): Unfoldings: Essays in Schenkerian Theory and Analysis, pp. 79–117. Oxford University Press, New York (1999)
Schenker, H.: Free Composition (2 vols.), translated by Ernst Oster. Longman, New York (1979)
Schenker, H.: Counterpoint (2 vols.), translated by John Rothgeb and Jürgen Thym. Schirmer Books, New York (1987)
Schenker, H.: The Masterwork in Music: A Yearbook (3 vols.), edited by William Drabkin. Cambridge University Press, Cambridge (1997)
Smoliar, S.: A Computer Aid for Schenkerian Analysis. Computer Music Journal 4(2), 41–59 (1980)
Stasheff, J. D.: Homotopy associativity of H-spaces I. Transactions of the American Mathematical Society 108, 275–292 (1963)
Yeston, M.: The Stratification of Musical Rhythm. Yale University Press, New Haven (1976)
Yust, J.: Formal Models of Prolongation. PhD Diss. University of Washington (2006)
Yust, J.: The Step-Class Automorphism Group in Tonal Analysis. In: Proceedings of the First International Conference on Mathematics and Computation in Music. Springer, Berlin (forthcoming 2007)

A Multi-tiered Approach for Analyzing Expressive Timing in Music Performance

Panayotis Mavromatis

Music and Audio Research Lab (MARL),
New York University,
35 West 4th St., Suite 777,
New York, NY 10012, USA
panos.mavromatis@nyu.edu
http://theory.smusic.nyu.edu/pm/

Abstract. This paper presents a method for analyzing expressive timing data from music performances. The goal is to uncover rules which explain a performer's systematic timing manipulations in terms of structural features of the music such as form, harmonic progression, texture, and rhythm. A multi-tiered approach is adopted, in which one first identifies a continuous *tempo curve* by performing non-linear regression on the durations of performed time spans at all levels in the metric hierarchy. Once the effect of tempo has been factored out, subsequent tiers of analysis examine how the performed subdivision of each metric layer (e.g., quarter note) typically deviates from an even rendering of the next lowest layer (e.g., two equal eighth notes) as a function of time. Structural features in the music are identified that contribute to a performer's tempo fluctuations and metric deviations.

1 Introduction

The study of expressive musical performance has been the subject of experimental as well as computational research [1,2]. It is generally acknowledged that expressive timing—a performer's deviations from an exact temporal rendering of the score—is an important component of musical expression. By manipulating timing, a performer is able to communicate musical structure and shape a listener's experience of the music. This paper presents a method for analyzing expressive timing data, extracted through audio analysis of recorded performances. The purpose is to uncover rules which explain a performer's systematic timing manipulations in terms of structural features such as form, harmonic progression, texture, and rhythm.

A fundamental assumption of this analysis is that a performer controls a hierarchically structured *metrical cycle* of measure, beat, and subdivision levels [3,4]. At each point in time, the performer's mental clock fires at a given tempo, which is evidenced by the *cumulative* effect of all the levels in the metrical cycle. The performer's clock rate as a function of time is represented by a *tempo curve*. Identifying this curve forms a natural first tier of analysis.

E. Chew, A. Childs, and C.-H. Chuan (Eds.): MCM 2009, CCIS 38, pp. 193–204, 2009.

Once the effect of tempo has been factored out, it is possible to examine how the performance rendering of subdivisions between adjacent metrical layers (e.g., the subdivision of a quarter note into two eighth notes) deviates from the corresponding exact duration ratios (e.g., 0.5 / 0.5). In the subsequent tiers of analysis, systematic deviations of this type are identified at each level in the metric hierarchy.

As justification for the proposed multi-tiered approach, we first note that it is supported by informal musical discourse: terms such as *ritardando* and *accelerando* typically refer to the first tier of expressive timing, whereas terms such as *rubato*, *notes inégales*, or "swing" most commonly represent deviations in the subsequent tiers.

More to the point, it appears that, in principle at least, a skilled performer can control each tier independently. For instance, a performer may be asked to manipulate the tempo of a performance, while maintaining even metric subdivisions. Conversely, the performer may be requested to perform at steady tempo, while producing various types of uneven metric subdivisions. Moreover, these uneven subdivisions can be executed independently at any particular metric level—up to a certain depth—while maintaining even timing at higher metrical levels.

One of the challenges of expressive performance research is to understand the cognitive mechanisms that underlie expert expressive rendering of a musical score. In line with current views on cognitive modeling, it is natural to seek *modular* rules that specialize in responding to specific features of the musical structure (such as metric accent) by shaping the expression in specific ways (such as lengthening the strong half of a subdivision). This modularity requirement poses challenges for any analytical approach to expressive performance data, which must identify and isolate the effect of individual rules from their surrounding context, where other rules may be simultaneously contributing to expressive deviations. It is in this spirit that the present analysis is offered; it represents work aiming towards a complete rule system for expressive timing. We believe that the multi-tiered analytical approach proposed in this paper can help identify and isolate the right ingredients in this complex multi-faceted manifestation of expert musical skill.

2 Related Previous Research

Several studies have focused on modeling specific aspects of expressive performance, such as *rubato* [5] or the final *ritardando* (see [6] for a review). In addition, some research groups have aimed for comprehensive models that integrate many different components of performance expression. As expected, timing plays a central role in such models.

An important early attempt at an integrated model was the work of Eric Clarke [7]. Clarke proposed nine generative rules to explain expressive deviations in terms of the performed piece's structural features, such as grouping and meter. These rules were derived from measurements of piano performances in experimental studies by Clarke and collaborators. Another important contribution was the KTH model by Sundberg and his group [8]. This model represents a synthetic approach, where expression rules were formulated by querying expert performers as to their expressive deviation practices.

The approach proposed in the present paper is inspired in part by the work of Gerhard Widmer and his collaborators [9,10,2], perhaps the most sophisticated proposal to date for an integrated model of expressive performance. Widmer's group applied machine learning techniques to analyze measurements of expressive performance by skilled musicians.

Most relevant to our approach is the fact that Widmer employed a *two-tiered* data analytic model, in which local note-to-note expressive deviations were separated from the more global expressive shaping of grouping units, such as phrases. Following earlier research [11], Widmer hypothesized that each grouping unit in the music contributes a parabolically shaped *accelerando-ritardando* component to the performance's tempo curve. The overall tempo curve is assumed to be the product of all such contributions coming from each grouping unit.

The first tier of Widmer's analysis consisted of identifying the parabolic coefficients corresponding to each unit of grouping. The process started from the highest grouping level and proceeded to the lowest. At each level, the coefficients were identified by least-squares fitting. After each level's contribution was factored out, the analysis was repeated at the next lowest level, until all levels of grouping were accounted for. The residual timing deviations were then attributed to local note-to-note expressive timing rules, which were extracted from the data via a machine learning algorithm.

The present work extends and modifies Widmer's approach in two different ways. First, we do not make the assumption that grouping is the only factor contributing to the shape of the tempo curve. Instead, we consider sources of additional contributions, such as texture and the tonal/formal function of phrases and sections. As we will see, there is indeed evidence that such factors come into play in determining a performance's tempo fluctuations.

The second difference between Widmer's approach and ours is that, rather than examining a single layer of low-level residual timing deviations, we analyze *separately* the deviations at each subdivision level in the metric hierarchy. As we will see, there is evidence that this separation could lead to simpler, more modular rules. At the same time, this allows us to develop rules that are specific to the absolute time scale of metric subdivision, as measured in seconds. Indeed, different time scales of pulsation can have different cognitive properties, as evidenced by several experimental studies, which are nicely summarized in [4] (see especially Chapter 2).

3 Tempo Curve Calculation Using a Non-parametric Regression Model

If we give up a specific functional dependence of the tempo curve on grouping, as implemented by the fitting of parabolic segments, we must consider the most general options for calculating the tempo curve from the timing data. This naturally leads to a non-parametric regression analysis, which does not assume a specific functional form for the tempo curve.

For the purposes of this study, we found it most flexible to use a non-linear regression model based on *radial basis functions*. The technique was first proposed in [12,13], and is a particular instance of density estimation using *Parzen windows* [14]. In its simplest form, the process can be illustrated as follows:

Let $\{x_i : i = 1 \ldots N\}$ be a set of values for the independent variable X, and let $\{y_i : i = 1 \ldots N\}$ be the corresponding values of the dependent variable Y, so that (x_i, y_i) are the coordinates of the i-th point in the data set. Then the regression curve $y(x)$ obtained from the above data set is given by

$$y(x) = \frac{\sum_{i=1}^{N} y_i \; exp[-(x - x_i)^2/2\sigma^2]}{\sum_{i=1}^{N} exp[-(x - x_i)^2/2\sigma^2]}$$

under the assumption of a Gaussian Parzen window. This expression, calculated through the Parzen density estimation formula, has a simple interpretation: it tells us that the predicted value of y at point x is equal to a weighted sum of the y_i observed at each x_i. The weights are determined by the distance of each x_i from x, and decay rapidly with that distance, according to a Gaussian function.

The variance σ of the Parzen window is also known as the *window width*, and can be viewed as a kind of smoothing parameter. Thus, the regression is formally equivalent to a (Gaussian-weighted) moving average filter. However, it should be noted that the window width is not set a priori, but it is inferred from the data. Indeed, a central problem of this regression analysis is to determine the right value of σ: If the latter is too large, the regression curve becomes too coarse to capture the meaningful fluctuations in the data. Conversely, when σ is too small, the regression curve displays over-fitting, i.e., it captures random noise fluctuations in the data and is a poor predictive model. A simple, yet effective way to determine the appropriate value of σ is through a form of *N-fold cross-validation*. This is effected by minimizing a cost function that represents a least-squares error on the cross-validation training sets (see [13] for more details).

The starting point for our analysis is a performance's set of *inter-onset durations*. These are extracted from the audio recording using Tristan Jehan's Echo Nest API. The latter is a programming toolkit for digital audio analysis that contains a tool for automatic note onset detection[1]. Depending on the value of a resolution parameter, the algorithm can miss a real note onset (if the resolution is too low), or detect a spurious one (e.g. caused by reverb, if the resolution is too high). There is no single optimal resolution, and so it is generally safest to perform onset detection using a relatively high value, to ensure that no notes have been missed. As a result, any spurious onsets detected by the algorithm must be filtered out manually by listening. The algorithm produces the time of each onset in seconds, correct to four decimal places, from which the inter-onset duration values can be calculated at the same precision.

---

[1] See *http://developer.echonest.com/pages/overview* (last visited March 2009)

Since each note's inter-onset duration reflects not only the local tempo, but also the note's *nominal* duration value (e.g., quarter-note, eighth note, etc.), we must normalize each of the raw inter-onset durations by dividing it by the corresponding note's nominal value, where a whole note equals 1.0, a quarter note equals 0.25, etc. This way, each normalized inter-onset duration is a consistent indicator of the local tempo: its value reflects the whole-note duration corresponding to the tempo at that specific point in time. Our solution is essentially equivalent to Widmer's representation of his timing data using percentage deviations instead of absolute durations, but has the added advantage that it keeps track of *absolute* tempo information, and not just its relation to some average. The normalized inter-onset durations for each performance were used as data presented to the non-linear regression model, in order to obtain that performance's tempo curve.

Figure 1 shows the application of the above analysis to a recording of Bach's F minor prelude, BWV 881, from the Well-Tempered Clavier, Book 2. The piece is performed on the harpsichord by an expert, and is recorded on a commercially available CD. This performance will be used as an illustration throughout the paper. In Figure 1, the data points corresponding to the normalized inter-onset durations are shown in grey. The tempo curve derived from the regression is shown in black.

The performances of three contrasting Bach preludes (BWV 845, 863, 881) were analyzed, each of them performed by two different harpsichordists. The most salient factors shaping the tempo curve appear to be

- an initial small *accelerando*;
- a pronounced final *ritardando*;
- less pronounced, but consistent *ritardandi* leading to important cadences, with magnitude usually reflecting the cadence's hierarchical depth in the Schenkerian sense;
- small but measurable contrasts in tempo to highlight sections marked off by distinctive texture or tonal function (e.g., extended dominant pedal).

Once the effect of tempo is factored out, one can examine the lengthening or shortening of individual measures with respect to their neighbors, in response to specific features of the music. This individual manipulation of measure lengths is distinct from overall tempo change, and can be represented in a graph such as that of Fig. 2. Identified variations of this type include lengthening a measure that

- begins a hypermetric pair;
- effects tonal arrival or resolution of a dissonant chord;
- contains unexpected material, such as a highly chromatic chord in a diatonic context.

One intriguing feature of the non-parametric regression analysis is that the optimal Parzen window width σ leading to each tempo curve emerges out of the regression analysis through the process of cross-validation. The absolute value of

σ is usually in the range of 2–4 seconds (2.0591 secs for the curve of Fig. 1). It is an open question whether this value may hold some special significance, either in terms of tempo, or the structure of the piece, or even in terms of psychological properties of time perception and production.

4 The Hierarchy of Metric Deviations

The performed subdivision of each metric layer (e.g., quarter note) typically deviates from an even rendering of the next lowest layer (e.g., two equal eighth notes) as a function of time. This information can be represented in a graph such as that of Figures 3 and 4. The nature of such deviations varies with metric depth. They are often embedded in a small amount of random noise, which reflects limits in the perception and production of exact rhythmic ratios [15].

However, some systematic variations are noteworthy. For instance, a consistent lengthening of the metrically strongest half in a two-fold subdivision highlights its stronger metric position through agogic accent. This is in line with findings reported in many other approaches [7,8,9]. In our analysis, such specialized rules are generally arrived at by inspection, and are subsequently confirmed using standard statistical tests. The possibility of employing some machine-learning classifier to uncover such rules algorithmically, in a manner akin to [9], is currently under investigation.

It is perhaps most remarkable that, even though deviation from exact subdivision is free to vary on a point-by-point basis, the deviations observed in performance often vary smoothly over extended time spans, which typically corresponding to formal units such as phrases (see Figs 3 and 4). This suggests that manipulation of subdivisions is not always controlled on a pulse-by-pulse basis, which might impose excessive demands on real-time processing. Instead, it is shaped by broader gestures in a performer's motor programs, coordinated so as to reinforce communication of musical structure. We would like to suggest that our multi-tiered analysis, which separates each layer of subdivision in the metric hierarchy, makes it easier for such patterns to be identified.

It should be added that the same non-parametric regression technique that is used to construct the tempo curve has been applied to subdivision timing data such as those of Figs 3 and 4, in order to extract the underlying envelope. Once that envelope is identified, one can seek rules that cause the subdivisions of *particular* pulses to deviate from the overall envelope. Such deviations can be typically attributed to the need to project some type of accent.

5 Conclusions and Future Directions

The present paper proposed a data analytic method that aims to uncover rules linking musical structure to specific expressive timing gestures in music performance. Several links were suggested between musical structure and expressive timing at one or several tiers in a hierarchy. The description of structure-to-timing associations remains to some extent qualitative at this stage. This could perhaps be partly attributed to an inevitable element of unpredictability that may exist from performance to performance, even for the same player under different circumstances. However, given the present analysis, there are many ways to explore the possibility of precise quantitative relations between musical structure and expressive timing deviations.

For instance, correlations between structural features of the music and specific expressive deviations can be established by (i) annotating the score with a large number of potentially relevant features, some of them objectively identifiable (e.g., location of cadences), and some requiring annotations by independent musical experts; (ii) seeking correlations between the above features and expressive deviation gestures such as peaks in the tempo curve, lengthened measures, or lengthened beats. Such features can be tabulated in contingency tables, to which standard statistical tests can be applied.

Another analytical approach might involve modeling the exact shape of the tempo curve, seeking a quantitative predictive model of tempo fluctuations as a function of specific musical features. This would entail (i) quantification of all the possibly relevant features as continuous functions of time [16], and (ii) complex regression analysis to identify features that are the best predictors of the tempo curve. We are currently exploring certain multivariate time-series models that could lead to such quantitative relations. As for the expressive subdivisions within each layer of the metric hierarchy, they can be effectively modeled as they unfold in time using the technique of Hidden Markov Models.

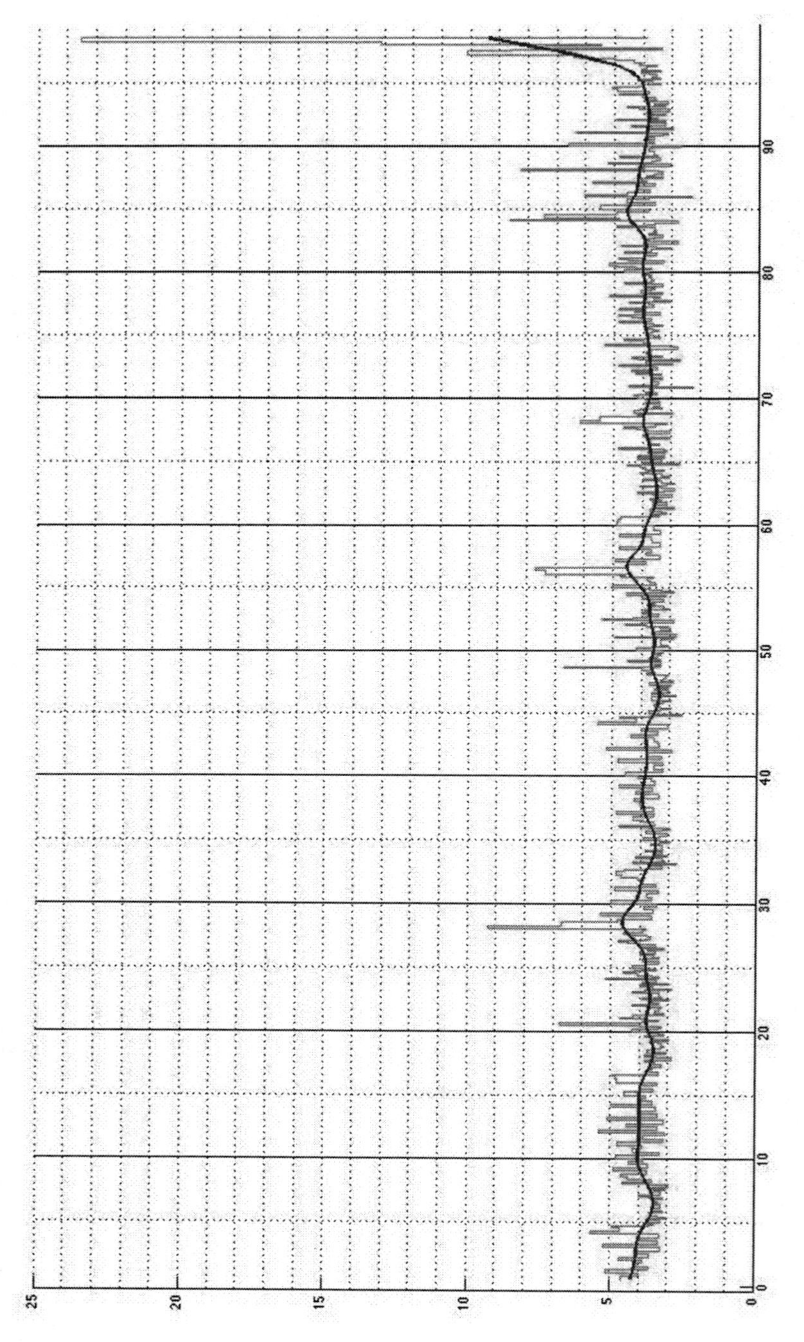

Fig. 1. Tempo curve for a recorded performance of Bach's F minor prelude, BWV 881 (WTC, Book 2), plotted against normalized inter-onset durations for each note in the piece. The x-axis represents position in the notated score using measure numbers (e.g., 2.5 is the middle of the second measure). The y-axis represents local tempo, measured by the duration of a whole note in seconds.

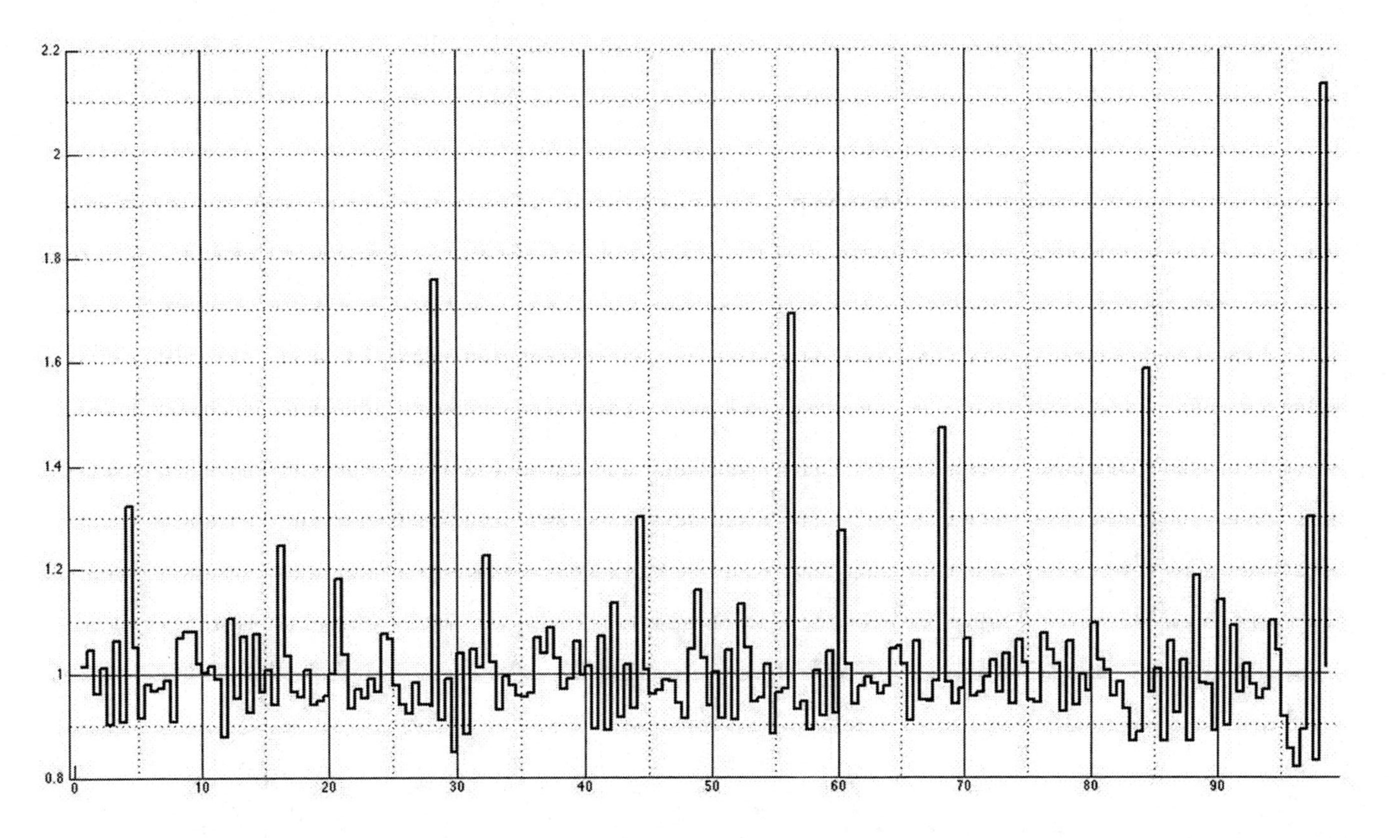

Fig. 2. Graph showing how individual measure durations deviate from the duration corresponding to performed tempo at each point in time. The x-axis represents position in the notated score using measure numbers (see Fig. 1). The y-axis represents individual measure durations in seconds. Therefore, a high spike represents a markedly lengthened measure. The graph comes from the same performance as that of Fig. 1.

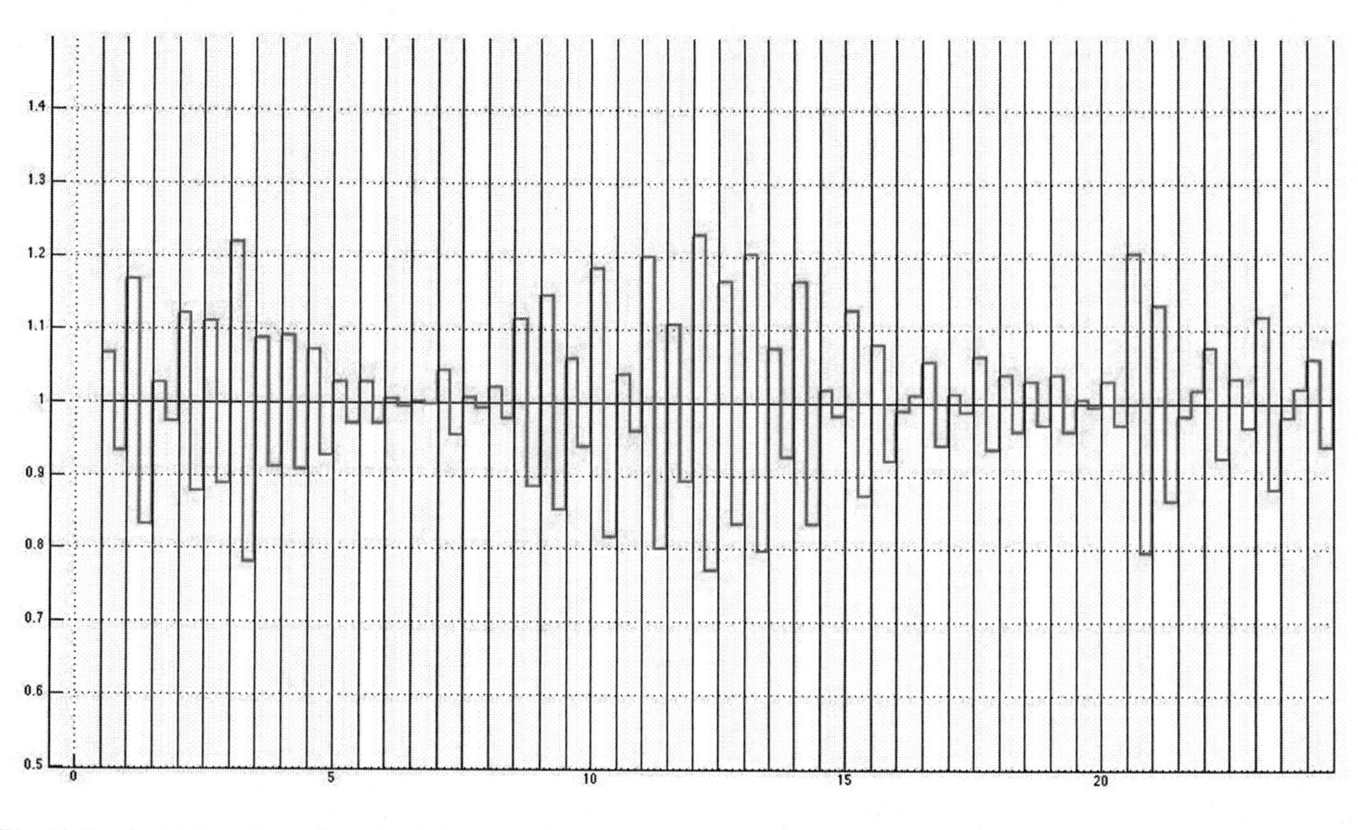

Fig. 3. Graph showing the performed subdivision of each quarter note into two eighth notes. As before, the x-axis represents position in the notated score using measure numbers (see Fig. 1). The y-axis represents the duration ratio of each subdivision. E.g. the first quarter note in the Figure is subdivided into two eighth-note time spans of duration ratio 1.06 : 0.94. The measurements are taken from the same performance as that of Figures 1–2. Figure 3 focuses on mm. 1–23. The complete piece is represented in Fig. 4.

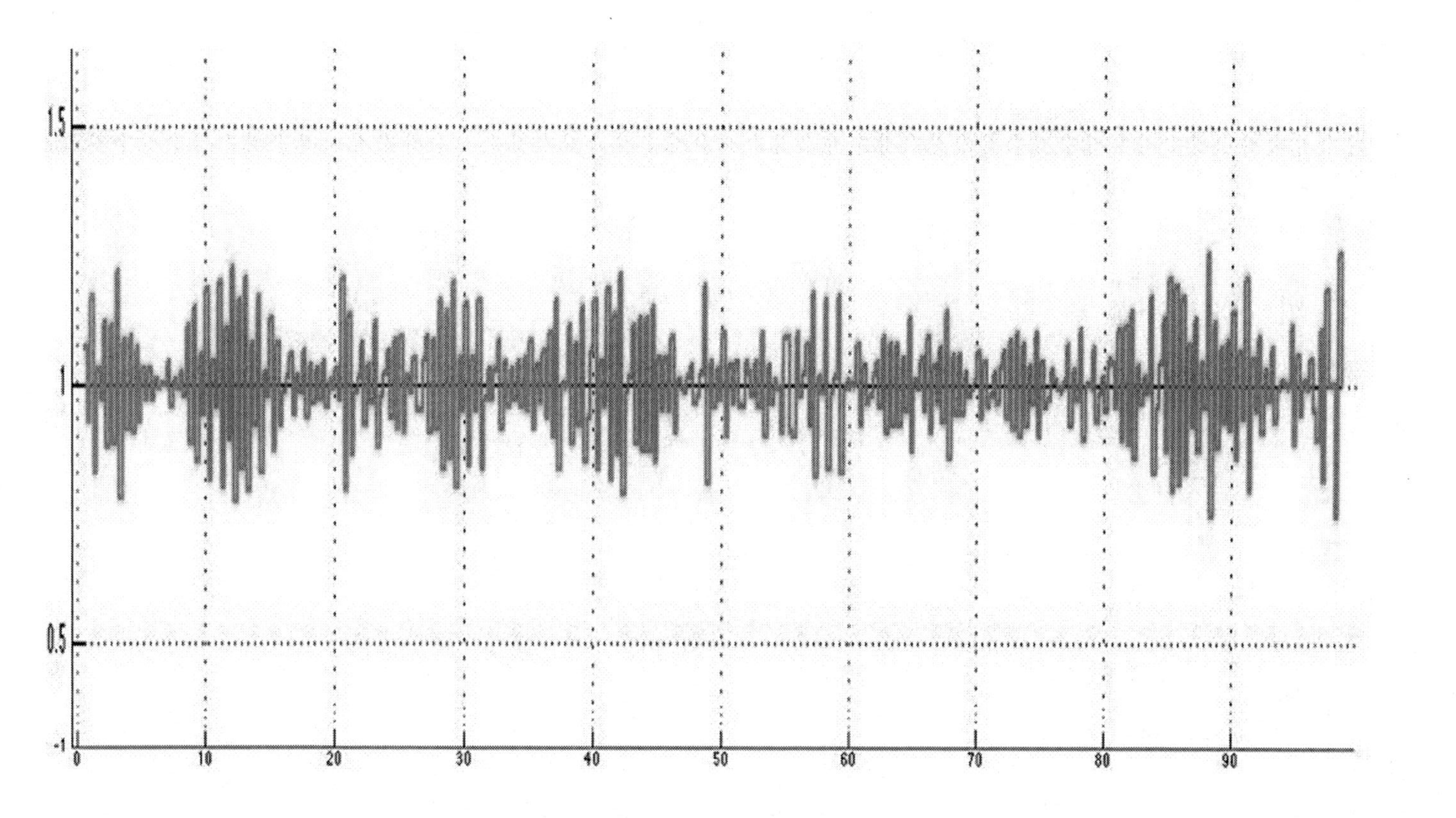

Fig. 4. The measurements presented in Fig. 3, now shown over a larger time scale that encompasses the entire piece. The axes carry the same meaning as those of Fig. 3. From this graph, one can clearly observe how the deviations from exact subdivisions vary smoothly in magnitude over time, as witnessed by the smooth envelope to the curve. These smooth variations are shaped by gestures (lumps in the envelope) corresponding to meaningful formal units such as phrases.

References

1. Gabrielsson, A.: Music Performance. In: Deutsch, D. (ed.) The Psychology of Music, 2nd edn. Academic Press, San Diego (1999)
2. Widmer, G., Goebl, W.: Computational Models of Expressive Music Performance: The State of the Art. Journal of New Music Research 33, 203–216 (2004)
3. Lerdahl, F., Jackendoff, R.S.: A Generative Theory of Tonal Music. MIT Press, Cambridge (1983)
4. London, J.: Hearing in Time: Psychological Aspects of Musical Meter. Oxford University Press, Oxford (2004)
5. Todd, N.P.M.: A Computational Model of Rubato. Contemporary Music Review 3, 69–88 (1989)
6. Honing, H.: Computational Modeling of Music Cognition: A Case Study on Model Selection. Music Perception 23, 365–376 (2006)
7. Clarke, E.F.: Generative Principles in Music Performance. In: Sloboda, J.A. (ed.) Generative Processes in Music: The Psychology of Performance, Improvisation, and Composition. Clarendon Press, Oxford (1988)
8. Friberg, A.: A Quantitative Rule System for Musical Performance. Doctoral dissertation, Royal Institute of Technology, Stockholm (1995)
9. Widmer, G.: Machine Discoveries: A Few Simple, Robust Local Expression Principles. Journal of New Music Research 31, 37–50 (2002)
10. Widmer, G., Tobudic, A.: Playing Mozart by Analogy: Learning Multi-Level Timing and Dynamics Strategies. Journal of New Music Research 32, 259–268 (2003)
11. Todd, N.P.M.: A Model of Expressive Timing in Tonal Music. Music Perception 3, 33–58 (1985)
12. Specht, D.F.: A General Regression Neural Network. IEEE Transactions on Neural Networks 2, 568–576 (1991)
13. Specht, D.F.: Probabilistic and General Regression Neural Networks. In: Chen, C.H. (ed.) Fuzzy Logic and Neural Network Handbook. McGraw-Hill, New York (1996)
14. Parzen, E.: On Estimation of a Probability Density Function and Mode. Annals of Mathematical Statistics 33, 1065–1076 (1962)
15. Clarke, E.F.: Rhythm and Timing in Music. In: Deutsch, D. (ed.) The Psychology of Music, 2nd edn. Academic Press, San Diego (1999)
16. Farbood, M.M.: A Quantitative, Parametric Model of Musical Tension. Doctoral dissertation. MIT, Cambridge, MA (2006)

HMM Analysis of Musical Structure: Identification of Latent Variables Through Topology-Sensitive Model Selection

Panayotis Mavromatis

Music and Audio Research Lab (MARL),
New York University,
35 West 4th St., Suite 777,
New York, NY 10012, USA
panos.mavromatis@nyu.edu
http://theory.smusic.nyu.edu/pm/

Abstract. Hidden Markov Models (HMMs) have been successfully employed in the exploration and modeling of musical structure, with applications in Music Information Retrieval. This paper focuses on an aspect of HMM training that remains relatively unexplored in musical applications, namely the determination of HMM topology. We demonstrate that this complex problem can be effectively addressed through search over model topology space, conducted by HMM state merging and/or splitting. Once successfully identified, the HMM topology that is optimal with respect to a given data set can help identify hidden (latent) variables that are important in shaping the data set's visible structure. These variables are identified by suitable interpretation of the HMM states for the selected topology. As an illustration, we present two case studies that successfully tackle two classic problems in music computation, namely (i) algorithmic statistical segmentation and (ii) meter induction from a sequence of durational patterns.

1 Introduction

Hidden Markov Models have been successfully employed in the exploration and modeling of musical structure [1,2], with applications in Music Information Retrieval [3].

Simply put, a Hidden Markov Model is a probabilistic version of a *Finite State Machine* (FSM), or formal specification of a finite state grammar. A FSM is formally defined by *states* and *transitions*, graphically represented by circles and arrows respectively. A FSM generates a symbolic sequence by traversing a path of states connected by transitions, following the direction of the arrows. The generated sequence is the string of output symbols encountered in the path. A FMS is a simple and flexible way to specify finite-memory constraints on the symbolic values of variables that characterize musical structure (e.g., pitch, duration, etc.) and as such offers useful formal characterizations of the structure of musical sequences.

E. Chew, A. Childs, and C.-H. Chuan (Eds.): MCM 2009, CCIS 38, pp. 205–217, 2009.

A *Hidden Markov Model* (HMM) is a FSM with probabilities attached to its transitions and output symbols [4,5]. The generation of a sequence through a specific HMM path has probability equal to the product of all transition and output probabilities encountered in traversing the generating path.

What gives the HMM technique its strength and flexibility is the fact that selecting the best HMM for a given data set can be generally accomplished through efficient algorithms. For instance, given a data set of symbolic sequences whose structure we wish to explore, it is customary to assume a HMM of fixed topology (i.e., number of states, and how they are connected by transitions) and identify the model parameters (i.e., transition and output probabilities) that best fit the data set, in the sense of Maximum Likelihood Estimation, using the so-called *Baum-Welch* algorithm.

This paper focuses on an aspect of HMM training that remains relatively unexplored in musical applications, namely the determination of HMM *topology*. Our aim is to algorithmically construct models whose topologies consist of states interpretable as values of latent ("hidden") variables that may play important role in the determination of musical structure. In a given application, one may wish to focus on a particular ("visible") musical variable, aiming to model syntactical constraints on its successive values (e.g., stylistically acceptable patterns of note durations). The states of a HMM obtained through topology-sensitive search should indicate which additional variables must be taken into consideration (e.g., metric position) in order to understand the syntax of the original "visible" variable that one set out to model. This can be accomplished by showing a close correspondence between HMM states and particular values of the candidate "hidden" variables.

For an HMM topology to be interpretable in the manner suggested in the preceding paragraph, special effort must be put in the topology selection algorithm. If one simply relies on Baum-Welch optimization of the HMM parameters, one will in most cases obtain HMMs whose states are not readily interpretable, however well these models may fit the data. Previous studies that attempted to address this complex problem have generally employed some form of search over model topology space, which was conducted by HMM state merging [6] or splitting [7]. In this paper, we use the same basic search procedure, except that we allow state merging and splitting to be combined in the same search. In addition, we evaluate each candidate model using a Bayesian approach, in which a HMM's prior probability is determined through the Minimum Description Length principle. This prior is optimal in that it leads to models that are neither too large nor too small, and has been found to provide a reliable termination criterion for the state merging/splitting search.

We will illustrate our method with the help of two case studies that successfully tackle two classic problems in music computation, namely (i) algorithmic statistical segmentation and (ii) meter induction from a sequence of durational patterns.

2 HMM Training and Topology Identification

The proposed method of topology identification takes place in the framework of Bayesian model selection. More specifically, given data set D, we seek the model M that maximizes the probability $P(M|D)$ of the model given the data. The latter is obtained through Bayes's Law as

$$P(M|D) = \frac{P(D|M)P(M)}{P(D)}$$

It is customary to use the simpler form

$$P(M|D) \propto P(D|M)P(M) \tag{1}$$

since $P(D)$ is constant over models M and therefore does not affect the maximization problem. $P(M)$ is known as the *model prior probability*, assigned to the model on general grounds before the data set is consulted. Likewise, $P(M|D)$ is known as the *model posterior probability*, and represents the probability of the model after the data has been taken into consideration.

Topology identification is achieved through a suitable choice of model prior $P(M)$, defined as a function of model topology alone, and designed to reward model simplicity. For a fixed topology, $P(M)$ is fixed, and so maximization of the model posterior amounts to maximizing the $P(D|M)$ part in eq. (1). This is achieved through the Baum-Welch (BW) algorithm, which chooses the model parameters maximizing the probability of the data set using the Expectation-Maximization principle. Overall, the maximization problem defined by eq. (1) is a concrete implementation of Occam's Razor, and achieves optimal balance between goodness-of-fit and model simplicity.

We have shown elsewhere [8] that an optimal choice for $P(M)$ is a *model complexity prior* given by

$$P(M) = Ke^{-D(M)} \tag{2}$$

where the function $D(M)$ is defined by

$$D(M) \equiv L(n_S) + L(d) + n_S \, log \frac{(d + n_S + 1)!}{d! n_S!} + n_T \, log \frac{(d + n_A + 1)!}{d! n_A!} \tag{3}$$

and $L(n)$ is the *universal prior for integers* [9, pp. 34–5], defined by

$$L(n - 1) = c + log(n) + log(log(n)) + log(log(log(n))) + \ldots \tag{4}$$

Here n_S is the number of HMM states, n_A is the number of distinct output symbols in the data sequences, and K and c are suitably chosen normalization constants. An additional integer d represents the decimal precision needed to express the real-valued HMM model parameters. The expression in eq. (3) was derived in [8] with the help of the Minimum Description Length principle [9,10].

The best way to tackle the problem of HMM topology selection is by systematizing the search over all possible HMM graphs. Such a search scheme typically begins with an extreme graph which is maximally simple or maximally complex. Incremental improvements are subsequently performed on each candidate graph by either (i) *splitting* one of its states, if the graph is too simple, or (ii) *merging* two of its states if the graph is too complex. As an illustration, the following procedure formalizes the state-splitting search:

1. Begin with a *one-state* HMM. This model has only one transition, namely the one from the single state to itself. The output probabilities on that transition can be determined by the BW algorithm.
2. For this and each subsequent candidate model,
   (a) Choose a state to split. Determine the new graph that results from the splitting.
   (b) Perform BW estimation of the new graph's parameters.
   (c) Evaluate the resulting HMM's posterior probability using eq. (1) with the model complexity prior (eqs 2–4).

   Continue Steps (a–c) until all the states have been tried for splitting. The split-state HMM with the best posterior becomes the next candidate model, and Step 2 is repeated for as long as the candidate models' posterior probability continues to improve.
3. The process terminates once the posterior probability of the candidate model begins to deteriorate, and the HMM with the highest overall posterior is identified as the optimal HMM for the given data set.

One can modify Step 2(a) above to replace state-splitting by state-merging. Alternatively, one can consider both possibilities at each step, choosing the option that maximizes the model posterior at that step.

The above HMM topology selection process will now be illustrated with the help of two case studies.

3 Case Study I: Statistical Segmentation of Symbolic Sequences

Statistical segmentation is used to refer to the process of identifying grouping boundaries in sequences based solely on the patterns of occurrences of symbol combinations, without relying on explicit cues or annotations for such boundaries.

The process can be illustrated with the help of a data set $D1$ based on a language that was artificially synthesized to investigate statistical learning of tone sequences by people in an experimental setting [11]. The set of symbols, or *alphabet*, for this artificial language consists of pitches of the chromatic scale, to be represented by the symbols {C, C♯, D . . . B}. The data sequences of $D1$ are built out of the following six three-symbol artificial segments ("words"):

A D B   D F E   G G♯ A   F C F♯   D♯ E D   C C♯ D

These words appear randomly with equal probability in the sequences of our data set $D1$. (Word combinations were more restricted in Saffran's stimuli, due

Table 1. Calculation of model posteriors for all the HMMs considered in the word segmentation example involving data set $D1$. Each model is obtained from the previous one by state splitting. The first column shows the HMM's number of states n_S. The second column shows the state split from which that model was obtained. Negative logarithms of probability values are used throughout. The selected model maximizes the model posterior or, equivalently, minimizes the value in Column 5. This model is marked with an asterisk in Column 1.

n_S	State Split	$-log_2 P(D\|M)$	$-log_2 P(M)$	$-log_2 P(M\|D)$
1	-	20365.8	84.2112	20450.1
2	0	15539.9	159.162	15699
3	0	13570.1	220.761	13790.9
4	1	12116.8	300.335	12417.2
5	3	11091.5	326.96	11418.4
6	2	10250.1	439.757	10689.8
7	0	9527.7	400.767	9928.47
8	1	8607.96	480.044	9088
9	4	7992.11	409.265	8401.38
10	7	7385.11	428.647	7813.76
11	6	6798.11	448.685	7246.8
12	5	6220.11	469.327	6689.44
13*	0	5653.91	624.18	6278.09
14	2	5653.91	670.633	6324.54

to the experimental design.) A typical sequence in $D1$ will therefore look like this:

$$\texttt{G G}\sharp\ \texttt{A A D B D}\sharp\ \texttt{E D A D B C C}\sharp\ \texttt{D D F E} \tag{5}$$

The output of the segmentation will be the same sequence annotated with word boundaries as follows:

$$\texttt{G G}\sharp\ \texttt{A / A D B / D}\sharp\ \texttt{E D / A D B / C C}\sharp\ \texttt{D / D F E}$$

Our HMM analysis was applied to a data set $D1$ constructed in the above manner, consisting of 200 randomly generated sequences with an average length of 27.21 symbols. A state-splitting search was performed to identify the best HMM topology. Each candidate split was followed by Baum-Welch estimation of the HMM parameters. The results of this search are summarized in Table 1. The model identified as the winner is the one that carries the maximum posterior probability. This model is marked with an asterisk in the first column of the table. The model's graph structure is given in Figure 1.

To illustrate how the HMM of Figure 1 performs segmentation on a data sequence, it is helpful to consider the *most likely* HMM path that generates the sequence in question, also known as the sequence's *Viterbi* path [4,5, pp. 331–3]. For the sequence of example (5), this path turns out to be the following:

$$\begin{array}{ccccccccccccccccccccc} & BEGIN & & \texttt{G} & & \texttt{G}\sharp & & \texttt{A} & & \texttt{A} & & \texttt{D} & & \texttt{B} & & \texttt{D}\sharp & & \texttt{E} & & \texttt{D} & \\ s_0 & \rightarrow & s_1 & \rightarrow & s_2 & \rightarrow & s_3 & \rightarrow & s_1 & \rightarrow & s_8 & \rightarrow & s_9 & \rightarrow & s_1 & \rightarrow & s_{12} & \rightarrow & s_4 & \rightarrow & s_1 \end{array}$$

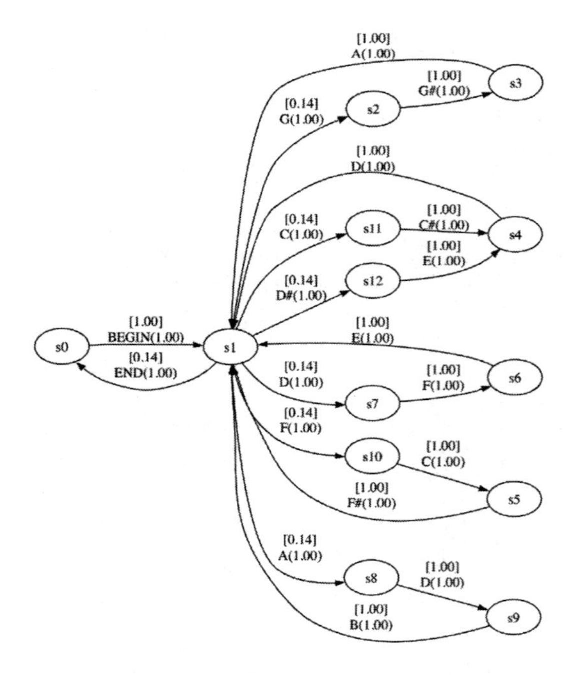

Fig. 1. The best HMM for data set $D1$, obtained through state-splitting

$$\begin{array}{ccccccccccccccccccccc} & \mathtt{A} & & \mathtt{D} & & \mathtt{B} & & \mathtt{C} & & \mathtt{C\sharp} & & \mathtt{D} & & \mathtt{D} & & \mathtt{F} & & \mathtt{E} & & END & \\ s_1 & \longrightarrow & s_8 & \longrightarrow & s_9 & \longrightarrow & s_1 & \longrightarrow & s_{11} & \longrightarrow & s_4 & \longrightarrow & s_1 & \longrightarrow & s_7 & \longrightarrow & s_6 & \longrightarrow & s_1 & \longrightarrow & s_0 \end{array} \tag{6}$$

With the help of this Viterbi path, all word boundaries in the sequence are clearly identified through the HMM state s_1. The significance of that state as a marker of word boundaries can also be confirmed by observing the graph structure of Figure 1 and following the derivation path of any sequence generated by that graph.

This simple example serves to illustrate that, just like the experimental subjects in the study by Saffran et al. [11], the HMM topology selection technique presented here can exploit the statistical structure of symbolic sequences to segment them into grouping units. This result is replicated with other similar data sets and suggests that—at least in certain cases—segmentation can be performed on the basis of statistical information alone, without recourse to other structure, such as Gestalt principles of grouping.

4 Case Study II: Meter Induction from Rhythmic Patterns

Meter induction refers to the inference of metrical structure from a pattern of note durations. Our second case study illustrates this process by analyzing patterns of durations found in Palestrina's vocal music. Table 2 lists all the possible note and rest durations employed in the style.

It should be noted that the goal of this application is not to do meter induction *per se*. Rather, we seek to model Renaissance rhythm by establishing a syntax of note durations. With the help of the HMM topology selection technique, we hope to identify any other variable(s) that may be most relevant in constraining and shaping the style's duration patterns. In this instance, the most crucial variable turns out to be metric placement, and is identified by the interpretation of HMM states as explained below.

The HMM analysis of the present case study was performed on a sample of melodies taken from the corpus of Palestrina's masses. The corpus was obtained from the Internet in *Humdrum*-encoded form.[1] The sample was constructed as follows:

1. The corpus of Palestrina masses was subdivided into movements, or sections of movements. Each such section was further subdivided into individual vocal lines. This processing was carried out using standard Humdrum tools. The result was a database of 5034 vocal lines covering the entire corpus.
2. Out of these 5034 vocal lines, fifty were chosen at random to form the sample, using a random number generator.
3. Each of the fifty lines was further subdivided into one or more data sequences. The divisions were made at places where there was a rest of one complete bar or longer. This subdivision was intended to ensure that the data sequences represented units close to the phrase level.
4. Finally, the durations of each data sequence were extracted and encoded using the symbols listed in the fourth column of Table 2.

An example of this encoding is shown above the staff in Figure 3. The resulting sample consisted of 190 such sequences with an average length of 34.48 symbols.

The results of HMM inference algorithm are shown on Table 3. The HMM with the highest posterior probability was a 6-state model, marked with an asterisk in the first column of the table. Figure 2 shows the model in graph form.

As in the previous case study, the model's structure will be easier to interpret with the help of the data sequences' Viterbi path. As an illustration, the Viterbi path for a typical melody in the data set is given in Figure 3.

Examination of the state sequences in the model's Viterbi paths reveals one striking property: there is a close correspondence between the HMM states and the various metric positions in the compositions' underlying 4/2 meter. As can be seen from the example of Figure 3, states s_1 and s_3 occur exclusively on strong beats (1 or 3), whereas state s_2 only occurs on weak beats (2 or 4); moreover, state s_4 only occurs on weak quarters, and the rare occurrence of state s_5 coincides with a weak eighth-note subdivision. In other words, the HMM appears to be "aware" of metric placement for each duration it generates. This

[1] URL: *http://csml.som.ohio-state.edu/HumdrumDatabases/classical/Renaissance/Palestrina/Masses/* (last visited March 2009).

Table 2. The note and rest durations available to the Renaissance vocal style. These are shown along with the corresponding symbolic value of the duration variable, as encoded for the HMM analysis of the present project. The rightmost column records the possible metric placements for each duration, as prescribed in counterpoint instruction.

Music symbol	Renaissance name	Modern name	Encoding	Metric position
	Longa		L	beats 1, 3
	Breve		B	beats 1, 3
	Semibreve	Whole note	W	beats 1, 2, 3, 4
	Minim	Half note	H	beats 1, 2, 3, 4
	Semiminim	Quarter note	Q	any quarter
	Fusa	Eighth note	E	pairs, weak quarter
	Dotted Longa		L.	beats 1, 3
	Dotted Breve		B.	beats 1, 3
	Dotted Semibreve	Dotted whole note	W.	beats 1, 3
	Dotted Minim	Dotted half note	H.	beats 1, 2, 3, 4
	Semibreve rest	Whole note rest	Rw	beats 1, 3
	Minim rest	Half note rest	Rh	beats 1, 3

awareness is embodied in the HMM states, whose job is to encapsulate the most decisive factors that determine the next output at each point in time. The fact that each HMM state has chosen to incorporate metric information should perhaps come as no surprise, given the generally acknowledged role of metric constraints in the style's rhythmic syntax. What is perhaps most remarkable is that metric position was not originally encoded explicitly in the data sequences. The HMM inference algorithm was able to detect the importance of this variable, based on statistical regularities in the sequential combinations of note durations.

Table 3. Calculation of model posteriors for all the HMMs considered in the analysis of Palestrina rhythm. As in the earlier example, each model is obtained from the previous one by state splitting. The columns of this table carry the same interpretation as those of Table 1.

n_S	State Split	$-log_2 P(D\|M)$	$-log_2 P(M)$	$-log_2 P(M\|D)$
3	-	14091.2	826.489	14917.7
3	0	14091.2	826.489	14917.7
4	1	12950.7	942.963	13893.7
5	0	12161.9	1051.110	13213.0
6*	4	12002.0	1204.640	13206.6
7	0	12002.0	1363.340	13365.3

Examination of the HMM states reveals a close correspondence between HMM states and the rules of metric placement found in standard Renaissance counterpoint textbooks [12,13], including the constraints on each duration's metric placement, and the general tendency to find longer note values near the beginnings and ends of phrases. The latter property is reflected in the differentiation between the two "strong beat" states s_1 and s_3; the former represents strong beats near the beginning and end of phrases, whereas the latter occurs in the phrases' interior positions.

5 Conclusions

The two case studies presented in this paper have demonstrated how topology-sensitive HMM training can successfully uncover hidden structure underlying the observable behavior of symbolic data sequences. Indeed, generic application of the Baum-Welch algorithm would not have resulted in readily interpretable graphs such as those of Figures 1 and 2. Only when HMM training incorporates model topology identification, in a way that is sensitive to the data set's statistical regularities, will the HMM states be readily interpretable in terms of the processes underlying the data sequence's generation. In such cases, we can interpret the different HMM states as representing the values of *hidden*, or *latent*, variables that are most crucial in shaping the structural constraints of the data sequences.

More specifically, one salient latent variable underlying Case Study I could be identified as "word completion status" with the two values 'yes' (corresponding to state s_2) and 'no' (corresponding to states s_1 s_3, and s_4); furthermore, a second latent variable of "word label" could account for the differences among the non-boundary states s_1 s_3, and s_4. For Case Study II, the most salient latent variable seemed to be "metric position" with most HMM states representing distinct values. A second latent variable representing "position in the phrase" was found to differentiate between states s_1 and s_3.

Of course, in both the above examples, identification of the relevant latent variables is relatively straightforward. This is because the HMM graphs are rather small, and so the correspondence between HMM states and latent variable values can be directly perceived. In more complicated situations, however, this need not be the case. We must have a way of interpreting HMM states that is more reliable than simple inspection. In general, the interpretation process could be systematized by compiling contingency tables that show how each HMM state aligns, or doesn't align, with the values of a set of candidate latent variables along the HMM paths that generate the data set (the Viterbi path offering the dominant contribution).

Finally, it should be noted that, as our experiments with various data sets indicate, our MDL prior of eq. (2–4) is an essential ingredient for the identification of the right model topology. Other priors that we have tried typically produce smaller graphs—e.g., caused by premature termination of state-splitting—whose states are not consistently interpretable. In general, whenever the data is abundant, it is found that the result is less sensitive to the choice of prior. However, that choice really matters when data is scarce, which is the case, for example, in historically delimited musical corpora (e.g. "all D-mode Gregorian tracts"). The MDL approach is a strongly motivated and principled way of choosing a prior, which in the majority of cases leads the topology search to discover interpretable graphs.

It should be also noted that a simple splitting/merging search over model topologies, unaided by other search heuristics, does not always yield readily interpretable graphs, especially in data sequences with rich alphabets of symbols. The problem is that the splitting/merging search is a form of "best first" search that guarantees an optimal next step in the search, leading to a local maximum of the model posterior; however, it cannot guarantee that the maximum reached in this way will be optimal in the global sense. This is of course a concern for any optimization problem. We have found that, in order to produce interpretable results in the most general cases, the search proposed in this paper has to be augmented with heuristics that determine an appropriate starting point for the splitting or merging. This issue is currently under investigation, and will be presented in a future work.

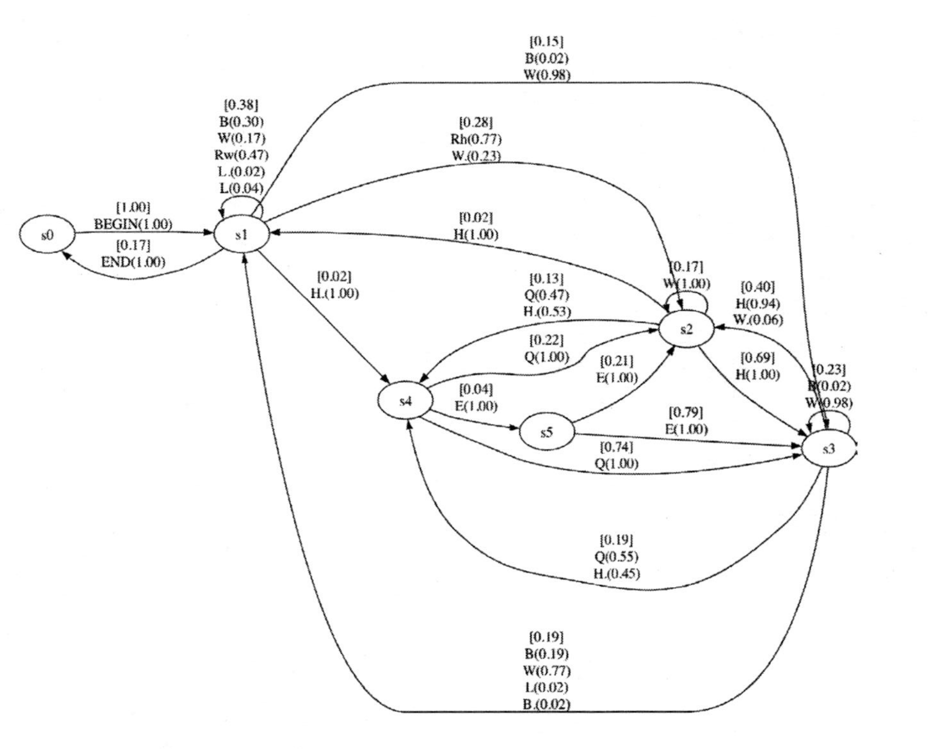

Fig. 2. The best HMM for the data sequences of durations in the Palestrina sample. The graph's output symbols are encoded using the symbols listed in the fourth column of Table 2.

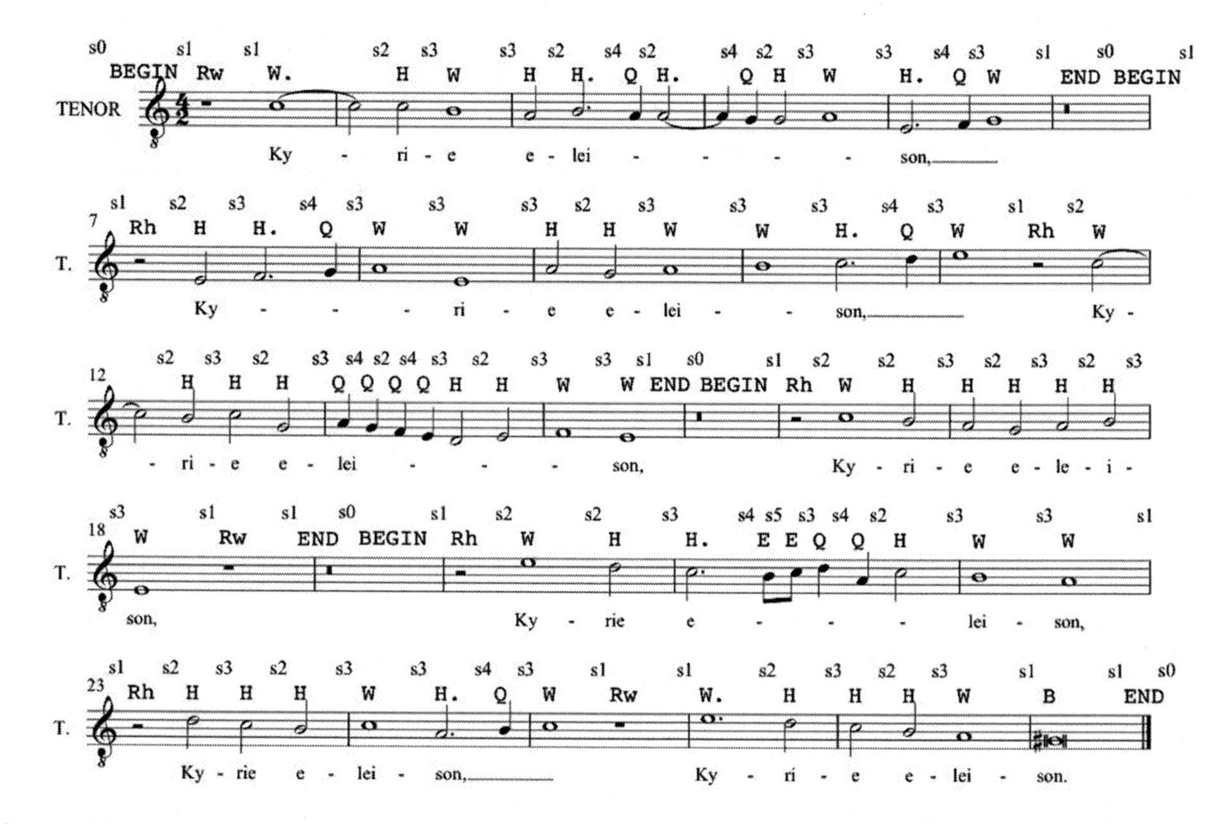

Fig. 3. A typical Palestrina vocal line (from *Missa Te Deum Laudamus*, *Kyrie* II, Tenor I) annotated with its Viterbi path. The state sequence corresponding to that path is marked with the symbols $s_0 - s_5$ above each staff. Arrows from one state to the next have been suppressed for visual clarity. The output symbols corresponding to the encoded durations appear below the state sequence and above the corresponding note or rest. Duration encodings are listed in column 4 of Table 2.

References

1. Raphael, C., Stoddard, J.: Functional Harmonic Analysis Using Probabilistic Models. Computer Music Journal 28, 45–52 (2004)
2. Mavromatis, P.: A Hidden Markov Model of Melody Production in Greek Church Chant. Computing in Musicology 14, 93–112
3. Bello, J.P., Pickens, J.: A Robust Mid-Level Representation for Harmonic Content in Music Signals. In: Proceedings of the 6th International Conference on Music Information Retrieval (ISMIR 2005), London, UK (September 2005)
4. Rabiner, L.R., Juang, B.-H.: An Introduction to Hidden Markov Models. IEEE ASSP Magazine 3, 4–16 (1986)
5. Manning, C.D., Schütze, H.: Foundations of Statistical Natural Language Processing. MIT Press, Cambridge (1999)
6. Stolcke, A., Omohundro, S.M.: Hidden Markov Model Induction by Bayesian Model Merging. In: Hanson, S.J., Cowan, J.D., Giles, C.L. (eds.) Advances in Neural Information Processing Systems, vol. 5, pp. 11–18. Morgan Kaufmann, San Mateo (1993)
7. Ostendorf, M., Singer, H.: HMM Topology Design Using Maximum Likelihood Successive State Splitting. Computer Speech and Language 11, 17–41 (1997)
8. Mavromatis, P.: Minimum Description Length Modeling of Musical Structure. Submitted to the Journal of Mathematics and Music (Under revision)
9. Rissanen, J.: Stochastic Complexity in Statistical Inquiry. Series in Computer Science, vol. 15. World Scientific, Singapore (1989)
10. Grünwald, P.D.: The Minimum Description Length Principle. Adaptive Computation and Machine Learning. MIT Press, Cambridge (2007)
11. Saffran, J.R., Johnson, E.K., Aslin, R.N., Newport, E.L.: Statistical Learning of Tone Sequences by Human Infants and Adults. Cognition 70, 27–52 (1999)
12. Jeppesen, K.: Counterpoint: The Polyphonic Vocal Style of the Sixteenth Century. Prentice Hall, New York (1939); reprinted by Dover (1992)
13. Gauldin, R.: A Practical Approach to Sixteenth-Century Counterpoint. Waveland Press, Long Grove (1995)

A Declarative Language for Dynamic Multimedia Interaction Systems*

Carlos Olarte[1,2] and Camilo Rueda[2,3]

[1] INRIA and LIX, École Polytechnique, France
{colarte}@lix.polytechnique.fr
[2] Pontificia Universidad Javeriana Cali, Colombia
{crueda}@cic.javerianacali.edu.co
[3] IRCAM, France

Abstract. Universal Timed Concurrent Constraint Programming (utcc) is a declarative model for concurrency tied to logic. It aims at specifying mobile reactive systems, i.e., systems that continuously interact with the environment and may change their communication structure. In this paper we argue for utcc as a declarative model for dynamic multimedia interaction systems. Firstly, we show that the notion of constraints as partial information allows us to neatly define temporal relations between interactive agents or events. Secondly, we show that mobility in utcc allows for the specification of more flexible and expressive systems. Thirdly, by relying on the underlying temporal logic in utcc, we show how non-trivial temporal properties of the model can be verified. We give two compelling applications of our approach. We propose a model for dynamic interactive scores where interactive points can be defined to adapt the hierarchical structure of the score depending on the information inferred from the environment. We then broaden the interaction mechanisms available for the composer in previous (more static) models. We also model a music improvisation system based on the factor oracle that scales up to situations involving several players, learners and improvisers.

1 Introduction

Process calculi provide a language in which the structure of terms represents the structure of processes together with an operational semantics to represent computational steps. Concurrent Constraint Programming (CCP) [13] has emerged as a declarative model for concurrency tied to logic. In CCP, concurrent systems are specified by means of constraints (e.g. $x + y \geq 10$) representing partial information about certain variables. This way, agents (or processes) interact with each other by telling and asking information represented as constraints in a global store: A process $\mathbf{tell}(c)$ adds the constraint c, thus making it available to other processes. A *positive ask* $\mathbf{when}\ c\ \mathbf{do}\ P$ remains blocked until the store is strong enough to entail c; if so, it behaves like P.

Interactivity in multimedia systems has become increasingly important. The aim is to devise ways for the machine to be an active partner in a collective behavior constructed

* This work has been partially supported by FORCES, an INRIA's *Equipe Associée* between the teams COMETE (INRIA), the Music Representation Research Group (IRCAM), and AVISPA.

E. Chew, A. Childs, and C.-H. Chuan (Eds.): MCM 2009, CCIS 38, pp. 218–227, 2009.

dynamically by many actors. In its simplest form, a musician signals the computer when processes should be launched or stopped. In more complex forms the machine is always actively adapting its behavior according to the information derived from the activity of the other partners. To be coherent these machine actions must be the result of a complex adaptive system composed of many agents that should be coordinated in precise ways. Constructing such systems is a challenging task. Moreover, ensuring their correctness poses a great burden to the usual test-based techniques. In this setting, CCP has much to offer: CCP calculi are explicitly designed for expressing complex coordination patterns in a very simple way by means of constraints. In addition, their declarative nature allows formally proving properties of systems modeled with them.

Interactive scores [3] are models for reactive music systems adapting their behavior to different types of intervention from a performer. Weakly defined temporal relations between components in an interactive score specifies loosely coupled music processes potentially changing their properties in reaction to stimulus from the environment (say, a performer). An interactive score defines a hierarchical structure of processes. Musical properties of a process depend on the context in which it is located. Although the hierarchical structure has been treated as static in previous works, there is no reason it should be so. A process, in reaction to a musician action, for example, could be programmed to move from one context to another or simply to disappear. Imagine, for instance, a particular set of musical materials within different contexts that should only be played when an expected information from the environment actually takes place. Modeling this kind of interactive score mobility in a coherent way is greatly simplified by using the calculus described in this paper.

Musical improvisation is another natural context for interacting agents. Improvisation is effective when agents behavior adapts to what has been learned in previous interactions. A music style-learning/improvisation scheme such as Factor Oracle (FO) [1,5] can be seen as a reactive system where several learning and improvising agents react to information provided by the environment or by other agents. In its simplest form three concurrent agents, a player, a learner and an improviser must be synchronized. Since only three independent processes are active, coordination can be implemented without major difficulties using traditional languages and tools. The question is whether such implementations would scale up to situations involving several concurrent agents. For an implementation using traditional languages the complexity of such systems would most likely impose many simplifications in coordination patterns if behavior is to be controlled in a significant way. A CCP model, as described here, provides a compact and simple model of the agents involved in the FO improvisation, one in which coordination is automatically provided by the blocking *ask* construct of the calculus. Moreover, additional agents could easily be incorporated in the system. As an extra bonus, fundamental properties of the constructed system can be formally verified in the model.

In this paper we argue for Universal Timed CCP (`utcc`) [10] as a declarative language for the modeling and verification of multimedia interaction systems. The `utcc` calculus is a timed extension of CCP with the ability to model *mobile reactive* system, i.e., systems that continuously interact with the environment and may change their communication structure.

After a brief introduction of utcc in Section 2, our contributions are as follows. In Section 3, we propose a utcc model for interactive scores where the interactive points allow the composer to dynamically change the hierarchical structure of the score. We then broaden the interaction mechanisms available for the user in previous (more static) models, e.g., [4], where temporal objects cannot be moved to different contexts according to the information derived from the environment. We also provide a framework based on the underlying linear temporal logic of utcc to formally verify fundamental properties of the constructed system. For instance, we can verify if certain musical structure is not played due to the absence of a stimulus from the environment. In Section 4 we model a music improvisation system based on the factor oracle that scales up to situations involving several agents and offers a more compact and efficient representation of the data structure wrt the model in [5]. Section 5 concludes the paper.

An extended version of this work, including further details, is available at [8].

2 Preliminaries

CCP-based languages are parametric in a constraint system [13] defining the kind of constraints that can be used in the program. Here, *constraints* $c, d, \ldots$ are understood as formulae in a first-order language. If the information of d can be *entailed* (or deduced) from the information represented by c we write $c \vdash d$ (e.g. $pitch > 64 \vdash pitch > 48$).

Universal timed CCP (utcc) [9] extends Timed CCP (tcc) [12] for mobile reactive systems. Time in utcc is conceptually divided into *time intervals* (or *time-units*). In a particular time-unit, a utcc process P gets an input c from the environment, it executes with this input as the initial *store*, and when it reaches its resting point, it *outputs* the resulting store d to the environment. Furthermore, the resting point determines a residual process, which is then executed in the next time interval.

Processes in utcc are built by the following syntax:

$$\begin{array}{ll} P, Q := & \mathbf{skip} \mid \mathbf{tell}(c) \mid (\mathbf{abs}\ \boldsymbol{x}; c)\, P \mid P \parallel Q \mid (\mathbf{local}\ \boldsymbol{x}; c)\, P \mid \\ & \mathbf{next}\ P \mid \mathbf{unless}\ c\ \mathbf{next}\ P \mid !\, P \end{array}$$

A process $\mathbf{skip}$ represents *inaction*. A process $\mathbf{tell}(c)$ adds c to the store in the current time interval, thus making it available to other processes.

In utcc, the CCP ask operator $\mathbf{when}\ c\ \mathbf{do}\ P$ (executing P if c can be deduced) is replaced by the *abstraction* operator $(\mathbf{abs}\ \boldsymbol{x}; c)\, P$. This construct is a parameterized ask where $P[\mathbf{t}/\boldsymbol{x}]$ is executed for *all the terms* $\mathbf{t}$ s.t $c[\mathbf{t}/\boldsymbol{x}]$ is entailed by the store.

A process $P \parallel Q$ denotes P and Q running in parallel possibly "communicating" via the common store. The process $(\mathbf{local}\ \boldsymbol{x}; c)\, P$ behaves like P but the information c about the variables in $\boldsymbol{x}$ is local to P. We shall omit c in $(\mathbf{local}\ \boldsymbol{x}; c)\, P$ when $c \equiv \mathtt{true}$.

From a programming language perspective, $\boldsymbol{x}$ in $(\mathbf{local}\ \boldsymbol{x}; c)\, P$ can be viewed as the local variables of P while $\boldsymbol{x}$ in $(\mathbf{abs}\ \boldsymbol{x}; c)\, P$ as the formal parameters of P. This way, abstractions can encode recursive definitions of the form $X(\boldsymbol{x}) \stackrel{\text{def}}{=} P$ (see [8]).

The unit-delay $\mathbf{next}\ P$ executes P in the next time interval. The (weak) time-out $\mathbf{unless}\ c\ \mathbf{next}\ P$ executes P in the next time-unit iff c *cannot* be entailed by the final store at the current time interval. The *replication* $!\, P$ means $P \parallel \mathbf{next}\ P \parallel \mathbf{next}^2 P$..., i.e. unboundedly many copies of P but one at a time.

We shall also use the derived operator $(\mathbf{wait}\ \boldsymbol{x}; c)\ \mathbf{do}\ P$ that *waits*, possibly for several time-units, until for some t, $c[\boldsymbol{t}/\boldsymbol{x}]$ holds and then it executes $P[\boldsymbol{t}/\boldsymbol{x}]$ (see [10]).

An Example. The abstraction operator allows us to communicate (local) names or variables between processes, i.e., mobility in the sense of the π-calculus [7]. Let us give a simple example of this situation. Let P be a process modeling a musician playing notes at different time-units, and Q be an improvisation system which after "reading" the note played by P performs some action R. Roughly, this scenario can be modeled as follows

$$P \stackrel{\text{def}}{=} \mathbf{tell}(\texttt{play}\,(A)) \parallel \mathbf{next}\,(\mathbf{tell}(\texttt{play}\,(G)) \parallel \mathbf{next}\,\mathbf{tell}(\texttt{play}\,(B)))\ldots$$
$$Q \stackrel{\text{def}}{=} !\,(\mathbf{abs}\,x; \texttt{play}(x))\,R$$

When executing $P \parallel Q$, we observe, e.g., $R[G/x]$ in the second time-unit. This means that P and Q synchronized on the constraint $\texttt{play}(\cdot)$ and the note played by P (i.e. G) was read by Q and then processed by R. See [8] for a more involved example defining synchronization of multiple agents.

Logic Characterization. The utcc calculus enjoys a declarative view of processes as first-order linear-time temporal logic (FLTL) formulae [6]. This means that processes can be seen, at the same time, as computing agents and as logic formulae.

Formulae in FLTL are built from the following syntax

$$F, G, \ldots := c \mid F \wedge G \mid \neg F \mid \exists x F \mid \circ F \mid \Box F.$$

where c is a constraint. The modalities $\circ F$ and $\Box F$ stand for resp., that F holds *next* and *always*. We use $\forall x F$ for $\neg \exists x \neg F$, and the *eventual* modality $\Diamond F$ as an abbreviation of $\neg \Box \neg F$. See [6] for further details on this logic.

Processes in utcc can be represented as FLTL formulae as follows:

$$\begin{array}{lll} [\![\mathbf{skip}]\!] = \texttt{true} & [\![\mathbf{tell}(c)]\!] = c & [\![P \parallel Q]\!] = [\![P]\!] \wedge [\![Q]\!] \\ [\![(\mathbf{abs}\,\boldsymbol{y}; c)\,P]\!] = \forall \boldsymbol{y}(c \Rightarrow [\![P]\!]) & [\![(\mathbf{local}\,\boldsymbol{x}; c)\,P]\!] = \exists \boldsymbol{x}(c \wedge [\![P]\!]) & \\ [\![\mathbf{next}\,P]\!] = \circ [\![P]\!] & [\![\mathbf{unless}\,c\,\mathbf{next}\,P]\!] = c \vee \circ [\![P]\!] & [\![!\,P]\!] = \Box [\![P]\!] \end{array}$$

Let $A = [\![P]\!]$. Roughly, $A \vdash \Diamond c$ (i.e., c eventually holds in A) iff the process P eventually outputs c (see [8,9] for further details).

3 A Model for Dynamic Interactive Scores

An interactive score [3] is a pair composed of *temporal objects* and Allen temporal relations [2]. In general, each object is comprised of a start-time, a duration, and a procedure. The first two can be partially specified by constraints, with different constraints giving rise to different types of temporal objects, so-called *events* (duration equals zero), *textures* (duration within some range), *intervals* (textures without procedures) or *control-points* (a temporal point occurring somewhere within an interval object). The procedure gives operational meaning to the action of the temporal object. It could just be playing a note or a chord, or any other action meaningful for the composer. Figure 1, based on one from [3], shows an interactive score where temporal objects are represented as *boxes*. Objects are T_i, durations D_i. Object T_4 is a control point, whereas T_0 and T_3 are intervals. Duration D_3 should be such that $D_s \leq D_3 \leq D_f$.

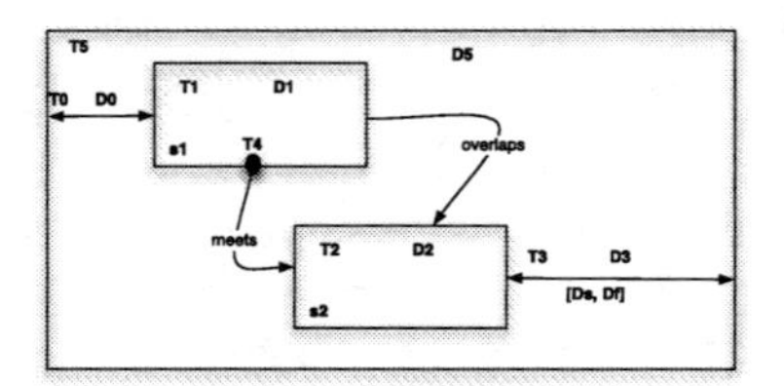

Fig. 1. Interactive score

The whole temporal structure is determined by the hierarchy of temporal objects. Suppose that, as a result of the information obtained by the occurrence of an event, object T_2 should no longer synchronize with a control-point inside T_1 but, say, with a similar point inside T_5. This very simple interaction cannot be modeled in the standard model of interactive scores [3]. Another example is an object waiting for some interaction from the performer within some temporal interval. If the interaction does not occur, the composer might then determine to probe the environment again later when a similar musical context has been defined. This amounts to moving the waiting interval from one box to another.

The model. Figure 2 shows our model for dynamic interactive scores. The process $BoxOperations$ may perform the following actions:

- $\texttt{mkbox}(id, d)$: defines a new box with id id and duration d. The start time is defined as a new (local) variable s whose value will be constrained by the other processes.
- $\texttt{destroy}(id)$: firstly, it retrieves the box sup which contains the box id. If the box id is not currently playing, in the next time-unit, it drops the boundaries of id by inserting all the boxes contained in id into sup.
- $\texttt{before}(x, y)$: checks if x and y are contained in the same box. If so, the constraint $\texttt{bf}(x, y)$ is added.
- $\texttt{into}(x, y)$: dictates that the box x is into the box y if x is not currently playing.
- $\texttt{out}(x, y)$: takes the box x out of the box y if x is not currently playing.

Process $Constraints$ adds the necessary constraints relating the start times of each temporal object to respect the hierarchical structure of the score. For each constraint of the form $\texttt{in}(x, y)$, this process dictates that the start time of x must be less than the one of y. Furthermore, the end time of y (i.e. $d_y + s_y$) must be greater than the end time of x. The case for $\texttt{bf}(x, y)$ can be explained similarly.

The process $Persistence$ transfers the information of the hierarchy (i.e. box declarations, $\texttt{in}$ and $\texttt{bf}$ relations) to the next time-unit.

The process $Clock$ defines a simple clock which binds the variable t to the value v in the current time-unit and to $v + 1$ in the next time-unit.

The process $Play(x, t)$ adds the constraint $\texttt{play}(x)$ during t time-units. This informs the environment that the box x is currently playing.

The process $Init(t)$ waits until the environment provides the constraint $init(x)$ for the outermost box x to start the execution of the system. Then, the *clock* is started and the start time of x is set to 0. The rest of the boxes wait until their start time is less or equal to the current time (t) to start playing.

$$
\begin{array}{rl}
BoxOperations \stackrel{\text{def}}{=} & (\textbf{abs}\ id, d; \texttt{mkbox}(id, d)) \\
& \quad (\textbf{local}\ s)\ \textbf{tell}(\texttt{box}(id, d, s)) \\
& \parallel (\textbf{abs}\ id; \texttt{destroy}(id)) \\
& \quad (\textbf{abs}\ x, sup; \texttt{in}(x, id) \wedge \texttt{in}(id, sup)) \\
& \qquad \textbf{unless}\ \texttt{play}(id)\ \textbf{next}\ \textbf{tell}(\texttt{in}(x, sup)) \\
& \parallel (\textbf{abs}\ x, y; \texttt{before}(x, y))\ \textbf{when}\ \exists_z(\texttt{in}(x, z) \wedge \texttt{in}(y, z))\ \textbf{do} \\
& \qquad \textbf{unless}\ \texttt{play}\,(y)\ \textbf{next}\ \textbf{tell}(\texttt{bf}(x, y)) \\
& \parallel (\textbf{abs}\ x, y; \texttt{into}(x, y))\ \textbf{unless}\ \texttt{play}\,(x)\ \textbf{next}\ \textbf{tell}(\texttt{in}(x, y)) \\
& \parallel (\textbf{abs}\ x, y; \texttt{out}(x, y))\ \textbf{when}\ \texttt{in}(x, y)\ \textbf{do} \\
& \qquad \textbf{unless}\ \texttt{play}\,(x)\ \textbf{next}\ (\textbf{abs}\ z, \texttt{in}(y, z); \textbf{tell}(\texttt{in}(x, z))) \\
Constraints \stackrel{\text{def}}{=} & (\textbf{abs}\ x, y; \texttt{in}(x, y))\ (\textbf{abs}\ d_x, s_x; \texttt{box}(x, d_x, s_x)) \\
& \quad (\textbf{abs}\ d_y, s_y; \texttt{box}(y, d_y, s_y)) \\
& \qquad \textbf{tell}(s_y \leq s_x) \parallel \textbf{tell}(d_x + s_x \leq d_y + s_y) \\
& \parallel (\textbf{abs}\ x, y; \texttt{bf}(x, y))\ (\textbf{abs}\ d_x, s_x; \texttt{box}(x, d_x, s_x)) \\
& \qquad (\textbf{abs}\ d_y, s_y; \texttt{box}(y, d_y, s_y))\ \textbf{tell}(s_x + d_x \leq s_y) \\
Persistence \stackrel{\text{def}}{=} & (\textbf{abs}\ x, y; \texttt{in}(x, y))\ \textbf{when}\ \texttt{play}(x)\ \textbf{do}\ \textbf{next}\ \textbf{tell}(\texttt{in}(x, y)) \\
& \qquad \parallel \textbf{unless}\ \texttt{out}(x, y) \vee \texttt{destroy}(x)\ \textbf{next}\ \textbf{tell}(\texttt{in}(x, y)) \\
& \parallel (\textbf{abs}\ x, y; \texttt{bf}(x, y))\ \textbf{when}\ \texttt{play}(y)\ \textbf{do}\ \textbf{next}\ \textbf{tell}(\texttt{bf}(x, y)) \\
& \qquad \parallel \textbf{unless}\ (\texttt{out}(x, y) \vee \texttt{destroy}(y)\ \textbf{next}\ \textbf{tell}(\texttt{bf}(x, y)) \\
& \parallel (\textbf{abs}\ x; \texttt{box}(x, d_x, s_x))\ \textbf{when}\ \texttt{play}(x)\ \textbf{do}\ \textbf{next}\ \textbf{tell}(\texttt{box}(x, d_x, s_x)) \\
& \qquad \parallel \textbf{unless}\ \texttt{destroy}(x)\ \textbf{next}\ \textbf{tell}(\texttt{box}(x, d_x, s_x)) \\
Clock(t, v) \stackrel{\text{def}}{=} & \textbf{tell}(t = v) \parallel \textbf{next}\ Clock(t, v + 1) \\
Play(x, t) \stackrel{\text{def}}{=} & \textbf{when}\ t \geq 1\ \textbf{do}\ \textbf{tell}(\texttt{play}(x)) \parallel \textbf{unless}\ t \leq 1\ \textbf{next}\ Play(x, t - 1) \\
Init(t) \stackrel{\text{def}}{=} & (\textbf{wait}\ x; init(x))\ \textbf{do} \\
& \quad (\textbf{abs}\ d_x, s_x; \texttt{box}(x, d_x, s_x)) \\
& \qquad Clock(t, 0) \parallel \textbf{tell}(s_x = t) \parallel \\
& \qquad !\,(\textbf{wait}\ y, d_y, s_y; \texttt{box}(y, d_y, s_y) \wedge s_y \leq t)\ \textbf{do}\ Play(y, d_y) \\
System \stackrel{\text{def}}{=} & (\textbf{local}\ t)\ Init(t) \parallel !\,Persistence \parallel !\,Constraints \parallel !\,BoxOperations \parallel UsrBoxes
\end{array}
$$

Fig. 2. A `utcc` model for Dynamic Interactive Scores

Finally, the whole system is the parallel composition between the previously defined processes and the specific user model, e.g.:

$$
\begin{array}{rl}
UsrBoxes \stackrel{\text{def}}{=} & \textbf{tell}(\texttt{mkbox}(a, 22) \wedge \texttt{mkbox}(b, 12) \wedge \texttt{mkbox}(c, 4)) \parallel \\
& \textbf{tell}(\texttt{mkbox}(d, 5) \wedge \texttt{mkbox}(e, 2)) \parallel \\
& \textbf{tell}(\texttt{into}(b, a) \wedge \texttt{into}(c, b) \wedge \texttt{into}(d, b) \wedge \texttt{into}(e, d)) \parallel \\
& \textbf{tell}(\texttt{before}(c, d)) \parallel \\
& \textbf{whenever}\ \texttt{play}(b)\ \textbf{do}\ \textbf{unless}\ \texttt{signal}\ \textbf{next} \\
& \qquad \textbf{tell}(\texttt{out}(d, b) \wedge \texttt{mkbox}(f, 2) \wedge \texttt{into}(f, a)) \parallel \\
& \qquad \textbf{tell}(\texttt{before}(b, f) \wedge \texttt{before}(f, d))
\end{array}
$$

This system defines the hierarchy in Figure 3(a). When b starts playing, the system asks if the signal `signal` is present (i.e., if it was provided by the environment). If it was not, the box d is taking out from the context b. Furthermore, a new box f is created such that b must be played before f and f before d as in Figure 3(b). Notice that when the box d is taken out from b, the internal box e is still into d preserving its structure.

Verification of the Model. The processes defined by the user may lead to situations where the final store is inconsistent as in $st < 5 \wedge st > 7$ where st is the start time of a given box. Take for example the process $UsrBoxes$ above. If the box f is defined with a duration greater than 5, the execution of f (and then that of d) will exceed the boundaries of the box a which contains both structures.

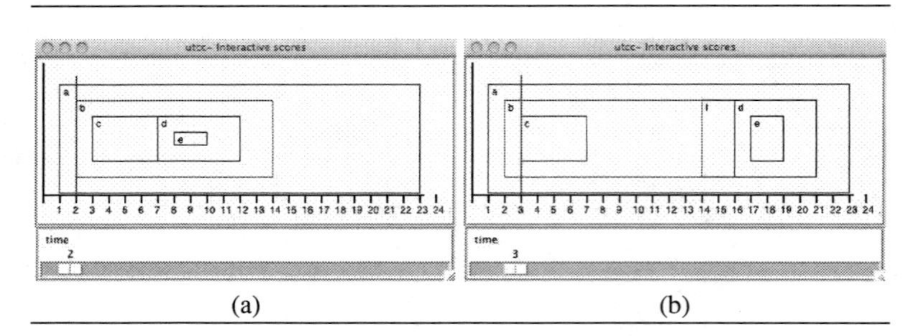

(a) (b)

Fig. 3. Example of an Interactive Score Execution

In this context, the declarative view of `utcc` processes as FLTL formulae provides a valuable tool for the verification of the model: The formula $A = [\![P]\!]$ allows us to verify whether the execution of P leads to an inconsistent store. Thus, we can detect pitfalls in the user model such as trying to place a bigger box into a smaller one or taking a box out of the outermost box.

In the following, we present some examples of temporal properties we could verify in an interactive score represented as the process P.

- $[\![P]\!] \vdash \Diamond \exists_{x,d_x,s_x,y,d_y,s_y}(\mathtt{box}(x,d_x,s_x) \wedge \mathtt{box}(y,d_y,s_y) \wedge \mathtt{in}(x,y) \wedge s_x + d_x > s_y + d_y)$: The end time of the box y is less than the end time of the inner box x. I.e., the box y cannot contain x.
- $[\![P]\!] \vdash \forall_x(\exists_{d_x,s_x}(\mathtt{box}(x,d_x,s_x) \Rightarrow \Diamond\,\mathtt{play}(x))$: All the musical structures are eventually played.
- $[\![P]\!] \vdash \Box \forall_{x,y}(\mathtt{in}(x,y) \wedge \mathtt{play}(x) \Rightarrow \mathtt{play}(y))$: The execution of the internal box implies the execution of the outer box.
- $[\![P]\!] \vdash \Box \forall_x(\exists_{d_x,s_x}\,\mathtt{box}(x,d_x,s_x) \Rightarrow \mathit{init}(x) \vee \exists_y(\mathtt{in}(x,y)))$: Every box is either the initial box or it is contained in another box.
- $[\![P]\!] \vdash \Diamond \forall_x(\exists_{d_x,s_x}(\mathtt{box}(x,d_x,s_x) \Rightarrow \mathtt{play}(x))$: At some point all the boxes are playing simultaneously.
- $[\![P]\!] \vdash \mathtt{signal} \vee \Diamond\,\mathtt{play}(x)$: The signal `signal` is present or else the box x must be played.

Remark. For the sake of presentation we only defined here the `before` relation. Our model can be straightforwardly extended to support all Allen temporal relations [2]. Making use of the `into` and `out` operations, we can define also the operation $\mathtt{move}(a,b)$ meaning, move the structure a into the structure b.

4 A Model for Music Improvisation

As described above, in interactive scores the actual musical output may change depending on interactions with a performer, but the framework is not meant for learning from those interactions, nor to change the score (i.e. improvise) accordingly.

Music improvisation provides a complex context of concurrent systems posing great challenges to modeling tools. In music improvisation, partners behave independently but are constantly interacting with others in controlled ways. The interactions allow building a complex global musical process collaboratively. Interactions become effective when each partner has somehow learned about the possible evolutions of each musical process launched by the others, i.e, their musical *style*. Getting the computer involved in the improvisation process requires learning the musical style of the human interpreter and then playing jointly in the same style. A *style* in this case means some set of meaningful sequences of musical material the interpreter has played. A graph structure called *factor oracle* (FO) is used to efficiently represent this set [1].

A FO is a finite state automaton constructed in an incremental fashion. A sequence of symbols $s = \sigma_1\sigma_2 \ldots \sigma_n$ is learned in such an automaton, which states are $0, 1, 2 \ldots n$. There is always a transition arrow (called factor link) labeled by the symbol σ_i going from state $i-1$ to state $i, 1 \leq i < n$. Depending on the structure of s, other arrows will be added. Some are directed from a state i to a state j, where $0 \leq i < j \leq n$. These also belong to the set of factor links and are labeled by symbol σ_j. Some are directed "backwards", going from a state i to a state j, where $0 \leq j < i \leq n$. They are called suffix links, and bear no label (represented as '$\star$' in our processes below). The factor links model a factor automaton, that is every factor p in s corresponds to a unique factor link path labeled by p, starting in 0 and ending in some other state. Suffix links have an important property : a suffix link goes from i to j iff the longest repeated suffix of $s[1..i]$ is recognized in j. Thus suffix links connect repeated patterns of s.

The oracle (see Figure 4) is learned on-line. For each new input symbol σ_i, a new state i is added and an arrow from $i-1$ to i is created with label σ_i. Starting from $i-1$, the suffix links are iteratively followed backward, until a state is reached where a factor link with label σ_i originates (going to some state j), or until there is no more suffix links to follow. For each state met during this iteration, a new factor link labeled by σ_i is added from this state to i. Finally, a suffix link is added from i to the state j or to state 0 depending on which condition terminated the iteration. Navigating the oracle in order to generate variants is straightforward : starting in any place, following factor links generates a sequence of labelling symbols that are repetitions of portions of the learned sequence; following one suffix link followed by a factor links creates a recombined pattern sharing a common suffix with an existing pattern in the original sequence. This common suffix is, in effect, the musical context at any given time.

In [5] a `tcc` model of FO is proposed. This model has three drawbacks. Firstly, it (informally) assumes the basic calculus has been extended with general recursion in order to correctly model suffix links traversal. Secondly, it assumes dynamic construction of new variables $\delta_{i\sigma}$ set to the state reached by following factor link labelled σ from state i. This construction cannot be expressed with the local variable primitive in basic `tcc`. Thirdly, the model assumes a constraint system over both finite domains and finite sets. We use below the expressive power of the abstraction construction in

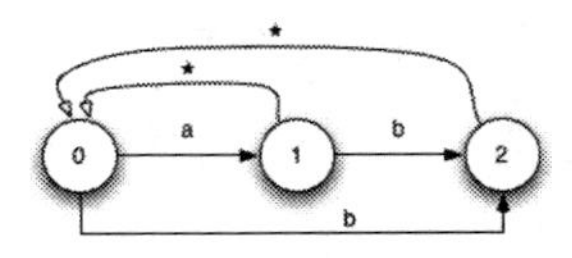

Fig. 4. A FO automaton for $s = ab$

$$
\begin{array}{lll}
\mathit{FO} & \stackrel{\text{def}}{=} & \mathit{Counter} \parallel \mathit{Persistence} \\
 & & \parallel ! \, (\mathbf{abs}\ \mathit{Note}; \mathtt{play}(\mathit{Note}))\ \mathbf{whenever}\ \mathit{ready}\ \mathbf{do}\ \mathit{Step}_1(\mathit{Note}) \\
\mathit{Counter} & \stackrel{\text{def}}{=} & \mathbf{tell}(i = 1) \parallel ! \, (\mathbf{abs}\ x; i = x)\ (\mathbf{when}\ \mathit{ready}\ \mathbf{do}\ \mathbf{next}\ \mathbf{tell}(i = x + 1) \\
 & & \qquad\qquad\qquad\qquad\qquad \parallel \mathbf{unless}\ \mathit{ready}\ \mathbf{next}\ \mathbf{tell}(i = x)) \\
\mathit{Persistence} & \stackrel{\text{def}}{=} & ! \, (\mathbf{abs}\ x, y, z; \mathsf{edge}(x, y, z))\ \mathbf{next}\ \mathbf{tell}(\mathsf{edge}(x, y, z)) \\
\mathit{Step}_1(\mathit{Note}) & \stackrel{\text{def}}{=} & \mathbf{tell}(\mathsf{edge}(i - 1, i, \mathit{Note})) \parallel \mathit{Step}_2(\mathit{Note}, i - 1) \\
\mathit{Step}_2(\mathit{Note}, E) & \stackrel{\text{def}}{=} & \mathbf{when}\ E = 0\ \mathbf{do} \\
 & & \quad (\mathbf{abs}\ k; \mathsf{edge}(E, k, \mathit{Note}))\ (\mathbf{tell}(\mathsf{edge}(i, k, \star)) \parallel \mathbf{next}\ \mathbf{tell}(\mathit{ready})) \\
 & & \quad \parallel \mathbf{unless}\ \exists_k\ \mathsf{edge}(E, K, \mathit{Note})\ \mathbf{next}\ (\mathbf{tell}(\mathit{ready}) \parallel \mathbf{tell}(\mathsf{edge}(i, 0, \star))) \\
 & & \mathbf{when}\ E \neq 0\ \mathbf{do} \\
 & & \quad (\mathbf{abs}\ j; \mathsf{edge}(E, j, \star)) \\
 & & \qquad \mathbf{when}\ \exists_k\ \mathsf{edge}(j, k, \mathit{Note})\ \mathbf{do} \\
 & & \qquad\quad (\mathbf{abs}\ k; \mathsf{edge}(j, k, \mathit{Note}))\ (\mathbf{tell}(\mathsf{edge}(i, k, \star)) \parallel \mathbf{next}\ \mathbf{tell}(\mathit{ready})) \\
 & & \qquad \parallel \mathbf{unless}\ \exists_k\ \mathsf{edge}(j, k, \mathit{Note})\ \mathbf{next}\ \mathbf{when}\ j \neq 0\ \mathbf{do}\ \mathbf{tell}(\mathsf{edge}(j, i, \mathit{Note})) \\
 & & \qquad\qquad\qquad\qquad\qquad\qquad \parallel \mathit{Step}_2(\mathit{Note}, j)
\end{array}
$$

Fig. 5. Implementing the FO into `utcc`

`utcc` to correct all these drawbacks (see Figure 5). Furthermore, our model leads to a compact representation of the data structure of the FO based on constraints of the form $\mathtt{edge}(x, y, N)$ representing an arc between node x and y labeled with N.

Process *Counter* signals when a new played note can be learned. It can be learned when all links for the previous note have already been added to the FO. Process *Persistence* transmits information about already constructed arcs (factor and suffix) to all future time-units. Process $Step_1$ adds a factor link from $i - 1$ to i labelled with a just played note and launches traversal of suffix links from $i - 1$. When state zero is reached by traversing suffix links, process $Step_2$ adds a suffix link from i to a state reached from 0 by a factor link labelled $Note$, if it exists, or from i to state zero, otherwise. For each state k different from zero reached in the suffix links traversal, process $Step_2$ adds factor links labelled $Note$ from k to i.

The inclusion of a new agent in our FO model (e.g. a learner agent for a second performer) entails a new process and new interactions, both with the new process and among the existing ones. In traditional models this usually means major changes in the synchronization scheme, which are difficult to localize and control. In `utcc`, all synchronization is done semantically, through the available information in the store. Each agent would thus have to be incremented with processes testing for the presence of new information (e.g. a factor link with some label in the other agent's FO graph). The new synchronization behavior that this demands is automatically provided by the blocking ask (abstraction) construct.

5 Concluding Remarks

Here we argued for `utcc` as a declarative framework for modeling and verifying dynamic multimedia interaction systems. We showed that the synchronization mechanism based on entailment of constraints leads to simpler models that scale up when more agents are added. Moreover, we showed that systems can be formally verified with the underlying temporal logic in `utcc`. We modeled two non trivial interacting systems. The model proposed for interactive scores in Section 3 improved considerably the expressivity of previous models such as [3]. It allows the composer to dynamically change the structure of the score according to the information derived from the environment.

The results presented here are so far encouraging although much remains to be done at the implementation level. Currently, to guarantee reliable responses in time, we are working on assessing the behavior of `utcc` processes in real-time contexts. We plan to provide a more principled notion of time where the duration of each time-unit can be related to the amount of computation involved in it. We also plan to enrich our FO model with probabilistic traversals of the graph in the lines of [11].

References

1. Allauzen, C., Crochemore, M., Raffinot, M.: Factor oracle: A new structure for pattern matching. In: Bartosek, M., Tel, G., Pavelka, J. (eds.) SOFSEM 1999. LNCS, vol. 1725, p. 295. Springer, Heidelberg (1999)
2. Allen, J.F.: Maintaining knowledge about temporal intervals. Commun. ACM 26(11) (1983)
3. Allombert, A., Assayag, G., Desainte-Catherine, M.: A system of interactive scores based on Petri nets. In: Proceedings of SMC 2007 (2007)
4. Allombert, A., Assayag, G., Desainte-Catherine, M., Rueda, C.: Concurrent constraints models for interactive scores. In: Proceedings of SMC 2006 (2006)
5. Assayag, G., Dubnov, S., Rueda, C.: A concurrent constraints factor oracle model for music improvisation. In: CLEI 2006 (2006)
6. Manna, Z., Pnueli, A.: The Temporal Logic of Reactive and Concurrent Systems: Specification. Springer, Heidelberg (1991)
7. Milner, R.: Communicating and Mobile Systems: the Pi-Calculus. Cambridge University Press, Cambridge (1999)
8. Olarte, C., Rueda, C.: A declarative language for dynamic multimedia interaction systems (April 2009), `http://www.lix.polytechnique.fr/~colarte/`
9. Olarte, C., Valencia, F.D.: The expressivity of universal timed CCP: Undecidability of monadic FLTL and closure operators for security. In: Proc. of PPDP 2008. ACM, New York (2008)
10. Olarte, C., Valencia, F.D.: Universal concurrent constraint programing: Symbolic semantics and applications to security. In: Proc. of SAC 2008. ACM Press, New York (2008)
11. Perez, J.A., Rueda, C.: Non-determinism and probabilities in timed concurrent constraint programming. In: ICLP 2008. LNCS (2008)
12. Saraswat, V., Jagadeesan, R., Gupta, V.: Foundations of timed concurrent constraint programming. In: Proc. of LICS 1994. IEEE CS, Los Alamitos (1994)
13. Saraswat, V.A.: Concurrent Constraint Programming. MIT Press, Cambridge (1993)

Generalized Voice Exchange

Robert Peck

School of Music, Louisiana State University
Baton Rouge, Louisiana 70803 USA
rpeck@lsu.edu

Abstract. The notion of voice exchange in ordered pitch-class space conforms closely to that of contextual inversion in neo-Riemannian theory: the melodic dyad (a, b) in one voice inverts in another voice, and we define an axis of inversion respectively for all such pairs. We may thus apply many of the transformational concepts of neo-Riemannian theory to a study of voice exchange. We draw our musical examples from the Prelude to Richard Wagner's *Tristan und Isolde*, for which a separate analytical thread exists that considers aspects of tonality in relation to the voice exchange in the resolution of the Tristan Chord.

Keywords: voice exchange, neo-Riemannian theory, contextual inversion, Wagner, Tristan chord.

1 Introduction

The notion of voice exchange in ordered pitch-class space conforms closely to that of contextual inversion in neo-Riemannian theory: the melodic dyad (a, b) in one voice inverts in another voice, and we define an axis of inversion respectively for all such pairs. A connection to the Parallel Exchange exists. Given a C major triad in a neo-Hauptmannian sense [1] with an *Einheit* C and a *Zweiheit* G, these pitch-classes invert about a contextual axis under the P operation, with G's assuming the *Einheit* function and C's assuming that of *Zweiheit* in the resulting C minor triad. We substitute order positions for chordal factors in translating this concept to voice exchange; therefore, "*Einheit*" becomes "first coordinate" and "*Zweiheit*" "second," and the ordered dyad (C, G) becomes (G, C). We may thus apply many of the transformational concepts of neo-Riemannian theory to a study of voice exchange.

We draw our musical examples from the Prelude to Richard Wagner's *Tristan und Isolde*, for which a separate analytical thread exists that considers aspects of tonality in relation to the voice exchange (or "interchange" [2]) in the resolution of the Tristan Chord [3,4,5]. Other recent research on the Prelude considers aspects of voice-leading efficiency, particularly in various resolutions of the Tristan Chord, wherein voice exchanges are viewed as surface embellishments of a more fundamental stepwise structure [6]. This notion follows from a more general conceptualization of voice exchanges as permutations, the attitude taken implicitly in [7].

We adopt here a transformational approach, wherein we regard voice exchanges as being aligned with contextual inversion. To highlight this connection, we model voice exchanges using pitch classes, hence in the integers modulo 12. The theory

E. Chew, A. Childs, and C.-H. Chuan (Eds.): MCM 2009, CCIS 38, pp. 228–235, 2009.

presented here, however, is easily extendable to other models of pitch in more robust number systems, including infinite pitch space using the integers, continuous pitch space using the reals, and so on. The transformational perspective allows us ultimately to relate voice exchanges—including various chromatic exchanges and those in differing harmonic contexts—to one another in terms of a transformational scheme by which we may describe networks that model processes in the music.

1.1 Connection to Contextual Inversion

We note four particular desiderata in the connection of voice exchange to contextual inversion. First, voice exchanges are defined in terms of the objects within the set on which the exchange acts, rather than abstractly [8]. Hence, the contextual axis of inversion in the above example is $(a + b)/2$ for any pair (a, b), not a fixed point in pitch or pitch-class space. Second, like contextual inversions, voice exchanges are involutions. Therefore, it does not matter if (C, G) appears in register above or below (G, C); both voicings represent the same exchange. Third, it is necessary that voice exchanges commute with the usual transposition and inversion operations, and are consequently preserved under their conjugation [9]. The voice exchange (C, G) (G, C) is accordingly equivalent to its images under T_n and I_n for all n. Hence, its image (C#, G#) (G#, C#) under T_1, (C, F) (F, C) under I_0, and so on, are all instances of the same operation. It follows as a consequent of this desideratum that if (a, b) exchanges with (c, d), then any dyad whose interval belongs to the same interval class as that of (a, b) will exchange with a dyad whose interval belongs to the interval class that includes (c, d). Fourth, if $0 < b - a \leq 5$ mod 12, then $6 < d - c \leq 11$ mod 12, or vice versa. This last point ensures a kind of contrary motion in pitch-class space, relating such operations further to inversion.

2 Generalized Voice Exchange

We define a preliminary operation X in which to model a voice exchange:

$$X := (a, b) \mapsto (b, a), \text{ for all } (a, b) \in \mathbb{Z}_{12} \times \mathbb{Z}_{12} \tag{1}$$

noting that X satisfies all four of the above criteria. That is: we define it contextually; it is an involution; it commutes with transposition and inversion; and it exhibits contrary motion.

Fig. 1. *Tristan* Prelude, mm. 2-3

We find two voice exchanges labeled in X in the first seven measures of the *Tristan* Prelude (Figures 1 and 2). In the first exchange, the minor third (G#, B) of the soprano voice inverts to (B, G#) in the tenor; the second presents a sequential image of the first under T_3. In both instances, we note that (a, b) and (b, a) as ordered intervals are inverses of one another and belong to the same interval class.

Fig. 2. *Tristan* Prelude, mm. 6-7

2.1 Generalized Chromatic Voice Exchange: The Variable $i \in \mathbb{Z}_{12}$

Another situation exists for chromatic voice exchanges: those in which (a, b) and its image have intervals that do not belong to the same interval class. Let us take the subsequent, altered leg of the above sequence in the *Tristan* Prelude as an example (Figure 3). Here, the previously ascending minor third in the soprano is extended to a major third (D, F#), while the tenor's descending minor third is diminished (F, D#)—a doubly chromatic exchange. We could define a transformation[1]

$$X' := (a, b) \mapsto (b - 1, a + 1), \text{ for all } (a, b) \in \mathbb{Z}_{12} \times \mathbb{Z}_{12} \tag{2}$$

in which to label this exchange, but as it does not commute with inversion, X' fails as a contextual inversion.

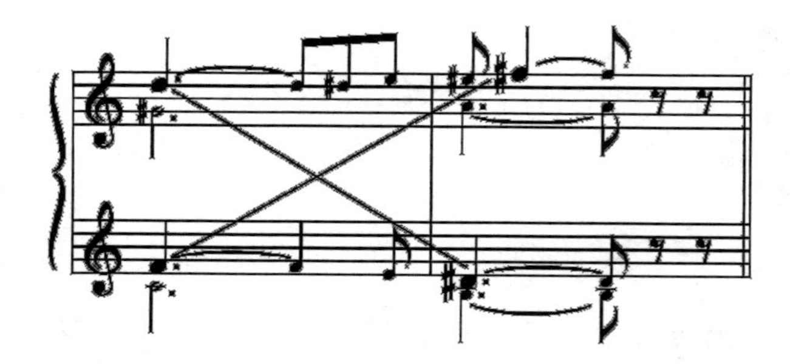

Fig. 3. *Tristan* Prelude, mm. 10-11

The centralizer of the *T/I* group's action on $\mathbb{Z}_{12} \times \mathbb{Z}_{12}$ in the symmetric group on the same set

[1] Throughout this article, we assume all arithmetic to be performed modulo 12, unless otherwise indicated.

$$C(T/I) = C_{Sym(\mathbb{Z}_{12} \times \mathbb{Z}_{12})} T/I \tag{3}$$

provides all 7,644,119,040 operations on $\mathbb{Z}_{12} \times \mathbb{Z}_{12}$ that commute with both transposition and inversion. $C(T/I)$ is the direct product of centralizers of two T/I orbit restrictions. One restriction is to the union of melodic dyads that belong to set classes [0, 1] through [0, 5], a wreath product of order $24^5 \cdot 5!$; the other is to the union of set classes [0, 0] and [0, 6], another wreath product of order $2^2 \cdot 2!$ [10]. We may use an involution from this centralizer, $X_{9,4}$, to model all three of the above voice exchanges, including the chromatic example. In the definition of such a function, we incorporate the interval between a and b in pitch-class space

$$i = b - a \bmod 12 \tag{4}$$

contextually, thereby facilitating the commutative property with both transposition and inversion.

$$X_{9,4} := (a,b) \mapsto \begin{cases} \big(b - 9 - i, a - 2 + (i + 4 \bmod 5)\big), & \text{for } 0 < i \leq 5 \\ \big(b + 9 - i, a - 2 + (i + 3 \bmod 5)\big), & \text{for } 6 < i \leq 11 \\ (b,a), & \text{for } i = 0 \bmod 6 \end{cases} \tag{5}$$

As we will see below, the subscript 9 in the label of this function associates with the initial harmonic interval in the exchange, whereas the subscript 4 relates to a particular permutation of T/I orbits.

We note the essential difference in the actions of $X_{9,4}$ on the T/I orbits of melodic unisons and tritones (those for which $i = 0 \bmod 6$) and on those of all other dyads. This variance is a consequence of the structure of the centralizer, as the action of T/I restricted to an orbit from among the former set is not permutation isomorphic to one from among the latter. Specifically, the orbits of melodic unisons and tritones have two fixed points each under I_0: (0, 0), (6, 6) and (0, 6), (6, 0), respectively, whereas the remaining orbits have no such fixed points.

As these three voice exchanges involve melodic thirds, we are particularly interested in the action of $X_{9,4}$ on dyads with intervals in interval classes 2 (as diminished thirds), 3 and 4. Table 1 summarizes this action; in particular, we note the two exchanges within interval class 3, and that within interval classes 2 and 4. More generally, interval class 3 is stabilized under $X_{9,4}$, and any melodic interval in interval class $3 + n$ will exchange with an interval in interval class $3 - n$ (where $n \leq 2$).

Table 1. Exchanges under $X_{9,4}$

$0 < i \leq 5$				
IC 1	IC 2	IC 3	IC 4	IC 5
		8 11 11 2 11 8 2 11	2 6 5 3	
IC 5	IC 4	IC 3	IC 2	IC 1
$6 < i \leq 11$				

2.2 Permutations of the Orbits: The Variable $p \in \mathbb{Z}_5$

Voice exchanges in which only one voice is altered chromatically require different, but related, operations. One such example derives from Rothstein's [11] reading of the *Tristan* Prelude's first two measures (Figure 4): here the bass voice's implied ascending minor third (D, F) inverts to a descending diminished third (F, D#) in the alto.

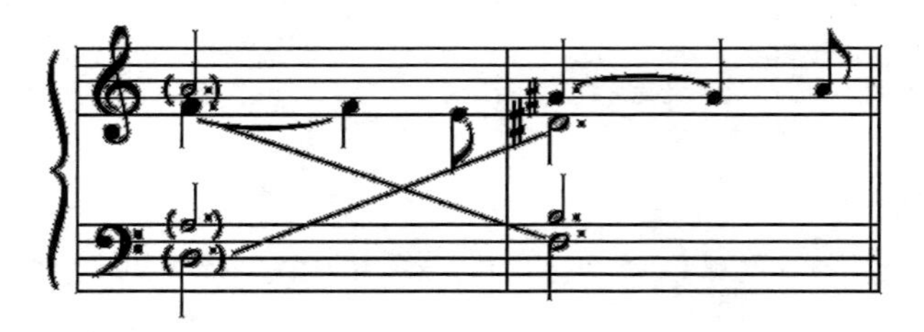

Fig. 4. *Tristan* Prelude, mm. 1-2 (after Rothstein [11])

We may model this exchange by using one such associated function:

$$X_{9,0} := (a,b) \mapsto \begin{cases} \big(b - 9 - i, a - 2 + (i + 0 \bmod 5)\big), & \text{for } 0 < i \leq 5 \\ \big(b + 9 - i, a - 2 + (i + 2 \bmod 5)\big), & \text{for } 6 < i \leq 11 \\ (b, a), & \text{for } i = 0 \bmod 6 \end{cases} \tag{6}$$

Whereas $X_{9,4}$ exchanges members of set classes [0, 1] with [0, 5], and [0, 2] with [0, 4] (as in Table 1), $X_{9,0}$ exchanges members of [0, 1] with [0, 4], and [0, 2] with [0, 3] (Table 2). Therefore, the ascending minor third in the implied bass of the example inverts to the descending diminished third in the alto.

Table 2. Exchanges under $X_{9,0}$

$6 < i \leq 11$				
IC 1	IC 2	IC 3	IC 4	IC 5
	5 ╳ 3 2 ╳ 5			
IC 4	IC 3	IC 2	IC 1	IC 5
$0 < i \leq 5$				

The resulting permutation of *T/I* orbits between functions (5) and (6) obtains from the variance in their respective modulo 5 components. In (5), we had ($i + 4$ mod 5) and ($i + 3$ mod 5), whereas in (6) we have ($i + 0$ mod 5) and ($i + 2$ mod 5). We note that for each pair, we may derive the latter addend by subtracting the former from 2 mod 5. That is, in (5), we may derive $i + 3$ from $i + 4$ by observing that $2 - 4 = 3$ mod 5. In the same way, in (6), we have $i + 0$ and $i + 2$. Again, the latter addend follows from the former: $i + (2 - 0 \bmod 5) = i + 2$. Henceforth in our generalization, let p represent the integer modulo 5 which is added to i whenever $0 < i \leq 5$, then ($2 - p$ mod 5) is added to i for $6 < i \leq 11$. Therefore, in (5), $p = 4$, and in (6), $p = 0$.

We observe that $X_{9,4}$ stabilizes set class [0, 3] as a set, while $X_{9,0}$ stabilizes [0, 5]. We may use p in determining these fixed sets, as well. In the former case, [0, 3] consists of all dyads with intervals in interval class 3; in the latter, [0, 5] consists of those in interval class 5. Then,

$$(p - (1 - p \bmod 5) \bmod 5) + 1 \tag{7}$$

yields the corresponding interval class. Hence, $p = 4$ for $X_{9,4}$; therefore, the interval class of intervals in the stabilized set class is $(4 - (1 - 4 \bmod 5) \bmod 5) + 1 = 3$. For $X_{9,0}$, $(0 - (1 - 0 \bmod 5) \bmod 5) + 1 = 5$.

2.3 Initial Harmonic Intervals: The Variable $q \in \mathbb{Z}_{12}$

All four of the examples above feature an initial harmonic interval in interval class 3: a major sixth in the first three examples, and a minor tenth in the last. Per our fourth desideratum, let (a, b) represent some melodic dyad in an exchange for which $0 < b - a \leq 5 \bmod 12$, and let (c, d) represent its image for which $6 < d - c \leq 11 \bmod 12$. Then, for our purposes, let

$$q = a - c \bmod 12 \tag{8}$$

represent this initial interval. In each of the previous examples, $q = 9$. It is possible, however, to describe voice exchanges with initial harmonic intervals in other interval classes by incorporating q into functions like (5) and (6) above.

For example, consider Figure 5, which presents m. 20 in the *Tristan* Prelude. Here we find a chromatic voice exchange for which $q = 8$ lies in interval class 4. We may model this exchange using the following function:

$$X_{8,3} := (a, b) \mapsto \begin{cases} \big(b - 8 - i, a - 1 + (i + 3 \bmod 5)\big), & \text{for } 0 < i \leq 5 \\ \big(b + 8 - i, a - 3 + (i + 4 \bmod 5)\big), & \text{for } 6 < i \leq 11 \\ (b, a), & \text{for } i = 0 \bmod 6 \end{cases} \tag{9}$$

noting that $b - 8$ and $b + 8$ in (9) now replace $b - 9$ and $b + 9$ in the former functions. In other words, $b - q$ and $b + q$ provide the generalization. Further, the previous instances of $a - 2$ in (5) and (6) now read in (9) as $a - 1$ and $a - 3$ for $0 < i \leq 5$ and $6 < i \leq 11$, respectively. These values may be derived from q, as well. For cases in which $0 < i \leq 5$, put $a - (q - 7)$; for those in which $6 < i \leq 11$, put $a - (11 - q)$. Then, as $q = 9$ in (5) and (6), we have $q - 7 = 2$ and $11 - q = 2$. In (9), $q = 8$; therefore, $q - 7 = 1$ and $11 - q = 3$.

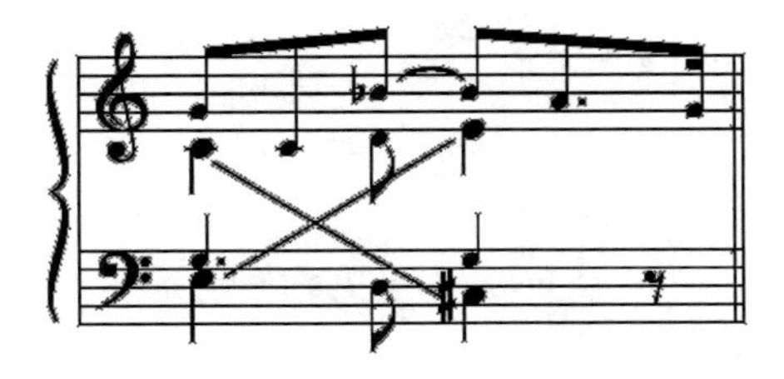

Fig. 5. *Tristan* Prelude, m. 20

Table 3 outlines the exchanges within interval classes 1 to 5 for $X_{8,3}$. In particular, we note the exchange within interval classes 4 and 3, for an initial harmonic interval $q = 8$, as seen in Figure 5. Moreover, as $p = 3$ for this example, we observe via (7) the stabilization of set class [0, 1], whose dyads have intervals in interval class (3 – (1 – 3 mod 5) mod 5) + 1 = 1.

Table 3. Exchanges under $X_{8,3}$

$0 < i \leq 5$				
IC 1	IC 2	IC 3	IC 4	IC 5
			0 4 ╳ 4 1	
IC 1	IC 5	IC 4	IC 3	IC 2
$6 < i \leq 11$				

3 Conclusions: The Group *R* and Transformational Networks

Thus far, we have described three operations on $\mathbb{Z}_{12} \times \mathbb{Z}_{12}$ in which to model voice exchanges with varying initial harmonic intervals (the variable q) and *T/I*-orbit permutations (the variable p). As q and p vary respectively within $\mathbb{Z}_{12}$ and $\mathbb{Z}_5$, we find 12 · 5 = 60 such operations in the following format:

$$X_{q,p} := (a,b) \mapsto \begin{cases} (b - q - i, a - (q - 7) + (i + p \bmod 5)), & \text{for } 0 < i \leq 5 \\ (b + q - i, a - (11 - q) + (i + (2 - p \bmod 5) \bmod 5)), & \text{for } 6 < i \leq 11 \\ (b, a), & \text{for } i = 0 \bmod 6 \end{cases} \quad (10)$$

Each of these sixty operations, then, represents twelve specific voice-leading patterns while i varies in $\mathbb{Z}_{12}$. Moreover, these operations are conjugate to each other in the centralizer of the transposition group's action on $\mathbb{Z}_{12} \times \mathbb{Z}_{12}$:

$$C(T) = C_{Sym(\mathbb{Z}_{12} \times \mathbb{Z}_{12})} T \quad (11)$$

a wreath product of order $12^{12} \cdot 12!$. Specifically, we may define an order 60 subgroup $R < C(T)$,

$$R := R_q, R_p \mid (R_q)^{12} = (R_p)^5 = 1 \quad (12)$$

where

$$R_q := (a,b) \mapsto \begin{cases} (a,b), & \text{for } i \leq 6 \\ (a+1, b+1), & \text{for } i > 6 \end{cases} \quad (13)$$

and

$$R_p := (a,b) \mapsto \begin{cases} (a,b), & \text{for } i \leq 6 \\ (a, a + 7 + (i - 1 \bmod 5)), & \text{for } i > 6 \end{cases} \quad (14)$$

isomorphic to $\mathbb{Z}_{12} \times \mathbb{Z}_5$. The set of conjugates of any $X_{q,\,p}$ under the members of R consists of all sixty operations in the form of (10). Using members of R, we may thus construct a network that relates the exchanges in all five of the previous examples (Figure 6).

$$X_{9,0} \xrightarrow{(R_p)^4} X_{9,4} \xrightarrow{R_q(R_p)^4} X_{8,3}$$

mm. 1-2 — mm. 2-3, mm. 6-7, mm. 10-11 — m. 20

Fig. 6. Network of exchanges in the *Tristan* Prelude, mm. 1-20

R provides us with a natural transformational scheme for relating these exchanges. For instance, $(R_p)^4$ gives us the specific *T/I*-orbit permutation that sends the exchange in mm. 1-2 into the three exchanges in mm. 2-11, and from those into the one in m. 20. Moreover, the first four exchanges have initial harmonic intervals in the same interval class. R_q, then, represents the particular shift in m. 20 to another initial interval. In this way, we may model additional instances of voice exchange within the Prelude, as well as in further repertoire.

References

1. Gollin, E.: Some Aspects of Three-Dimensional Tonnetze. Journal of Music Theory 42(2), 195–206 (1998)
2. Mitchell, W.J.: The Tristan Prelude: Techniques and Structure. The Music Forum 1, 162–203 (1967)
3. Harrison, D.: Supplement to the Theory of Augmented-Sixth Chords. Music Theory Spectrum 17(2), 170–195 (1995)
4. Rothgeb, J.: The Tristan Chord: Identity and Origin. Music Theory Online 1(1) (1995)
5. Rothstein, W.: The Tristan Chord in Historical Context: A Response to John Rothgeb. Music Theory Online 1(1) (1995)
6. Tymoczko, D.: Scale Theory, Serial Theory, and Voice Leading. Music Analysis 27(1), 1–49 (2008)
7. Callender, C., Quinn, I., Tymoczko, D.: Generalized Voice-Leading Spaces. Science 320, 346–348 (2008)
8. Kochavi, J.: Contextually Defined Musical Transformations. Ph.D. dissertation, State University of New York, Buffalo (2002)
9. Peck, R.: Transformational Preservation and Set-Multiclasses. In: The Thirty-first Annual Meeting of the Society for Music Theory, Nashville, Tennessee (2008)
10. Hook, J .: Uniform Triadic Transformations. Ph.D . dissertation. Indiana University-Bloomington (2002)
11. Rothstein, W.: The Tristan Chord in Historical Context: A Response to John Rothgeb. Music Theory Online 1(1) (1995)

Representing and Estimating Musical Expression in Melody*

Christopher Raphael

School of Informatics, Indiana University, Bloomington

Abstract. A method for expressive melody synthesis is presented seeking to capture the prosodic (stress, direction, and grouping) element of musical interpretation. An expressive performance is represented as a note-level annotation, classifying each note according to a small alphabet of symbols describing the role of the note within a larger context. An audio performance of the melody is represented in terms of two functions describing the time-evolving frequency and intensity. A method is presented that transforms the expressive annotation into the frequency and intensity functions, thus giving the audio performance. The problem of expressive rendering is then cast as estimation of the most likely sequence of hidden variables corresponding to the prosodic annotation. Examples are presented on a dataset of around 50 folk-like melodies, realized both from hand-marked and estimated annotations.

1 Introduction

A traditional musical score represents music *symbolically* in terms of notes, formed from a discrete alphabet of possible pitches and durations. Human performance of music often deviates substantially from the score's cartoon-like recipe, by inflecting, stretching and coloring the music in ways that bring it to life. *Expressive music synthesis* seeks algorithmic approaches to this expressive rendering task, so natural to humans.

There is really a great deal of past work on expressive synthesis — more than can be summarized here, though some of the leading authors give an overview of several important lines of work in [1]. Most past work, for example [2], [3], [4], as well as the many RENCON piano competition entries, for example [5] [6], has concentrated on piano music. The piano is attractive for one simple reason: a piano performance can be described by giving the onset time, damping time, and initial loudness of each note. Since a piano performance is easy to represent, it is easy to define the task of expressive piano synthesis as an estimation problem: one must simply estimate these three numbers for each note. In contrast, we treat here the synthesis of *melody*, which finds its richest form with "continuously controlled" instruments, such as the violin, saxophone or voice. This area has been treated by a handful of authors, including the KTH group [7], [8], as well as a

* This work supported by NSF grants IIS-0739563 and IIS-0812244.

E. Chew, A. Childs, and C.-H. Chuan (Eds.): MCM 2009, CCIS 38, pp. 236–244, 2009.

number others, including a commercial singing voice system. Continuously controlled instruments simultaneously modulate many different parameters, leading to wide variety of tone color, articulation, dynamics, vibrato, and other musical elements, making it difficult to represent the performance of a melody. However, it is not necessary to replicate any of these familiar instruments to effectively address the heart of the melody synthesis problem. We will propose a minimal audio representation we call the theremin, due to its obvious connection with the early electronic instrument by the same name [9]. Our theremin controls only time-varying pitch and intensity, thus giving a relatively simple, yet capable, representation of a melody performance.

The efforts cited above include some of the most successful attempts to date. All of these approaches map observable elements in the musical score, such as note length and pitch, to aspects of the performance, such as tempo and dynamics. One example the rule-based KTH system, which represents several decades of focused effort. In this system, each rule maps various musical contexts into performance decisions, which can be layered, so that many rules can be simultaneously applied. The rules were chosen, and iteratively refined, by a music expert seeking to articulate and generalize a wealth of experience into performance principles, in conjunction with the KTH group. In contrast, the work of Widmer [2], [4] takes a machine learning perspective by *automatically* learning rules from actual piano performances. We share the perspective of machine learning. In [4], phrase-level tempo and dynamic curve estimates are combined with the rule-based prescriptions through a case-based reasoning paradigm. That is, this approach seeks musical phrases in a training set that are "close" to the phrase being synthesized, using the tempo and dynamic curves from the closest training example. As with the KTH work, the performance parameters are computed directly from the observable score attributes with no real attempt to describe any *interpretive* goals such as repose, passing tone, local climax, surprise, etc.

Our work differs significantly from these, and all other past work we know of, by explicitly trying to represent aspects of the interpretation *itself*. Previous work does not represent the interpretation, but rather treats the *consequences* of this interpretation, such as dynamic and timing changes. We introduce a hidden sequence of variables representing the prosodic interpretation (stress and grouping) itself by annotating the role of each note in the larger prosodic context. We believe this hidden sequence is naturally positioned between the musical score and the observable aspects of the interpretation. Thus the separate problems of estimating the hidden annotation and generating the actual performance from the annotation require shorter leaps, and are therefore easier, than directly bridging the chasm that separates score and performance.

Once we have a representation of interpretation, it is possible to *estimate* the interpretation for a new melody. Thus, we pose the expressive synthesis problem as one of statistical estimation and accomplish this using familiar methodology from the statistician's toolbox. We present a deterministic transformation from our interpretation to the actual theremin parameters, allowing us to *hear* both hand labeled and estimated interpretations. We present a data set of about 50 hand-annotated melodies, as well as expressive renderings derived from both the hand-labeled and estimated annotations. A brief user study helps to contextualize the results, though we hope readers will reach independent judgments.

2 The Theremin

Our goal of expressive melody synthesis must, in the end, produce actual sound. We introduce here an audio representation we believe provides a good trade-off between expressive power and simplicity.

Consider the case of a sine wave in which both frequency, $f(t)$, and amplitude, $a(t)$, are modulated over time:

$$s(t) = a(t)\sin(2\pi \int_0^t f(\tau)d\tau). \tag{1}$$

These two time-varying parameters are the ones controlled in the early electronic instrument known as the *theremin*. Continuous control of these parameters can produce a variety of musical effects such as expressive timing, vibrato, glissando, variety of attack and dynamics. Thus, the theremin is capable of producing a rich range of expression. One significant aspect of musical expression which the theremin *cannot* capture is tone color — as a time varying sine wave, the timbre of the theremin is always the same. Partly because of this weakness, we have modified the above representation to allow tone color to change as a function of amplitude. Thus our sound is still parametrized by $f(t)$ and $a(t)$, while we increase the perceived dynamic range.

3 Representing Musical Interpretation

There a number of aspects to musical interpretation which we cannot hope to do justice to here. Palmer [10] gives a very nice overview of current thinking on this subject from the the Psychology perspective.

Our focus here is on *musical prosody* — the placing, avoidance, and foreshadowing of local (note-level) stress and the associated low-level groupings that follow. Clearly this is only a piece of the larger interpretive picture. We make this choice because we believe the notion of "correctness" is more meaningful with prosody than with other aspects of interpretation, in addition to the fact that musical prosody is somewhat easy to isolate. The music we treat consists of simple melodies of slow to moderate tempo where *legato* phrasing is appropriate. Thus the range of affect or emotional state has been intentionally restricted, though still allowing for much diversity.

We introduce now a way of *representing* the desired musicality in a manner that makes clear interpretive choices and conveys these unambiguously. Our representation labels each melody note with a symbol from a small alphabet,

$$A = \{l^-, l^\times, l^+, l^\rightarrow, l^\leftarrow, l^*\}$$

Fig. 1. *Amazing Grace* (**top**) and *Danny Boy* (**bot**) showing the note-level labeling of the music using symbols from our alphabet

describing the role the note plays in the larger context. These labels, to some extent, borrow from the familiar vocabulary of symbols musicians use to notate phrasing in printed music. The symbols $\{l^-, l^\times, l^+\}$ all denote stresses or points of "arrival." The variety of stress symbols allows for some distinction among the kinds of arrivals we can represent: l^- is the most direct and assertive stress; $l^\times$ is the "soft landing" stress in which we relax into repose; l^+ denotes a stress that continues *forward* in anticipation of future unfolding, as with some phrases that end in the dominant chord. Examples of the use of these stresses, as well as the other symbols are given in Figure 1. The symbols $\{l^\rightarrow, l^*\}$ are used to represent notes that move *forward* towards a future goal (stress). Thus these are usually shorter notes we pass through without significant event. Of these, $l^\rightarrow$ is the "garden-variety" passing tone, while l^* is reserved for the passing stress, as in a brief dissonance, or to highlight a recurring beat-level emphasis, still within the context of forward motion. Finally, the $l^\leftarrow$ symbol denotes receding movement as when a note is connected to the stress that precedes it. This commonly occurs when relaxing out of a strong-beat dissonance *en route* to harmonic stability. We will write $x = x_1, \ldots, x_N$ with $x_n \in A$ for the prosodic labeling of the notes.

These concepts are illustrated with the examples of *Amazing Grace* and *Danny Boy* in Figure 1. Of course, there may be several reasonable choices in a given musical scenario, however, we also believe that most labellings do *not* make interpretive sense and offer evidence of this is Section 7. Our entire musical collection is marked in this manner and available for scrutiny at

http://www.music.informatics.indiana.edu/papers/mcm09

4 From Labeling to Audio

Ultimately, the prosodic labeling of a melody, using symbols from A, must be translated into the amplitude and frequency functions we use for sound synthesis. We have devised a deterministic mapping from our prosodically-labeled score to the actual audio parameter outlined here.

Our synthesis of frequency begins with a literal interpretation of frequency, $f(t)$ as given by the score. To this we add vibrato, as indicated by the length of notes and by the score annotation (the prosodic labels), and pitch bending to encourage a sense of legato. Figure 2 shows a short piece of this pitch function over the transition between two notes.

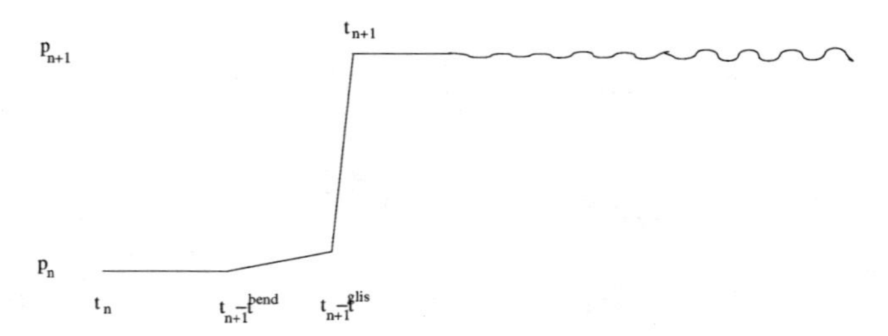

Fig. 2. A graph of the frequency function, $f(t)$, between two notes. Pitches are bent in the direction of the next pitch and make small *glissandi* in transition.

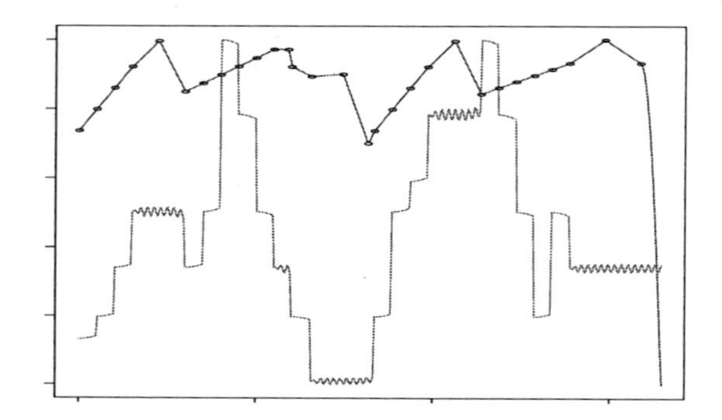

Fig. 3. The functions $f(t)$ (green) and $a(t)$ (red) for the first phrase of *Danny Boy*. These functions have different units so their ranges have been scaled to 0-1 to facilitate comparison.

The heart of the transformation, however, is in the construction of the amplitude function $a(t)$. This function is constructed through a series of soft constraints that are placed on the amplitude, constructed from the annotation and score. Through a quadratic penalty function, we encourage increasing amplitude through forward-moving notes and decreasing amplitude through receding notes. We also encourage fixed high values for the various kinds of stresses, and fixed low values for the beginning of phrases (changes in direction of the prosodic labels). While the details are many, we construct such an objective function and solve for the value of the amplitude at "knot" locations by minimizing our quadratic objective function. We then interpolate to produce a continuous $a(t)$ function. An example of both the $a(t)$ and $f(t)$ functions for a familiar examples are given in Figure 3.

5 Does the Labeling Capture Musicality?

The theremin parameters, $f(t), a(t)$, and hence the audio signal, $s(t)$, depend entirely on our prosodic labeling, x, and the musical score, through the mapping described in Section 4. We want to understand the degree to which x captures musically important interpretive notions. To this end, we have constructed a dataset of about 50 simple melodies containing a combination of genuine folk songs, folk-like songs, Christmas carols, and examples from popular and art music of various eras. The melodies were chosen to have simple chords, simple phrase structure, all at moderate to slow tempo, and appropriate for *legato* phrasing, and to be widely known. Examples include *Danny Boy*, *Away in a Manger*, *Loch Lomond*, *By the Waters of Babylon*, etc. These melodies were painstakingly hand-annotated by the author.

We rendered these melodies into audio according to our hand-marked annotations and the process of Section 4. For each of these audio files we provide harmonic context by superimposing sustained chords, as indicated in the scores. The entire collection of symbolic melodies along with audio files describing this synthesis is available at the aforementioned web site.

We do observe some aspects of musical interpretation that are not captured by our representation. For example, the interpretation of *Danny Boy* clearly requires a climax at the highest note, as do a number of the musical examples. We currently do not represent such an event through our markup. It is possible that we could add a new category of stress corresponding to such a highpoint, though we suspect that the degree of emphasis is continuous, thus not well captured by a discrete alphabet of symbols. Another occasional shortcoming is the failure to distinguish contrasting material, as in *O Come O Come Emmanuel.* This melody has a Gregorian chant-like feel and should mostly be rendered with deliberate calmness. However, the short outburst corresponding to the word "Rejoice" takes on a more declarative affect. Our prosodically-oriented markup simply has no way to represent such a contrast of styles. There are, perhaps some other general shortcomings of the interpretations, though we believe there is quite a bit that is "right" in them, especially considering the simplicity of our representation of interpretation.

6 Estimating the Interpretation

The essential goal of this work is to *algorithmically* generate expressive renderings of melody. Having formally represented our notion of musical interpretation, we can generate an expressive rendering by *estimating* the hidden sequence of note-level annotations, $x_1, \ldots, x_N$. Our estimation of this unobserved sequence will be a function of various observables, $y_1, \ldots, y_N$, where the feature vector $y_n = y_n^1, \ldots, y_n^J$ measures various attributes of the musical score at the nth note.

The features we consider come mostly from surface-level attributes of the musical score. While a great many possibilities were considered, we ultimately culled the set to the metric strength of the onset position, the first difference of note length in seconds, the first difference of pitch.

Our fundamental modeling assumption is that our label sequence has a Markov structure, given the data:

$$p(x|y) = p(x_1|y_1) \prod_{n=2}^{N} p(x_n|x_{n-1}, y_n, y_{n-1}) \tag{2}$$
$$= p(x_1|y_1) \prod_{n=2}^{N} p(x_n|x_{n-1}, z_n)$$

where $z_n = (y_n, y_{n-1})$. The intuition behind this assumption is the observation (or opinion) that much of phrasing results from a cyclic alternation between forward moving notes, $\{l^{\rightarrow}, l^*\}$, stressed notes, $\{l^-, l^+, l^{\times}\}$, and optional receding notes $\{l^{\leftarrow}\}$. Usually a phrase boundary is present as we move from either stressed or receding states to forward moving states. Thus the notion of *state*, as in a Markov chain, seems to be relevant. However, it is, of course, true that music has hierarchical structure *not* expressible through the regular grammar of a Markov chain.

We estimate the conditional distributions $p(x_n|x_{n-1}, z_n)$ for each choice of $x_{n-1} \in A$, as well as $p(x_1|y_1)$, using our labeled data. We will use the notation

$$p_l(x|z) = p(x_n = x|x_{n-1} = l, z_n = z)$$

for $l \in A$. In training these distributions we split our score data into $|A|$ groups, $D_l = \{(x_{li}, z_{li})\}$, where D_l is the collection of all (class label, feature vector) pairs over all notes that immediately follow a note of class l.

Our estimation method makes no prior simplifying assumptions and follows the familiar classification tree methodology of CART [11]. That is, for each D_l we begin with a "split," $z^j > c$ separating D_l into two sets: $D_l^0 = \{(x_{li}, z_{li}) : z_{li}^j > c\}$ and $D_l^1 = \{(x_{li}, z_{li}) : z_{li}^j \leq c\}$. We choose the feature, j, and cutoff, c, to achieve maximal "purity" in the sets D_l^0 and D_l^1 as measured by the average entropy over the class labels. We continue to split the sets D_l^0 and D_l^1, splitting their "offspring," etc., in a greedy manner, until the number of examples at a tree node is less than some minimum value. $p_l(x|z)$ is then represented by finding the terminal tree node associated with z and using the empirical label distribution over the class labels $\{x_{li}\}$ whose associated $\{z_{li}\}$ fall to the same terminal tree node.

Given a piece of music with feature vector $z_1, \ldots, z_N$,we can compute the optimizing labeling

$$\hat{x}_1 \ldots, \hat{x}_N = \arg \max_{x_1,\ldots,x_N} p(x_1|y_1) \prod_{n=2}^{N} p(x_n|x_{n-1}, z_n)$$

using dynamic programming.

7 Results

We estimated a labeling for each of the $M = 50$ pieces in our corpus by training our model on the remaining $M - 1$ pieces and finding the most likely labeling, $\hat{x}_1, \ldots, \hat{x}_N$, as described above. When computing the most likely labeling for each melody in our corpus we found a total of 678/2674 errors (25.3%) with detailed results as presented in Figure 4.

	l^*	$l^{\rightarrow}$	$l^{\leftarrow}$	l^-	$l^\times$	l^+	total
l^*	135	112	0	18	2	0	267
$l^{\rightarrow}$	62	1683	8	17	0	0	1770
$l^{\leftarrow}$	3	210	45	6	2	0	266
l^-	49	48	4	103	15	0	219
$l^\times$	5	32	2	65	30	0	134
l^+	0	3	0	12	3	0	18
total	254	2088	59	221	52	0	2674

Fig. 4. Confusion matrix of errors over the various classes. The rows represent the true labels while the columns represent the predicted labels. The block structure indicated in the table shows the confusion on the coarser categories of stress, forward movement, and receding movement.

The notion of "error" is somewhat ambiguous, however, since there really is no correct labeling. In particular, the choices among the forward-moving labels: $\{l^*, l^{\rightarrow}\}$, and stress labels: $\{l^-, l^\times, l^+\}$ are especially subject to interpretation. If we compute an error rate using these categories, as indicated in the table, the error rate is reduced to 15.3%.

One should note a mismatch between our evaluation metric of recognition errors with our estimation strategy. Using a forward-backward-like algorithm it is possible to compute $p(x_n|y_1, \ldots, y_N)$. Thus if we choose

$$\bar{x}_n = \arg\max_{x_n \in A} p(x_n|y_1, \ldots, y_N),$$

then the sequence $\bar{x}_1, \ldots, \bar{x}_N$ minimizes the expected number of estimation errors

$$E(\text{errors}|y_1, \ldots, y_N) = \sum_n p(x_n \neq \bar{x}_n|y_1, \ldots, y_N)$$

We have not chosen this latter metric because we want a *sequence* that behaves reasonably. It the sequential nature of the labeling that captures the prosodic interpretation, so the most likely sequence $\hat{x}_1, \ldots, \hat{x}_n$ seems like a more reasonable choice.

In an effort to measure what we believe to be *most* important — the perceived musicality of the performances — we performed a small user study. We took a subset of the most well-known melodies of the dataset and created audio files from the random, hand, and estimated annotations. We presented all three versions of each melody to a collection of 23 subjects who were students in the Jacobs School of Music at Indiana University, as well as some other comparably educated listeners. We regard the cohort as highly educated and sophisticated, musically speaking. The subjects were presented with random orderings of the three versions, with different orderings for each user, and asked to respond to the statement: "The performance sounds musical and expressive" with the Likert-style ratings 1=strongly disagree, 2=disagree, 3=neutral, 4=agree, 5=strongly agree, as well as to rank the three performances in terms of musicality (the ranking does not always follow from the Likert ratings). Out of a total of 244 triples that were evaluated in this way, the randomly-generated annotation received a mean score of 2.96 while the hand and estimated annotations received mean scores of 3.48 and 3.46. The rankings showed no preference for the hand annotations over the estimated annotations ($p = .64$), while both the hand and estimated annotations were clearly preferred to the random annotations ($p = .0002$, $p = .0003$).

Perhaps the most surprising aspect of these results is the high score of the random labellings — in spite of the meaningless nature of these labellings, the listeners were, in aggregate, "neutral" in judging the musicality of the examples. We believe the reason for this is that musical prosody, the focus of this research, accounts for only a portion of what listeners respond to. All of our examples were rendered with the same sound engine of Section 4 which tries to create a sense of smoothness in the delivery with appropriate use of vibrato and timbral variation. We imagine that the listeners were partly swayed by this appropriate *affect*, even when the use of stress was not satisfactory. The results also show that our estimation produced annotations that were, essentially, as good as the hand-labeled annotations. This demonstrates, to some extent, a success of our research, though it is possible that this also reflects a limit in the expressive range of our interpretation representation. Finally, while the computer-generated interpretations clearly demonstrate some musicality, the listener rating of 3.46 — halfway between "neutral" and "agree" — show there is considerable room for improvement.

References

1. Goebl, W., Dixon, S., De Poli, G., Friberg, A., Bresin, R., Widmer, G.: Sense in expressive music performance: Data acquisition, computational studies, and models, ch. 5, pp. 195–242. Logos Verlag, Berlin (2008)
2. Widmer, G., Goebl, W.: Computational models for expressive music performance: The state of the art. Journal of New Music Research 33(3), 203–216 (2004)
3. Todd, N.P.M.: The kinematics of musical expression. Journal of the Acoustical Society of America 97(3), 1940–1949 (1995)
4. Widmer, G., Tobudic, A.: Playing Mozart by analogy: Learning multi-level timing and dynamics strategies. Journal of New Music Research 33(3), 203–216 (2003)
5. Hiraga, R., Bresin, R., Hirata, K., Katayose, H.: Rencon 2004: Turing Test for musical expression. In: Proceedings of the 2004 Conference on New Interfaces for Musical Expression (NIME 2004), pp. 120–123 (2004)
6. Hashida, Y., Nakra, T., Katayose, H., Murao, Y.: Rencon: Performance Rendering Contest for Automated Music Systems. In: Proceedings of the 10th Int. Conf. on Music Perception and Cognition (ICMPC 10), Sapporo, Japan, pp. 53–57 (2008)
7. Sundberg, J.: The KTH synthesis of singing. Advances in Cognitive Psychology. Special issue on Music Performance 2(2-3), 131–143 (2006)
8. Friberg, A., Bresin, R., Sundberg, J.: Overview of the KTH rule system for musical performance. Advances in Cognitive Psychology 2(2-3), 145–161 (2006)
9. Roads, C.: The Computer Music Tutorial. MIT Press, Cambridge (1996)
10. Palmer, C.: Music Performance. Annual Review Psychology 48, 115–138 (1997)
11. Breiman, L., Friedman, J., Olshen, R., Stone, C.: Classification and Regression Trees. Wadsworth and Brooks, Monterey (1984)

Evaluating Tonal Distances between Pitch-Class Sets and Predicting Their Tonal Centres by Computational Models

Atte Tenkanen

Abstract. The pitch-class set belongs to the core concepts within musical set theory. The mathematical properties of pitch-class sets (in terms of interval-class content, evenness, etc.) as well as their mutual relations to other sets have been widely studied. In this paper, we concentrate on investigating them as carriers of tonal implications. Results provided by four algorithmic models, which propose hypothetical tonal centres for pitch-class sets, are compared. In addition to finding reference pitch class(es) for each set class of cardinality 3-9, the models are used for evaluating tonal distances between pitch-class sets. They are applied as 'similarity measures' in conjunction with an automated, computer-aided analysis method called comparison set analysis.

1 Introduction

In its narrowest sense, the concept of tonality is encapsulated within the concept of tonal centre (TC). Following Huovinen [1, xvii], the tonal centre is defined as a 'reference pitch class that attains the greatest stability in a musical passage or in a tonally perceived local musical object'. A pitch-class set (PCS) may be seen as such an object. Traditionally, music theorists have not conceived of pitch-class sets primarily as carriers of tonal implications. Instead, the discussions have centred on their symmetrical properties, interval-class contents and other features that are easily verified. However, all PCSs except the empty set and the chromatic aggregate are –at least in theory– able to induce tonal implications [2].

There are two main aims in this paper: four different algorithmic models are applied 1) to predict the tonal centre(s) for any unordered PCS and 2) to evaluate a 'tonal distance' or 'tonal stability' between two PCSs. Our models take a PCS as an input vector and produce a 12-dimensional vector, which includes resulting weights related to each pitch class (0-11). For the first aim, one pitch-class set is entered into the model (see Figure 1a). An index (0-11), which attains the greatest value in the resulting vector, is a hypothetical TC of the PCS. In order to evaluate the tonal distance between two PCSs, their 12-dimensional resulting vectors are compared using the correlation distance. Both procedures have been exploited using similar algorithms (c.f.[3], [4]). The latter approach is used in conjunction with *comparison set analysis* (CSA) [5] in Section 4. The main aim in CSA is to create representations of extensive musical surfaces that, for their part, expose the prevalence of a particular *comparison set* throughout

E. Chew, A. Childs, and C.-H. Chuan (Eds.): MCM 2009, CCIS 38, pp. 245–257, 2009.

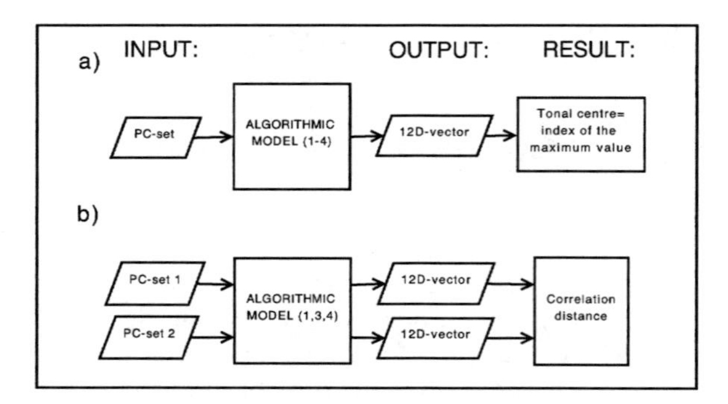

Fig. 1. Two applications that are based on the algorithmic models. The first application (a) produces a hypothetical tonal centre or centres for any pitch-class set. Another application (b) evaluates the tonal distance between two pitch-class sets.

a lengthy musical piece[1]. Instead of using a set class as a comparison set, a pitch-class set (containing transposition information) is applied in the present context.

Unlike in other well-known cases of key-finding algorithms, like those of Krumhansl and Schmuckler [4] and Chew [6], which are especially taken to reflect cognitive pitch-class (pc) hierarchies in tonal contexts, we consider PCSs independently of the key contexts.

Our algorithmic models, presented in Sections 2.1-2.4, consist of two parts: 1) *hypotheses*, and 2) a function that *generalizes* the tonal properties of the hypotheses across all the pitch-class sets. The hypotheses are either intuitively selected 'constraints' (Figure 2) or based on empirical results. They are used in conjunction with three linear models (in algorithms 1, 3 and 4) and with one function based on the Circle-Of-Fifths (2). The first three algorithms utilize methods applied in the field of artificial intelligence: these models are *trained* using the hypotheses. The first and fourth algorithms are modified versions of models introduced by Parncutt [3,2]. The second and the third algorithmic model are developed by the authors. The information presented above is compiled in Table 1.

Table 1. The basic properties of the four algorithmic models presented in Sections 2.1-2.4

Section	Hypotheses	Math. model	Is it trained?	Origin
2.1	'Tonal' constraints (Fig. 2)	Linear function	Yes	[3]
2.2	'Tonal' constraints (Fig. 2)	'COF-relation'	Yes	[7]
2.3	Empirical results [1]	Linear function	Yes (MLP)	New
2.4	(K-K profiles)	Linear function	No	[2]

[1] This might be, for example, a pitch-class collection or a numerical string of rhythmic units.

By using several different models, we aim to obtain more generic results than is possible when using only a couple of them. In fact, the hypotheses and models could be combined in other ways as well in order to produce even more alternatives. Regarding the tonal distance applications, if the results significantly agree they may be safely applied to CSA.

However, we do not suggest that there are 'correct' and unambiguous answers as far as TCs are concerned. On the contrary, we see alternative solutions as being equally plausible, in the same way that different listeners may have diverse strategies for selecting a TC for a chord [1, x]. The degree to which the results are reliable is an issue that requires further comparisons between the model predictions and the results of perceptual and cognitive experiments.

In Section 3, we discuss differences between the results provided by the different methods and explain our findings in Table 4 (Appendix C) which presents TC candidates for some sample set classes[2] along with some other information in condensed form. Although all SCs of cardinality ranging from 3 to 9 are used to make comparisons between all the approaches (Fig. 5) we have included only some results involved in the SCs of cardinality 3 and 6 for demonstration purposes. Finally in Section 4, we apply the distance-based method to tonal analysis of the *Intermezzo* from the opera *Wozzeck* by Alban Berg in order to show that the methods are intended to apply to all kinds of pitch class sets, not only to sets familiar from tonal contexts, which has often been the case in studies on tonality induction.

2 Algorithmic Models

2.1 Training the Weights of a Linear Polynomial with Tonal Constraints

Parncutt [3,2] uses a first-order polynomial $W(t) = pw^{(t)}$ for predicting the perceptual root(s) of a PCS. Vector w contains 'root-support weights'. In [8] he proposes w as {10,0,1,0,3,0,0,5,0,0,2,0}, which are estimations of the root-support weights for the ordered intervals between the fundamental of a harmonic complex tone and its first eight harmonics. p denotes the components of a PCS as a binary vector: for example, in the case of C-major triad $p = \{1,0,0,0,1,0,0,1,0,0,0,0\}$. t $(t = 0...11)$ indicates a cyclic permutation of w. Thus, the result vector W is a series of pc-weights. For the C-major triad $W = \{18,0,3,3,10,6,2,10,3,7,1,0\}$, which means that when a C-major triad is sounded, C $(W = 18)$ is more salient than, for example, E or G $(W = 10)$[3].

We used Parncutt's model in our first algorithm, but instead of estimating the weights using the overtone approximations, the procedure was turned around: the weights were 'tuned' anew by defining the desired TCs for a limited amount of PCSs and by requiring that the function satisfies all these *constraints*. This

[2] I.e., prime-form PCSs.

[3] E.g. for the C-major triad, $W(0) = pw^{(0)}$:
$\{1,0,0,0,1,0,0,1,0,0,0,0\} * \{10,0,1,0,3,0,0,5,0,0,2,0\} = 10 + 3 + 5 = 18$.

Fig. 2. The selected constraints used in the first algorithm

kind of procedure is called a *constraint satisfaction problem* (CSP)[4]. We defined our constraints by using the C-major scale, the harmonic a-minor scale, and some of their subsets as referential sets: the most common chord types in tonal music (major, minor, diminished and augmented triads, plus a dominant 7th chord) are represented together with their 'fundamentals,' or the degree to which they most probably resolve themselves (Figure 2). There may be other alternatives for the constraints.

After training the model $5 * 10^6$ times by assigning random values between 0 and 100 to each weight $w(0)...w(11)$ and selecting only those alternatives that fulfil all seven of the constraints, the normalized vector of averaged weights (i.e. w) converged to {0.144, 0.052, 0.106, 0.095, 0.094, 0.125, 0.043, 0.129, 0.033, 0.037, 0.036, 0.106}[5]. The new weighting vector seems to emphasize the pure fifths between pc5-pc0 (i.e. F and C) and p0-pc7 (C and G), which form the first pitches in the diatonic chain of fifths and, at the same time, the rarest interval class of the diatonic scale, i.e. the tritone between pc5 and pc11, which can be interpreted as an interval between the outermost pitch classes of the diatonic chain of fifths.

2.2 Circle-of-Fifths-Based Algorithm

Our second algorithm utilizes the Circle-of-Fifths (COF), which has been seen modelling tonal distances between pitch classes, TCs or keys [4]. If the COF-distance between two pitch classes a fifth apart is defined as 1, it follows that the most distant pitch classes (e.g. C and F#) would have a COF-distance of 6[6]. All COF-distances between pairs can be represented in a *COF-distance profile*: $(0, 5, 2, 3, 4, 1, 6, 1, 4, 3, 2, 5)$.

The COF-based relation between any two PCSs A and B, proposed by Tenkanen and Gualda [7], is defined as

$$cofrel(A, B) = \sqrt{\sum_{i \in A, j \in B} (c_{ij})^2} / \sqrt{(|A| * |B|)}$$

[4] "Basically, a CSP is a problem composed of a finite set of variables, each of which is associated with a finite domain, and a set of constraints that restricts the values the variables can simultaneously take. The task is to assign a value to each variable satisfying all the constraints." [9, 1] A more precise and formal explanation can be found in [10, 137].

[5] The resulting vector of a Monte Carlo process was normalized by the sum of the vector components.

[6] I.e., by using the generator $\langle 7 \rangle$ of a cyclic group Z_{12} under addition modulo 12, all pitch classes can be obtained from any given pitch class. This allows for distance measurements between any two pitch classes.

where c_{ij} includes all the COF-distances between the PCSs A and B. To demonstrate, the *cofrel*-value between the PCSs
$A = \{0, 4, 7, 10\}$ and $B = \{0, 5, 9\}$ is

$$(\sqrt{0^2 + 1^2 + 3^2 + ... + 2^2 + 1^2 + 5^2})/\sqrt{(4 * 3)} \approx 2.75.$$[7]

If a single COF-distance is interpreted as representing the degree of tonal stability between the two pitch classes, the *cofrel*-value may be interpreted as the normalized sum of stability values between the two PCSs: the smaller the value, the greater the tonal stability between the PCSs. It must be noted that the COF-relation does not obey any metric axioms other than symmetry and non-negativity.

In order to apply the COF-relation for evaluating reference pitch classes in PCSs, either A or B has to be maintained as a *constant reference set* (but allowing for all 12 transpositions in calculating hypothetical TCs), i.e. our TC-algorithm consists of the cofrel-function and a suitable reference set. A valid set was found by utilizing our constraints introduced in the previous section. This time, we entered all Tn-type SCs and their transpositions to the COF-relation function and asked the procedure to investigate in each case whether or not all of the seven constraints were fulfilled. Both PCS {0,2,3,4} and {2,3,4,7,8,9,10} fulfilled them all[8].

In our previous model (Section 2.1), the learning information was condensed into the vector of weights, but in the present case it is included in the pitch classes of the reference set. The first reference set {0,2,3,4} will be used in comparisons in Section 3 because it is in closer agreement with the results achieved by the other three models. Figure 3 represents the system in its entirety.

2.3 Training a Neural Network with Empirical Results

The previous algorithms were based on the constraint satisfaction problem, in which the constraints were selected intuitively. On the other hand, why not use empirical results that directly reflect intuitive knowledge related to the TCs of PCSs?

For his studies concerning the perception of tonal centricity, Huovinen [1] carried out six experiments using specially-composed samples of pitch strings derived from particularly selected SCs. He tested his subjects individually, and asked them to sing or hum what they felt to be the most stable tone in relation to the tones they heard [1, 108]. He noticed that there is no single cognitive structural basis for the perceptual phenomenon that we call tonality, even if the principal importance of P4 and P5 intervals was confirmed.[9]

[7] I.e., the COF-distances between pc:s (0,0)=0, (0,5)=1, (0,9)=3,..., (10,0)=2, (10,5)=1 and (10,9)=5.

[8] Thus, pitch classes 1, 5, 6, and 11 were missing.

[9] Nearly half of Huovinen's subjects were students and colleagues at Turku University's Department of Musicology, and the rest ranged from music amateurs to professional musicians. [1, p. 249, 283].

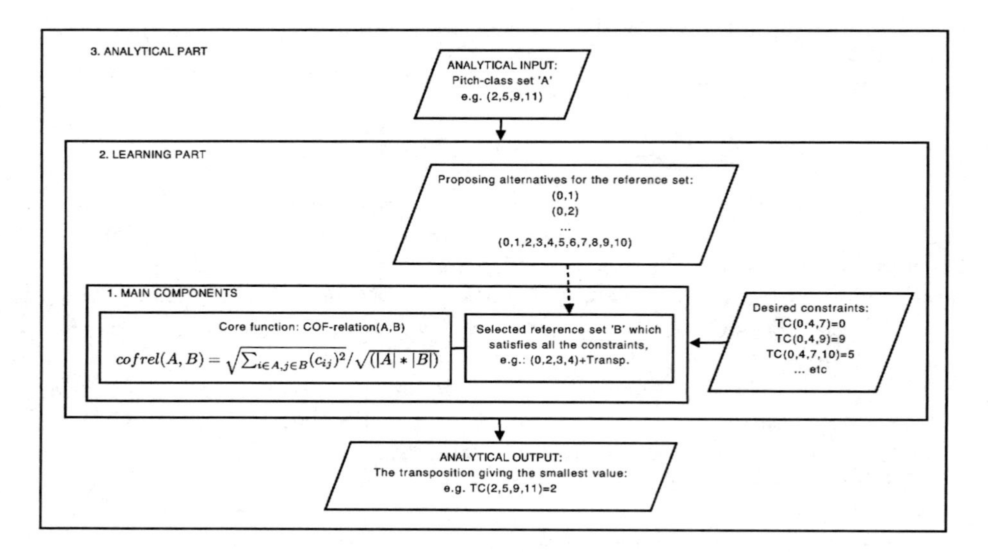

Fig. 3. The COF-based algorithm in its details. In the example, the algorithm proposes pc2 (d) as a TC for the d-minor triad with an added major sixth {2,5,9,11}.

We selected three response distributions[10] from Huovinen's study [1, 258, 259, 290] (table 9.4.1 from experiment 5 is included in Table 2, Appendix A) and used them to train a multilayer perceptron (MLP) with one hidden layer and a back-propagation algorithm.

Each row in the response distribution tables now represented an *input pattern (pc-set components) – desired output pattern (response values) pair*. Only the three greatest values from the rows were taken into consideration to reduce 'noise'. The components of an input pattern were entered as a binary vector. For an example of pcset–response values pair, look at Figure 4. The PCS 6-Z19A, i.e. {0,1,3,4,7,8} and the three greatest values associated with it are found in Table 2. These pairs were entered in random order 5000 times[11] into the network during one training period. The results given by the other three models were used as a reference set[12] for finding the optimal parameters for the network[13].

---

[10] These were tables 9.4.1, 9.4.2 and 10.3.1.

[11] We found that 5000 epochs was enough to attain optimal fitting.

[12] Such a procedure forms a kind of 'circular argument' in this context, but, since there are no 'right' extrinsic answers as far as the tonal centres in pitch-class sets are concerned, we did not have any other alternatives.

[13] In the MLP, the number of input and output neurons is determined by the data. In this case the number of input and output neurons was 12 (see Figure 4). The number of hidden neurons, which determines the fitting of data accuracy, was set randomly to 20-40 in each training period. A sigmoid $f(x) = \frac{1}{1+exp(-x)}$ was used as an activation function in all the mlp-neurons. A *learning coefficient* was initially set to 0.7 but its value was decreased smoothly towards 0 during each iteration.

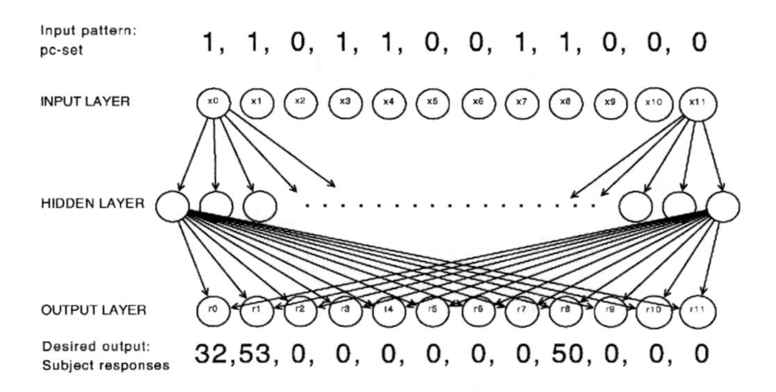

Fig. 4. Training the MLP, used in the third TC-algorithm. Binary vector {1,1,0,1,1,0,0,1,1,0,0,0} is associated with the SC 6-Z19A, cf. the first row of Table 2, Appendix A.

After training the network weights, all of the SCs of cardinality 3 to 9 were entered into the network and for each of them a vector of TC-weights was calculated. This procedure was repeated 300 times and, finally, a normalized mean vector was assigned for each SC. These ('tonal profile-') vectors were used as the basis for the information in column 5 of Table 4 in Appendix C.

2.4 The Tonal Profile of a PCS as a Weighted Mean of KK-Profiles

Our last algorithm also originates from studies by Parncutt [2]. It is based on the Krumhansl-Kessler (K-K) key profiles, which are taken to reflect important cognitive pitch-class hierarchies in a tonal context: in all major and minor K-K profiles, each pitch class is assigned a relative numerical value [4]. The idea of the algorithm is that a PCS can be heard in any key, but with different probabilities, and that K-K profile values are seen as probability values. The tonal profile of a PCS is calculated as a *weighted mean of all 24 K-K profiles.* The tonal profile *tp* for a PCS *pcset* is

$$tp = \frac{v_{pcset} M}{|pcset|}$$

where the PCS is represented as a row matrix $v_{pcset} = (v_0, v_1, ...v_{11})$. The value of v_i is 1 if there is a pitch class i in the PCS, otherwise the entry is 0. For example, for PCS {2,5,9,11} $v_{pcset} = (0, 0, 1, 0, 0, 1, 0, 0, 0, 1, 0, 1)$. The columns of the matrix M consist of all 24 K-K profiles from the C-major profile to the b-minor profile[14]. The values are standardized by the length of the pcset-vector. The d-minor triad with added major sixth {2,5,9,11} produces a tonal profile

[14] The *key profile* for C major is (6.35, 2.23, 3.48, 2.33, 4.38, 4.09, 2.52, 5.19, 2.39, 3.66, 2.29, 2.88) and for C minor (6.33, 2.68, 3.52, 5.38, 2.60, 3.53, 2.54, 4.75, 3.98, 2.69, 3.34, 3.17) [11].

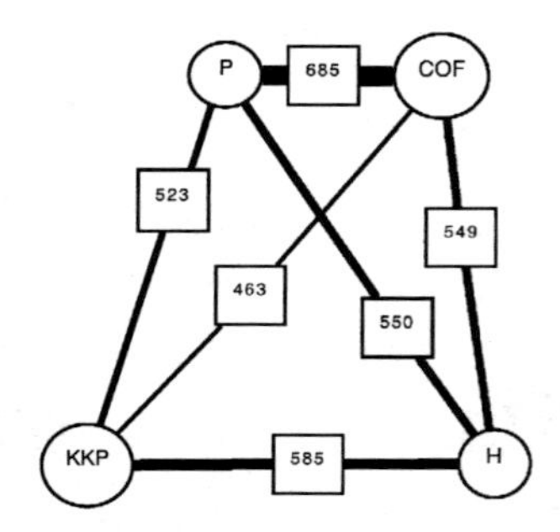

Fig. 5. The number of TC candidates in common between the four algorithms. P refers to the first, COF to the second, H to the third and KKP to the fourth algorithm, as presented in Section 2.

(3.53, 2.82, 4.38, 2.82, 3.45, 4.23, 2.92, 3.84, 2.68, 4.08, 3.67, 3.37, 3.23, 3.15, 4.79, 3.30, 3.58, 3.54, 4.01, 3.55, 3.32, 4.34, 3.30, 4.40). Hence, the greatest probability values are assigned to d minor (4.79).

3 Comparing and Combining Predictions

In order to compare our models, we entered all the SCs of cardinality 3-9 (336 pcs) into the four TC-algorithms (henceforth abbreviated as P, COF, H and KKP) and selected the three most probable TCs[15] predicted by each algorithm for each SC. The resulting values concerning some selected trichordal and hexachordal SCs are represented in Table 4, columns 3-6. The leftmost value always denotes the most probable TC according to the algorithm, and so forth.

The number of TC candidates in common between each pair of algorithms is given in Figure 5. The first two models have more candidates in common ($\frac{685}{1008}$ = 68%). The *order* of three TC candidates is not taken into account. Although the models are quite different, using the same constraints in both algorithms seems to produce similar results. The model based on KKP-profiles seems to differ most from the others. All of the algorithms propose the same three TC candidates for four SCs (for SCs 4-7, 6Z-19B, 6-20 and 7-6A).[16]

[15] That means 3*336=1008 TC candidates per algorithm, except that the last algorithm might introduce the same TC two times, i.e. for the same major and minor tonality, e.g. F major and f minor in the case of SC 3-4A. See Table 4 in Appendix C, the KKP column. It produced 948 different TC candidates altogether.

[16] If we think about the consistency of the results across four algorithms, four cases out of 336 (1.1%) appear to be quite insignificant. On the other hand, if all of the TC-combinations consisting of three candidates were equally probable in the results for each algorithm, the theoretical probability that they all would give the same three candidates would be as small as $1/\binom{12}{3}^3 = 9.4*10^{-8}$. It has to be remembered, however, that the algorithms are based on different underlying assumptions and, thus, they 'speak with a different voice'.

Which pitch classes are then the best TC candidates for a SC according to the results? For example, for SC 3-1 there are 6 candidates altogether (see Appendix C, the last 'dispersion'-column refers to the number of different TC candidates predicted by these models): 9, 0, 1, 7, 2 and 6. To answer the question about the best TC candidates for a SC, we ranked all the candidates in each cell and counted 'points': the best candidates for SC 3-1 are thus pc0 and pc2 (see the 'Rank' column), because pc0 is ranked once in the first position (3 points) and twice in the second position (2*2 points), which gives 7 out of a maximum 12 points (2+3+0+2). The second candidate, pc2, also scores 7 points and the third candidate, pc7, scores 4 points.

In addition, we checked if the winning candidates form pure fifths between them. There is a p5 between pc0 and pc7 as well as between pc7, and pc2 for SC 3-1. Thus, both of their 'bottom pitch classes', pc0 and pc7, are marked as 'tonic' in Tonic-column. Finally, we observed the difference between the tonic candidates according to their position in the SC: tonic pitch classes that are not members of that SC are underlined. A SC that exhibit such a quality is paralleled by the dominant chord.

To present a more challenging example, we considered the TC candidates of the *Promethean* SC 6-34A. The first and second algorithms rank pc10 as the strongest TC but the third and fourth algorithms rank pc9 and pc0 instead. Jim Samson [12, 156-7] has pointed out that in Scriabin's music this chord may take on a dominant quality of Eb or A (i.e. pc3 or pc9, Samson's sample chord is in this case in prime form transposition). Nevertheless, we can accept the pc10-interpretation given by the models as well, by playing the *Promethean* chord and the Bb-note or Bb-major triad successively.[17] The authors are not aware of whether or not Scriabin used his *Promethean* chord in this way: that would be an issue for further study.

4 Is Alban Berg's *Invention on a Key* in D Minor?

In the previous two sections, we have concentrated on investigating tonality within the context of the most local harmonic objects, pitch-class sets. TCs were calculated according to resulting 12-dimensional profiles produced by the algorithms. Such profiles can also be used to evaluate the tonal similarity between two PCSs, as represented in Figure 1. This is done most easily by calculating a correlation between two profiles. We thus define a *correlation distance* between profiles x and y as $d(x,y) = 1 - corr(x,y)$, where *corr* stands for the Pearson correlation. The *cofrel*-value of the second algorithm was defined as the stability between two PCSs and, as such, it can be applied to a similar task. Instead of finding hypothetical TCs for separate PCSs in a musical piece, the correlation

[17] In theory, this may however lead to a voice-leading problem: if the sample chord is followed by the Bb-major triad, C# has to progress to D (?) and thus, C, for its part, to Bb. Since G, most probably, would move to F, this may lead to the parallel fifths (C-Bb, G-F) depending on the arrangement of the pitches in the Promethean chord. On the other hand, this may be avoided by using a suspension.

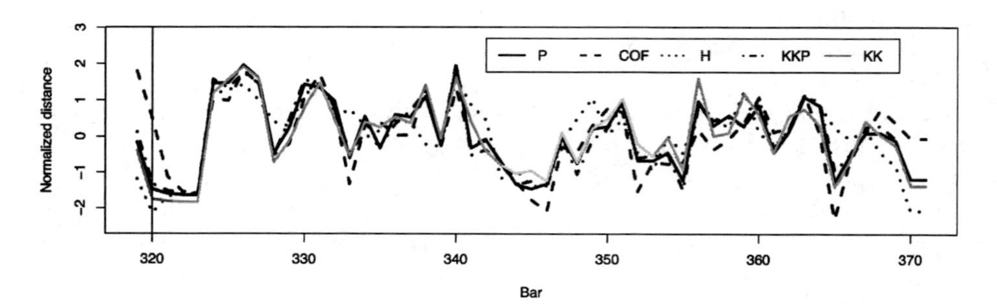

Fig. 6. Alban Berg: *Intermezzo* from *Wozzeck*. Correlation distances (P, H, KKP) and tonal stability values (COF) calculated using the harmonic d-minor collection {1,2,4,5,7,9,10} as a comparison set. The comparison curve produced by using the K-K d-minor profile is also added to the figure (KK, gray line).

distance and the cofrel-stability offer a more flexible way to evaluate tonalities in the piece. A suitable *comparison set* is required against which musical segments, derived from the piece, are compared.

For analysis, we chose the movement *Intermezzo* ('Invention on a Key (d minor)') from the opera *Wozzeck* by Alban Berg. It is found in the fourth Scene of the third Act, bars 320-371. We wanted to assess to what extent the d-minor tonality actually occurs in the section.

To begin with, all of the note onsets of the *Intermezzo* were clustered into PCSs of cardinality 7, according to their temporal proximity to their nearest neighbours. In practice, that means that a heptachord consisting of the nearest pitch classes was assigned to each *unique note onset time* (we follow here the segmentation method introduced by in [13]). These heptachordal segments were compared with the harmonic d-minor scale {1,2,4,5,7,9,10} using the correlation distance along with the first (P), the third (H), the fourth (KKP) algorithm and the cofrel-stability function (COF) as such. Musical properties like duration, voicing and loudness were not taken into consideration. Thereafter, the correlation distance values assigned to individual segments were averaged over each bar and the resulting values were normalized to a zero mean and unit variance for the purpose of comparability. To facilitate comparison, we added a curve by using a K-K d-minor profile. For each pitch-class in *each heptachordal segment*[18] was assigned a value according to the d-minor key profile and these values were averaged over each bar. Thus, five different curves, seen in Figure 6, were produced.

D minor seems to dominate the tonality in the beginning and the end of the section as well as in bars 345 and 365. All five approaches –even the COF-relation, which is based on a purely mathematical model– seem to correlate strongly with each other. The result is confirmed numerically through correlation estimates, calculated between the different curves (Table 3, Appendix B).

Another question is: How would the average distance-values look like if some other transposition of the harmonic minor scale or another type of reference

[18] Thus, the basis of calculations was equal to all approaches.

set is used instead? After using every transposition of the harmonic minor scale and calculating the mean of bar averages (without normalization), the d-minor collection was found to produce the lowest mean value. Furthermore, when all heptachordal PCSs were entered into the procedure, the lowest mean value was attained by the PCS {0,1,2,5,7,9,10}, which differs from the harmonic d-minor scale only in that pc4 is replaced by pc0. Although the music is quite chromatic[19] throughout, the d-minor key thus not only articulates it at the beginning and the end but also dominates it in terms of statistical significance.

5 Conclusions

We have considered PCSs as carriers of tonal implications, independent of musical context. By using four different algorithmic models, we predicted TCs for all SCs of cardinality 3 to 9 and compared the results noticing that they agree quite strongly. Additionally, we showed that our algorithmic models are useful in music analysis. The results obtained through CSA in Section 4 encourage us to apply the tonality models in a similar way as CSA is used with abstract set-classes (c.f. [5]).

However, as the experiments with the perception of reference pitch classes has shown, it is a highly subjective process, in which there are no right answers concerning the TCs [1]. Thus, TCs might be better viewed from the perspective of probability (c.f. the distribution table by Huovinen in Appendix A). The results concerning the hypothetical TCs may thus serve as a point of departure for further empirical studies. In computer-aided analysis, one interesting issue might be to compare the relationships between hypothetical TCs and chordal bass notes.

References

1. Huovinen, E.: Pitch-Class Constellations: Studies in the Perception of Tonal Centricity. Acta Musicologica Fennica 23, Turku. The Finnish Musicological Society (2002)
2. Parncutt, R.: Tonal implications of harmonic and melodic Tn-types. In: Proceedings of Mathematics and Computation in Music, MCM 2007. Springer, Berlin (in press) (2007)
3. Parncutt, R.: Revision of Terhardt's psychoacoustical model of the root(s) of a musical chord. Music Perception 6, 65–94 (1988)
4. Krumhansl, C.L.: Cognitive Foundations of Musical Pitch. Oxford University Press, Oxford (1990)
5. Huovinen, E., Tenkanen, A.: Bird's Eye Views of the Musical Surface - Methods for Systematic Pitch-Class Set Analysis. Music Analysis 26(1-2), 159–214 (2007)
6. Chew, E.: Towards a Mathematical Model of Tonality. PhD Dissertation. MIT (2000)

[19] The distribution of pitch classes is quite plain. The numbers of summed occurrences for the twelve pitch classes are 41,38,43,34,43,48,29,38,32,36,37, and 29 when their occurrences are counted at a maximum of once per bar.

7. Tenkanen, A., Gualda, F.: Detecting Changes in Musical Texture. In: Extended Abstracts of International Workshop on Machine Learning and Music 2008 (2008), `http://www.iua.upf.es/~rramirez/MML08/abstracts.pdf`
8. Parncutt, R.: A model of the perceptual root(s) of a chord accounting for voicing and prevailing tonality. In: Leman, M. (ed.) JIC 1996. LNCS, vol. 1317, pp. 181–199. Springer, Heidelberg (1997)
9. Tsang, E.P.K.: Foundations of Constraint Satisfaction. Academic Press, London (1993)
10. Russell, S., Norvig, P.: Artificial Intelligence: A Modern Approach, 2nd edn. Prentice Hall Series in Artificial Intelligence (2003)
11. Krumhansl, C.L., Kessler, E.J.: Tracing the dynamic changes in perceived tonal organization in a spatial representation of musical keys. Psychological Review 89(4), 334–368 (1982)
12. Samson, J.: Music in Transition: A Study of Tonal Expansion and Atonality, 1900-1920. W.W. Norton & Company, New York (1977)
13. Tenkanen, A., Gualda, F.: Multiple Approaches to Comparison Set Analysis. Springer, Heidelberg (in press)

APPENDIX A

Table 2. A response distribution for the trials in Experiment 5 by Huovinen, table 9.4.1 in [1, 258]. The SCs are shown in their prime forms with the set members underlined.

6-Z19A	32	53	1	11	7	5	2	10	50	0	0	1
6-Z19B	52	35	1	1	13	41	0	11	13	1	3	1
6-20	14	36	6	2	13	27	3	2	15	46	3	5
6-Z26	40	25	1	12	0	12	3	15	55	2	4	3
6-Z29	17	29	0	8	4	0	26	1	64	20	2	1
6-32	59	2	28	1	10	15	1	17	2	33	1	3
6-33A	38	1	18	14	3	55	0	12	0	20	10	1
6-33B	24	1	55	2	16	2	12	17	1	38	0	4
6-Z49	14	8	5	10	29	1	4	7	5	84	1	4
6-Z50	16	20	4	3	12	0	25	10	2	75	2	3

APPENDIX B

Table 3. Correlation estimates between the different approaches, explained in Section 4

	P	COF	H	KKP	KK
P	1	0.81	0.81	0.98	0.94
COF	0.81	1	0.49	0.79	0.75
H	0.81	0.49	1	0.79	0.82
KKP	0.98	0.79	0.79	1	0.92
KK	0.94	0.75	0.82	0.92	1
Average	0.91	0.77	0.78	0.90	0.89

APPENDIX C

Table 4. Predicted TCs for some set classes of cardinality 3 and 6

SC	Pitch classes	P	COF	H	KKP	Rank	Tonics	Disp
3-1	●●●○○○○○○○○○	9,0,1	0,7,2	2,6,7	2,0,1	0,2,7	0,$\underline{7}$	6
3-2A	●●○●○○○○○○○○	1,8,10	8,1,10	1,8,6	0,8,1	1,8,0	1	5
3-2B	●○●●○○○○○○○○	0,10,7	10,0,3	8,0,6	0,3,2	0,10,3	3	7
3-3B	●○○●●○○○○○○○	0,1,4	0,5,10	0,8,9	0,4,0	0,4,1,5	$\underline{5}$	7
3-4A	●●○○○●○○○○○○	1,10,0	10,1,5	5,1,0	5,5,10	1,5,10	$\underline{10}$	4
3-4B	●○○○●●○○○○○○	5,0,1	5,0,10	0,5,9	0,5,5	0,5,1,9,10	5,$\underline{10}$	5
3-5A	●●○○○○●○○○○○	1,6,11	1,6,11	1,8,6	6,6,1	1,6,8,11	6,1,$\underline{11}$	4
3-7A	●○●○○●○○○○○○	0,10,5	0,5,10	0,5,9	5,2,0	0,5,10	5,$\underline{10}$	5
3-7B	●○○●○●○○○○○○	0,10,5	10,3,5	0,5,8	0,5,5	0,5,10	5,$\underline{10}$	5
3-8A	●○●○○○●○○○○○	7,0,1	7,0,2	2,0,6	2,6,11	2,0,7	$\underline{7}$,0	6
3-10	●○○●○○●○○○○○	1,0,3	1,8,3	8,6,1	3,0,6	1,3,8	$\underline{1,8}$	5
3-11A	●○○●○○○●○○○○	0,8,7	0,5,10	0,8,5	0,3,0	0,8,5	$\underline{5}$	6
3-11B	●○○○●○○●○○○○	0,5,7	0,5,7	0,2,5	0,4,7	0,5,7	$\underline{5}$,0	5
3-12	●○○○●○○○●○○○	1,5,9	1,5,9	0,1,8	1,5,9	1,5,0	$\underline{5}$	5
6-1	●●●●●●○○○○○○	0,1,10	0,10,1	0,5,9	2,5,0	0,5,1	5	6
6-2A	●●●●●○●○○○○○	1,11,2	11,1,2	6,0,8	1,6,3	1,6,11	6,$\underline{11}$	7
6-9A	●●●●○●○●○○○○	0,10,2	0,10,3	0,5,8	0,5,7	0,5,10	5,$\underline{10}$	7
6-Z13	●●○●●○●●○○○○	1,0,11	1,11,0	0,1,8	4,1,0	1,0,4	–	5
6-15B	●○○●●○●●●○○○	1,4,3	1,8,3	8,0,1	0,4,8	1,8,0	$\underline{1}$	5
6-Z19B	●●○○●●○●●○○○	5,1,0	5,0,1	0,5,1	5,1,0	5,0,1	5	3
6-20	●●○○●●○○●●○○	1,5,9	1,5,9	1,5,9	1,5,9	1,5,9	–	3
6-21A	●○●●●○●○●○○○	1,3,9	1,11,4	8,0,6	0,3,8	1,0,3,8	$\underline{1}$	8
6-27B	●○●●○●●○○●○○	10,2,0	10,5,0	9,6,8	2,9,5	10,2,9	2	7
6-Z28	●●○●○●●○○●○○	1,10,6	1,6,8	9,1,8	6,10,3	1,6,10	6	6
6-30A	●●○●○○●●○●○○	1,7,2	2,8,1	8,1,9	0,6,1	1,8,2	1	7
6-30B	●○●●○○●○●●○○	1,7,3	1,7,2	8,6,9	0,6,3	1,6,7	6	8
6-32	●○●○●●○●○●○○	0,2,5	0,5,7	0,9,2	9,0,2	0,9,2	2	5
6-33A	●○●●○●○●○●○○	0,10,2	0,5,10	5,0,9	0,2,7	0,5,2,10	5,$\underline{10}$	6
6-34A	●●○●○●○●○●○○	10,0,2	10,5,0	9,5,0	0,5,9	0,5,10	5,$\underline{10}$	5
6-34B	●○●○●○●○●●○○	9,7,2	9,2,7	9,6,2	9,6,9	9,2,6	2	4
6-35	●○●○●○●○●○●○	1,3,5	1,3,5	0,8,4	0,2,4	0,1,3	–	7
6-Z39B	●○○●●●●○●○○○	1,5,3	1,3,8	8,0,1	5,0,8	1,5,8	$\underline{1}$	5
6-Z42	●●●●○○●○○●○○	1,2,10	2,1,7	6,8,9	6,3,9	6,1,2	6	8
6-Z44A	●●●○○●●○○●○○	2,10,1	2,7,0	9,6,5	6,2,9	2,6,9	2	8
6-Z44B	●●○○●●●○○●○○	1,2,5	2,9,4	9,5,1	9,6,1	9,1,2	$\underline{2}$	6
6-Z46B	●○●○●●●○○●○○	2,9,5	2,7,0	9,2,0	9,2,6	2,9,0,7	2,$\underline{7}$	6
6-Z49	●●○●●○○●○●○○	2,0,9	0,2,5	9,4,0	4,9,1	0,9,2	$\underline{2}$	6
6-Z50	●○●●○●○○●●○○	10,3,9	10,3,5	9,6,1	0,5,2	10,3,9	3	8

Three Conceptions of Musical Distance

Dmitri Tymoczko

310 Woolworth Center, Princeton University, Princeton, NJ 08544

Abstract. This paper considers three conceptions of musical distance (or inverse "similarity") that produce three different musico-geometrical spaces: the first, based on voice leading, yields a collection of continuous quotient spaces or orbifolds; the second, based on acoustics, gives rise to the *Tonnetz* and related "tuning lattices"; while the third, based on the total interval content of a group of notes, generates a six-dimensional "quality space" first described by Ian Quinn. I will show that although these three measures are in principle quite distinct, they are in practice surprisingly interrelated. This produces the challenge of determining which model is appropriate to a given music-theoretical circumstance. Since the different models can yield comparable results, unwary theorists could potentially find themselves using one type of structure (such as a tuning lattice) to investigate properties more perspicuously represented by another (for instance, voice-leading relationships).

Keywords: Voice leading, orbifold, tuning lattice, *Tonnetz,* Fourier transform.

1 Introduction

We begin with voice-leading spaces that make use of the *log-frequency* metric.[1] Pitches here are represented by the logarithms of their fundamental frequencies, with distance measured according to the usual metric on $\mathbb{R}$; pitches are therefore "close" if they are near each other on the piano keyboard. A point in $\mathbb{R}^n$ represents an ordered series of pitch classes. Distance in this higher-dimensional space can be interpreted as the aggregate distance moved by a collection of musical "voices" in passing from one chord to another. (We can think of this, roughly, as the aggregate physical distance traveled by the fingers on the piano keyboard.) By disregarding information—such as the octave or order of a group of notes—we "fold" $\mathbb{R}^n$ into an non-Euclidean quotient space or orbifold. (For example, imposing octave equivalence transforms $\mathbb{R}^n$ into the n-torus $\mathbb{T}^n$, while transpositional equivalence transforms $\mathbb{R}^n$ into $\mathbb{R}^{n-1}$, orthogonally projecting points onto the hyperplane whose coordinates sum to zero.) Points in the resulting orbifolds represent equivalence classes of musical objects—such as chords or set classes—while "generalized line segments" represent equivalence classes of voice leadings.[2] For example, Figure 1, from Tymoczko 2006,

[1] For more on these spaces, see Callender 2004, Tymoczko 2006, and Callender, Quinn, and Tymoczko 2008.

[2] The adjective "generalized" indicates that these "line segments" may pass through one of the space's singular points, giving rise to mathematical complications.

E. Chew, A. Childs, and C.-H. Chuan (Eds.): MCM 2009, CCIS 38, pp. 258–272, 2009.

represents the space of two-note chords, while Figure 2, from Callender, Quinn, and Tymoczko 2008, represents the space of three-note transpositional set classes. In both spaces, the distance between two points represents the size of the smallest voice leading between the objects they represent.

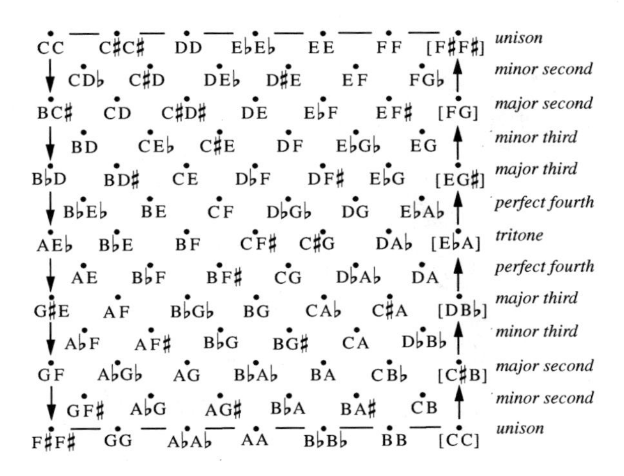

Fig. 1. The Möbius strip representing voice-leading relations among two-note chords

Let's now turn to a very different sort of model, the *Tonnetz* and related structures, which I will describe generically as "tuning lattices." These models are typically discrete, with adjacent points on a particular axis being separated by the same interval. The leftmost lattice in Figure 3 shows the most familiar of these structures, with the two axes representing acoustically pure perfect fifths and major thirds. (One can imagine a third axis, representing either the octave or the acoustical seventh, projecting outward from the paper.) The model asserts that the pitch G4 has an acoustic affinity to both C4 (its "underfifth") and D5 (its "overfifth"), as well as to E♭4 and B4 (its "underthird" and "overthird," respectively). The lattice thus encodes a fundamentally different notion of musical distance than the earlier voice leading models: whereas A3 and A♭3 are very close in log-frequency space, they are four steps apart our tuning lattice. Furthermore, where chords (or more generally "musical objects") are represented by *points* in the voice leadings spaces, they are represented by polytopes in the lattices.[3]

Finally, there are measures of musical distance that rely on chords' shared interval content. From this point of view, the chords {C, C♯, E, F♯} and {C, D♭, E♭, G} resemble one another, since they are "nontrivially homometric" or "Z-related": that is, they share the same collection of pairwise distances between their notes. (For instance, both contain exactly one pair that is one semitone apart, exactly one pair that is two semitones apart, and so on.) However, these chords are *not* particularly close

[3] For a modern introduction to the *Tonnetz*, see Cohn 1997, 1998, and 1999.

in either of the two models considered previously. It is not intuitively obvious that this notion of "similarity" produces any particular geometrical space. But Ian Quinn has shown that one can use the discrete Fourier transform to generate (in the familiar equal-tempered case) a six-dimensional "quality space" in which chords that share the same interval content are represented by the same point.[4] We will explore the details shortly.

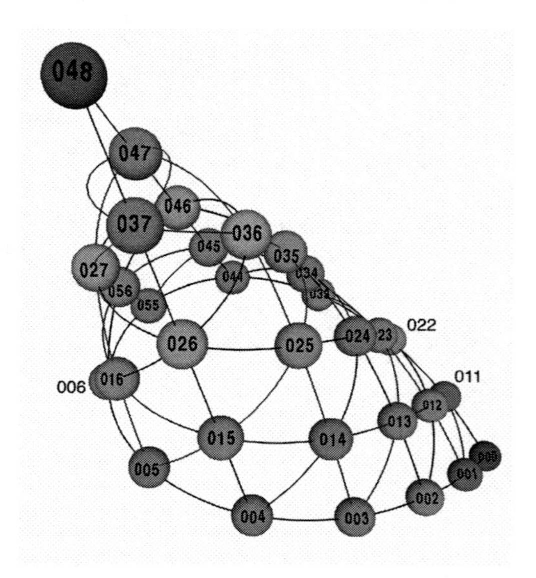

Fig. 2. The cone representing voice-leading relations among three-note transpositional set classes

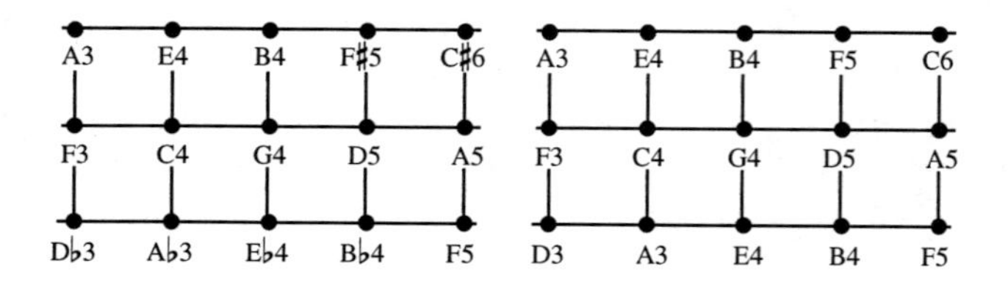

Fig. 3. Two discrete tuning lattices. On the left, the chromatic *Tonnetz*, where horizontally adjacent notes are linked by acoustically pure fifths, while vertically adjacent notes are linked by acoustically pure major thirds. On the right, a version of the structure that uses diatonic intervals.

Clearly, these three musical models are very different, and it would be somewhat surprising if there were to be close connections between them. But we will soon see that this is in fact this case.

---

[4] See Lewin 1959, 2001, Quinn 2006, 2007, Callender 2007.

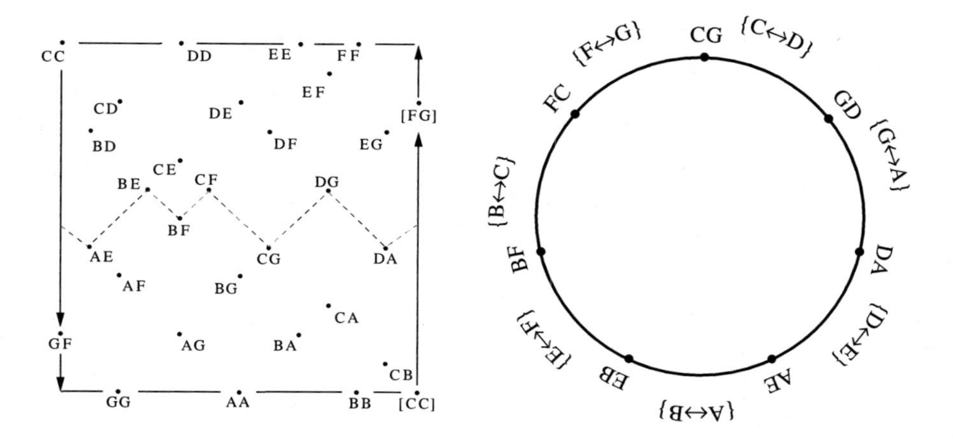

Fig. 4. (*left*) Most efficient voice-leadings between diatonic fifths form a chain that runs through the center of the Möbius strip from Figure 1. (*right*) These voice leadings form an abstract circle, in which adjacent dyads are related by three-step diatonic transposition, and are linked by single-step voice leading.

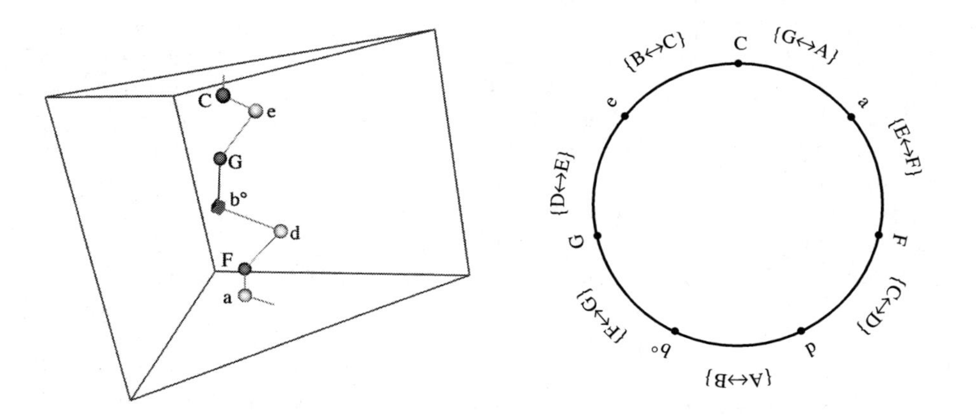

Fig. 5. (*left*) Most efficient voice-leadings between diatonic triads form a chain that runs through the center of the orbifold representing three-note chords. (*right*) These voice leadings form an abstract circle, in which adjacent triads are linked by single-step voice leading. Note that here, adjacent triads are related by transposition by two diatonic steps.

2 Voice-Leading Lattices and Acoustic Affinity

Voice-leading and acoustics seem to privilege fundamentally different conceptions of pitch distance: from a voice-leading perspective, the semitone is smaller than the perfect fifth, whereas from the acoustical perspective the perfect fifth is smaller than the semitone. Intuitively, this would seem to be a fundamental gap that cannot be bridged.

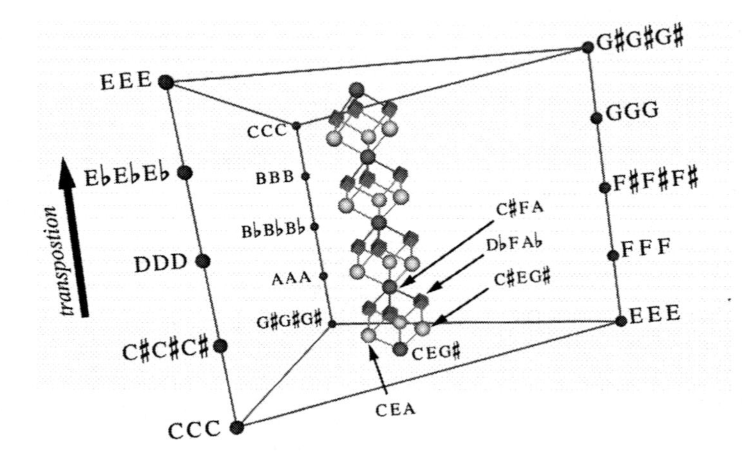

Fig. 6. Major, minor, and augmented triads as they appear in the orbifold representing three-note chords. Here, triads are particularly close to their major-third transpositions.

Things become somewhat more complicated, however, when we consider the discrete lattices that represent voice-leading relationships among familiar diatonic or chromatic chords. For example, Figure 4 records the most efficient voice leadings among diatonic fifths—which can be represented using an irregular, one-dimensional zig-zag near the center of the Möbius strip $\mathbb{T}^2/\mathcal{S}_2$. (The zig-zag seems to be irregular because the figure is drawn using the chromatic semitone as a unit; were we to use the diatonic step, it would be regular.) Abstractly, these voice leadings form the circle shown on the right of Figure 4. The figure demonstrates that there are purely contrapuntal reasons to associate fifth-related diatonic fifths: from this perspective {C, G} is close to {G, D}, not because of acoustics, but because the first dyad can be transformed into the second by moving the note C up by one diatonic step. One fascinating possibility—which we unfortunately can not pursue here—is that acoustic affinities actually *derive from* voice-leading facts, at least in part: it is possible that the ear associates the third harmonic of a complex tone with the second harmonic of another tone a fifth above it, and the fourth harmonic of the lower note with the third of the upper, in effect tracking voice-leading relationships among the partials.

Figures 5-7 present three analogous structures: Figure 5 connects triads in the C diatonic scale by efficient voice leading, and depicts third-related triads as being particularly close; Figure 6 shows the position of major, minor, and augmented triads in three-note chromatic chord space, where major-third-related triads are close[5]; Figure 7 shows (symbolically) that fifth-related diatonic scales are close in twelve-note chromatic space. Once again, we see that there are purely contrapuntal reasons to associate fifth-related diatonic scales and third-related triads.

[5] This graph was first discovered by Douthett and Steinbach (1998).

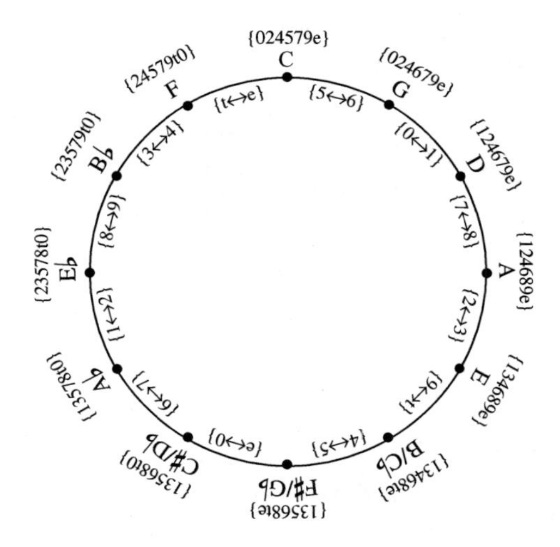

Fig. 7. Fifth-related diatonic scales form a chain that runs through the center of the seven-dimensional orbifold representing seven-note chords. It is structurally analogous to the circles in Figures 4 and 5.

		Correlation
MAJOR	**Bach**	.96
	Haydn	.93
	Mozart	.91
	Beethoven	.96
MINOR	**Bach**	.95
	Haydn	.91
	Mozart	.91
	Beethoven	.96

Fig. 8. Correlations between modulation frequency and voice-leading distances among scales, in Bach's *Well-Tempered Clavier,* and the piano sonatas of Haydn, Mozart, and Beethoven. The very high correlations suggest that composers typically modulate between keys whose associated scales can be linked by efficient voice leading.

This observation, in turn, raises a number of theoretical questions. For instance: should we attribute the prevalence of modulations between fifth-related keys to the acoustic affinity between fifth-related pitches, or to the voice-leading relationships between fifth-related diatonic scales? One way to study this question would be to compare the frequency of modulations in classical pieces to the voice-leading distances among their associated scales. Preliminary investigations, summarized in Figure 8, suggest that voice-leading distances are in fact very closely correlated to modulation frequencies. Surprising as it may seem, the acoustic affinity of perfect fifth-related notes may be superfluous when it comes to explaining classical modulatory practice.[6]

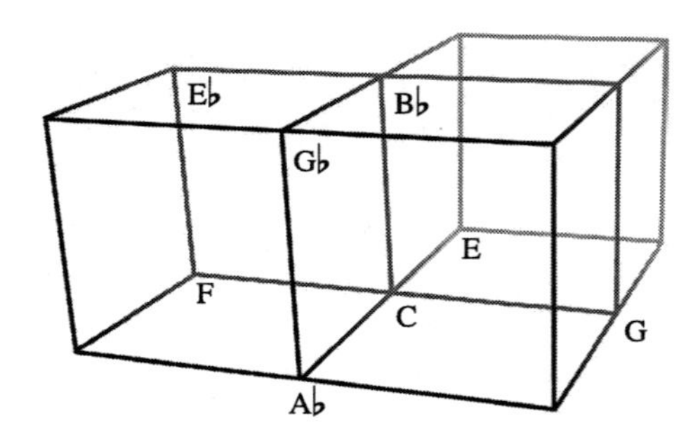

Fig. 9. On this three-dimensional *Tonnetz,* the C^7 chord is represented by the tetrahedron whose vertices are C, E, G, and B♭. The $C^{ø7}$ chord is represented by the nearby tetrahedron C, E♭, G♭, B♭, which shares the C-B♭ edge.

3 Tuning Lattices as Approximate Models of Voice Leading

We will now investigate the way tuning lattices like the *Tonnetz* represent voice-leading relationships among familiar sonorities. Here my argumentative strategy will by somewhat different, since it is widely recognized that the *Tonnetz* has something to do with voice leading. (This is largely due to the important work of Richard Cohn, who has used the *Tonnetz* to study what he calls "parsimonious" voice leading.[7]) My goal will therefore be to explain why tuning lattices are only an *approximate* model of contrapuntal relationships, and only for certain chords.

The first point to note is that inversionally related chords on a tuning lattice are near each other when they share common tones.[8] For example, the *Tonnetz* represents perfect fifths by line segments; fifth-related perfect fifths, such as {C, G} and {G, D} are related by inversion around their common note, and are adjacent on the lattice (Figure 3). Similarly, major and minor triads on the *Tonnetz* are represented by triangles; inversionally related triads that share an interval, such as {C, E, G} and {C, E, A}, are joined by a common edge. (On the standard *Tonnetz,* the more common tones, the closer the chords will be: C major and A minor, which share two notes, are closer than C major and F minor, which share only one.) In the three-dimensional *Tonnetz* shown in Figure 9, where the *z* ax is represents the seventh, C^7 is near its

[6] Similar points could potentially be made about the prevalence, in functionally tonal music, of root-progressions by perfect fifth. It may be that the diatonic circle of thirds shown in Figure 5 provides a more perspicuous model of functional harmony than do more traditional fifth-based representations.

[7] See Cohn 1997.

[8] This is not true of the voice leading spaces considered earlier: for example, in three-note chord space {C, D, F} is not particularly close {F, A♭, B♭}.

inversion $C^{\varnothing 7}$. The point is reasonably general, and does not depend on the particular structure of the *Tonnetz* or on the chords involved: on tuning lattices, inversionally related chords are close when they share common tones.[9]

The second point is that acoustically consonant chords often divide the octave relatively evenly; such chords can be linked by efficient voice leading to those inversions with which they share common notes.[10] It follows that proximity on a tuning lattice will indicate the potential for efficient voice leading *when the chords in question are nearly even and are related by inversion.* Thus {C, G} and {G, D} can be linked by the stepwise voice leading (C, G)→(D, G), in which C moves up by two semitones. Similarly, the C major and A minor triads can be linked by the single-step voice leading (C, E, G) →(C, E, A), and C^7 can be linked to $C^{\varnothing 7}$ by the two semitone voice-leading (C, E, G, B♭)→(C, E♭, G♭, B♭). In each case the chords are also close on the relevant tuning lattice. (Interestingly, triadic distances on the diatonic *Tonnetz* in Fig. 3 exactly reproduce the circle-of-thirds distances from Fig. 5.) This will not be true for uneven chords: {C, E} and {E, G♯} are close on the *Tonnetz,* but cannot be linked by particularly efficient voice leading; the same holds for {C, G, A♭} and {G, A♭, D♭}. Tuning lattices are approximate models of voice-leading only when one is concerned with the nearly-even sonorities that are fundamental to Western tonality.

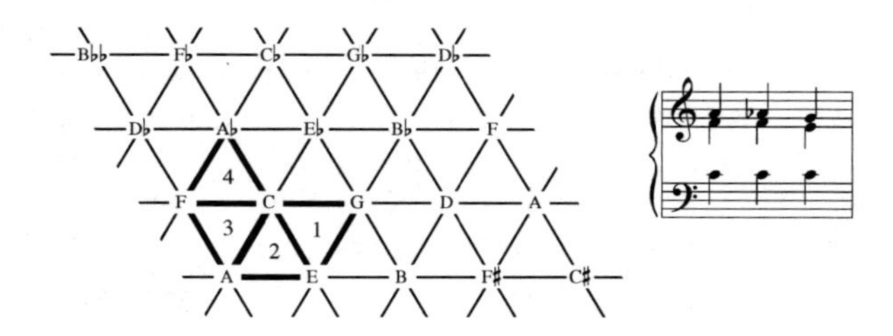

Fig. 10. On the *Tonnetz,* F major (triangle 3) is closer to C major (triangle 1) than F minor (triangle 4) is. In actual music, however, F minor frequently appears as a passing chord between F major and C major. Note that, unlike in Figure 3, I have here used a *Tonnetz* in which the axes are not orthogonal; this difference is merely orthographical, however.

Furthermore, on closer inspection *Tonnetz*-distances diverge from voice-leading distances even for these chords. Some counterexamples are obvious: for instance, {C, G} and {C♯, F♯} can be linked by semitonal voice leading, but are fairly far apart on the *Tonnetz*. Slightly more subtle, but more musically pertinent, is the following example: on the *Tonnetz,* C major is two units away from F major but *three* units from

[9] In the general case, the notion of "closeness" needs to be spelled out carefully, since chords can contain notes that are very far apart on the lattice. In the applications we are concerned with, chords occupy a small region of the tuning lattice, and the notion of "closeness" is fairly straightforward.

[10] See Tymoczko 2006 and 2008a. The point is relatively obvious when one thinks geometrically: the two chords divide the pitch-class circle nearly evenly into the same number of pieces; hence, if any two of their notes are close, then each note of one chord is near some note of the other.

F minor (Figure 10). (Here I measure distance in accordance with "neo-Riemannian" theory, which considers triangles sharing an edge to be one unit apart and which decomposes larger distances into sequences of one-unit moves.) Yet it takes only two semitones of total motion to move from C major to F minor, and three to move from C major to F major. (This is precisely why F minor often appears as a passing chord between F major and C major.) The *Tonnetz* thus depicts F major as being closer to C major than F minor is, even though contrapuntally the opposite is true. This means we cannot use the figure to explain the ubiquitous nineteenth-century IV-iv-I progression, in which the two-semitone motion $\hat{6}\rightarrow\hat{5}$ is broken into a pair of single-semitone steps $\hat{6}\rightarrow\flat\hat{6}\rightarrow\hat{5}$.

One way to put the point is that while adjacencies on the *Tonnetz* reflect voice-leading facts, other relationships do not. As Cohn has emphasized, two major or minor triads share an edge if they can be linked by "parsimonious" voice-leading in which a single voice moves by one or two semitones. If we are interested in this particular kind of voice leading then the *Tonnetz* provides an accurate and useful model. However, there is no analogous characterization of larger distances in the space. In other words, we do not get a recognizable notion of voice-leading distance by "decomposing" voice leadings into sequences of parsimonious moves: as we have seen, (F, A C)→(E, G, C) can be decomposed into *two* parsimonious moves, while it takes *three* to represent (F, A♭, C)→(E, G, C); yet intuitively the first voice leading is larger than the second. The deep issue here is that it is problematic to assert that "parsimonious" voice leadings are always smaller than non-parsimonious voice-leadings: by asserting that (C, E, A)→(C, E, G) is smaller than (C, F, A♭)→(C, E, G), the theorist runs afoul what Tymoczko calls "the distribution constraint," known to mathematicians as the submajorization partial order.[11] Tymoczko argues that violations of the distribution constraint invariably produce distance measures that violate intuitions about voice leading; the problem with larger distances on the *Tonnetz* is an illustration of this general point.

Nevertheless, the fact remains that the two kinds of distance are roughly consistent: for major and minor triads, the correlation between *Tonnetz* distance and voice-leading distance is a reasonably high .79.[12] Furthermore, since Tymoczko's "distribution constraint" is not intuitively obvious, unwary theorists might well think that they could declare the "parsimonious" voice leading (C, E, G)→(C, E, A) to be smaller than the non-parsimonious (C, E, G)→(C♯, E, G♯). (Indeed, the very meaning of the term "parsimonious" would seem to suggest that some theorists have done so.) Consequently, *Tonnetz*-distances might well appear, at first or even second blush, to reflect some reasonable notion of "voice-leading distance"; and this in turn could lead the theorist to conclude that the *Tonnetz* provides a generally applicable tool for

[11] See Tymoczko 2006, and Hall and Tymoczko 2007. Metrics that violate the distribution constraint have counterintuitive consequences, such as preferring "crossed" voice leadings to their uncrossed alternatives. Here, the claim that A minor is closer to C major than F minor leads to the F minor/F major problem discussed in Figure 10.

[12] Here I use the L^1 or "taxicab" metric. The correlation between *Tonnetz* distances and the number of shared common tones is an even-higher .9; however, "number of shared common tones" is not interpretable as a voice-leading metric.

investigating triadic voice-leading. I have argued that we should resist this conclusion: if we use the *Tonnetz* to model chromatic music, than Schubert's major-third juxtapositions will seem very different from his habit of interposing F minor between F major and C major, since the first can be readily explained using the *Tonnetz* whereas the second cannot.[13] The danger, therefore, is that we might find ourselves drawing unnecessary distinctions between these two cases—particularly if we mistakenly assume the *Tonnetz* is a fully faithful model of voice-leading relationships.

4 Voice Leading, "Quality Space," and the Fourier Transform

We conclude by investigating the relation between voice leading and the Fourier-based perspective.[14] The mechanics of the Fourier transform are relatively simple: for any number n from 1 to 6, and every pitch-class p in a chord, the transform assigns a two-dimensional vector whose components are:

$$V_{p,n} = (\cos(2\pi pn/12), \sin(2\pi pn/12))$$

Adding these vectors together, for one particular n and all the pitch-classes p in the chord, produces a composite vector representing the chord as a whole—its "nth Fourier component." The length (or "magnitude") of this vector, Quinn observes, reveals something about the chord's harmonic character: in particular, chords saturated with $(12/n)$-semitone intervals, or intervals approximately equal to $12/n$, tend to score highly on this index of chord quality.[15] The Fourier transform thus seems to quantify the intuitive sense that chords can be more-or-less diminished-seventh-like, perfect-fifthy, or whole-toneish. Interestingly, "Z-related" chords—or chords with the same interval content—always score identically on this measure of chord-quality. In this sense, Fourier space (the six-dimensional hypercube whose coordinates are the Fourier magnitudes) seems to model a conception of similarity that emphasizes interval content, rather than voice leading or acoustic consonance.

However, there is again a subtle connection to voice leading: it turns out that the magnitude of a chord's nth Fourier component is approximately linearly related to the (Euclidean) size of the minimal voice leading to *the nearest subset* of any perfectly even n-note chord.[16] For instance, a chord's first Fourier component (FC_1) is approximately related to the size of the minimal voice leading to any transposition of {0}; the second Fourier component is approximately related to the size of the minimal voice leading to any transposition of either {0} or {0, 6}; the third component is approximately related to the size of the minimal voice leading to any transposition of either {0}, {0, 4} or {0, 4, 8}, and so on. Figure 11 shows the location of the subsets

[13] See Cohn 1999.

[14] This material in this section appears in Tymoczko 2008b. It is influenced by Robinson (2006), Hoffman (2007), and Callender (2007).

[15] Here I use continuous pitch-class notation where the octave always has size 12, no matter how it is divided. Thus the equal-tempered five-note scale is labeled {0, 2.4, 4.8, 7.2, 9.6}.

[16] Here I measure voice-leading using the Euclidean metric, following Callender 2004. See Tymoczko 2006 and 2008a for more on measures of voice-leading size.

of the n-note perfectly even chord, as they appear in the orbifold representing three-note set-classes, for values of n ranging from 1 to 6.[17] Associated to each graph is one of the six Fourier components. For any three-note set class, the magnitude of its nth Fourier component is a decreasing function of the distance to the nearest of these marked points: for instance, the magnitude of the third Fourier component (FC_3) decreases, the farther one is from the nearest of {0}, {0, 4} and {0, 4, 8}. Thus, chords in the shaded region of Figure 12 will tend to have a relatively large FC_3, while those in the unshaded region will have a smaller FC_3. Figure 13 shows that this relationship is very-nearly linear for twelve-tone equal-tempered trichords.

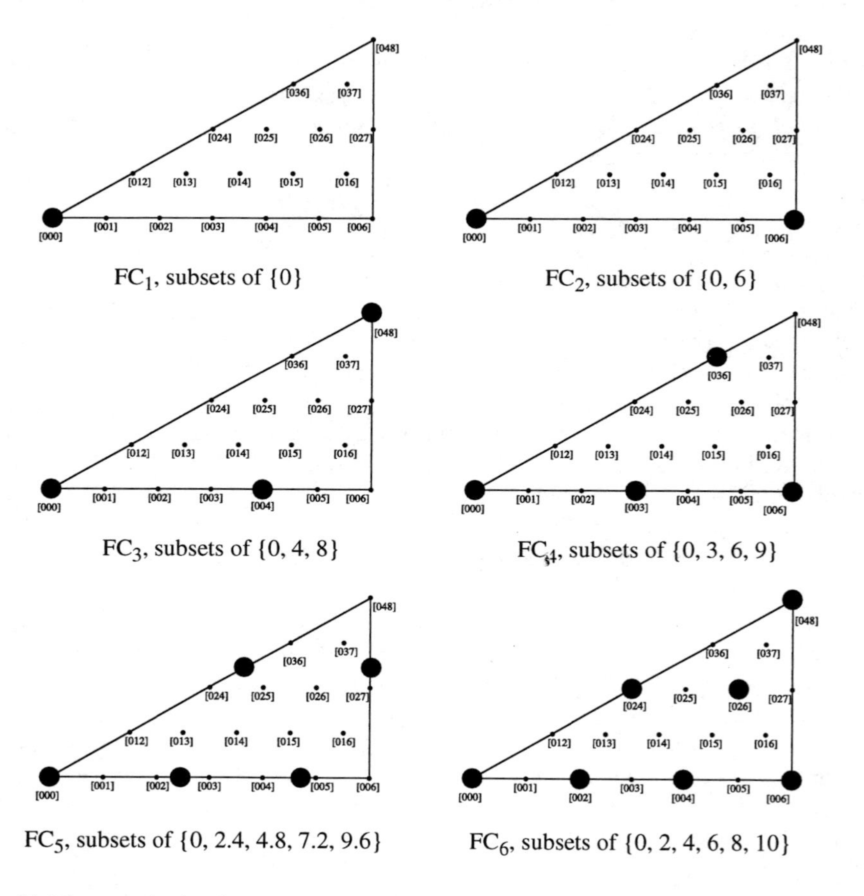

FC_1, subsets of {0} FC_2, subsets of {0, 6}

FC_3, subsets of {0, 4, 8} FC_4, subsets of {0, 3, 6, 9}

FC_5, subsets of {0, 2.4, 4.8, 7.2, 9.6} FC_6, subsets of {0, 2, 4, 6, 8, 10}

Fig. 11. The magnitude of a set class's nth Fourier component is approximately linearly related to the size of the minimal voice leading to the nearest subset of the perfectly even n-note chord, shown here as dark spheres.

[17] See Callender 2004, Tymoczko 2006, Callender, Quinn, and Tymoczko, 2008. These triangles result from bisecting the cone in Figure 2. Every point represents a set class, while every line segment represents an equivalence class of voice leadings.

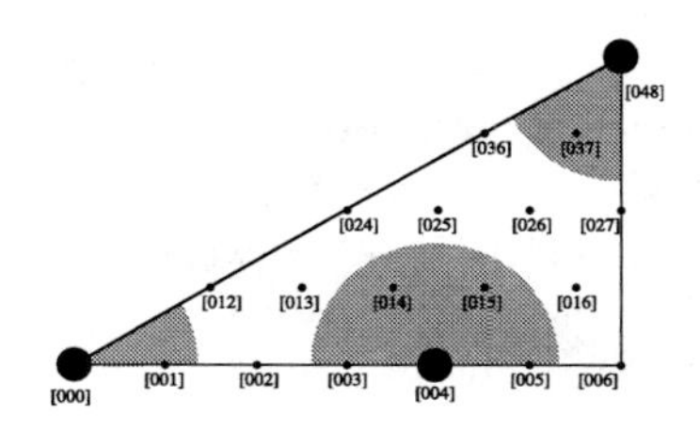

Fig. 12. Chords in the shaded region will have a large FC_3 component, since they are near subsets of {0, 4, 8}. Those in the unshaded region will have a smaller FC_3 component.

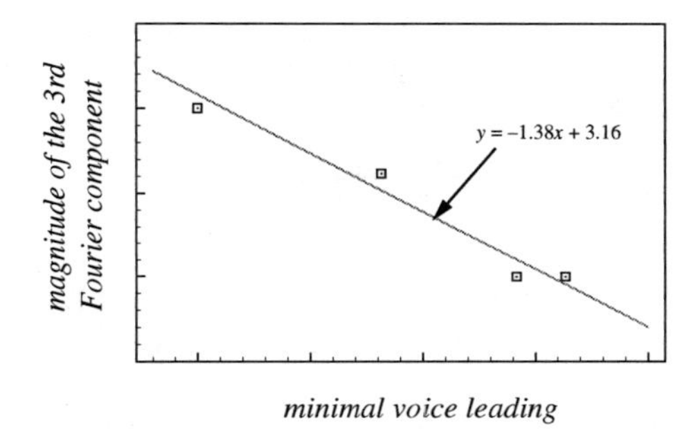

Fig. 13. For trichords, the equation $FC_3 = -1.38VL + 3.16$ relates the third Fourier component to the Euclidean size of the minimal voice leading to the nearest subset of {0, 4, 8}

Table 1. Correlations between voice-leading distances and Fourier magnitudes

	FC_1	FC_2	FC_3	FC_4	FC_5	FC_6
Dyads	-.97	-.96	-.97	-1	-.97	-1*
Trichords	-.98	-.97	-.97	-.98	-.98	-1*
Tetrachords	-.96	-.96	-.95	-.98	-.96	-1*
Pentachords	-.96	-.96	-.95	-.98	-.96	-1*
Hexachords	-.96	-.96	-.95	-.96	-.96	-1*
Septachords	-.96	-.96	-.96	-.97	-.96	-1*
Octachords	-.96	-.96	-.95	-.98	-.96	-1*
Nonachords	-.96	-.96	-.96	-.98	-.96	-1*
Decachords	-.96	-.96	-.96	-.98	-.96	-1*

* Voice leading calculated using L^1 (taxicab) distance rather than L^2 (Euclidean).

Table 1 uses the Pearson correlation coefficient to estimate the relationship between the voice-leading distances and Fourier components, for twelve-tone equal-tempered multisets of various cardinalities. The strong anti-correlations indicate that one variable predicts the other with a very high degree of accuracy. Table 2 calculates the correlation coefficients for three-to-six-note chords in 48-tone equal temperament.

These strong anticorrelations, very similar to those in Table 1, show that there continues to be a very close relation between Fourier magnitudes and voice-leading size in very finely quantized pitch-class space. Since 48-tone equal temperament is so finely quantized, these numbers are approximately valid for continuous, unquantized pitch-class space.[18]

Table 2. Correlations between voice-leading distances and Fourier magnitudes in 48-tone equal temperament

	$\mathbf{FC_1}$
Trichords -	.99
Tetrachords -	.97
Pentachords -	.97
Hexachords -	.96

Explaining these correlations, though not very difficult, is beyond the scope of this paper. From our perspective, the important question is whether we should measure chord quality using the Fourier transform or voice leading.[19] In particular, the issue is whether the Fourier components model the musical intuitions we want to model: as we have seen, the Fourier transform requires us to measure a chord's "harmonic quality" in terms of its distance from *all* the subsets of the perfectly even *n*-note chord. But we might sometimes wish to employ a different set of harmonic prototypes. For instance, Figure 14 uses a chord's distance from the augmented triad to measure the

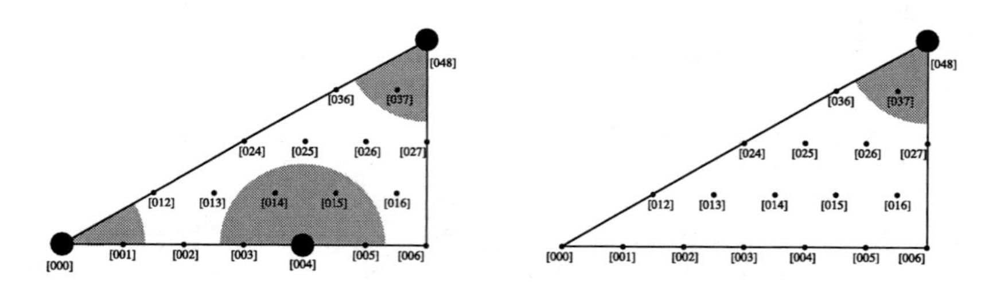

Fig. 14. The mathematics of the Fourier transform requires that we conceive of "chord quality" in terms of the distance to *all subsets* of the perfectly even *n*-note chord (*left*). Purely voice-leading-based conceptions instead allow us to choose our harmonic prototypes freely (*right*). Thus we can voice leading to model a chord's "augmentedness" in terms of its distance from the augmented triad, but not the tripled unison {0, 0, 0} or the doubled major third {0, 0, 4}.

[18] It would be possible, though beyond the scope of this paper, to calculate this correlation analytically. It is also possible to use statistical methods for higher-cardinality chords. A large collection of randomly generated 24- and 100-tone chords in continuous space produced correlations of .95 and .94, respectively.

[19] See Robinson 2006 and Straus 2007 for related discussion.

trichordal set classes' "augmentedness." Unlike Fourier analysis, this purely voice-leading-based method does not consider the triple unison or doubled major third to be particularly "augmented-like"; hence, set classes like {0, 1, 4} do not score particularly highly on this index of "augmentedness." This example dramatizes the fact that, when using voice leading, we are free to choose any set of harmonic prototypes, rather than accepting those the Fourier transform imposes on us.

5 Conclusion

The approximate consistency between our three models is in one sense good news: since they are closely related, it may not matter much—at least in practical terms—which we choose. We can perhaps use a tuning lattice such as the *Tonnetz* to represent voice-leading, as long as we are interested in gross contrasts ("near" vs. "far") rather than fine quantitative differences ("3 steps away" vs. "2 steps away"). Similarly, we can perhaps use voice-leading spaces to approximate the results of the Fourier analysis, as long as we are interested in modeling generic harmonic intuitions ("very fifthy" vs. "not very fifthy") rather than exploring very fine differences among Fourier magnitudes.

However, if we want to be more principled, then we need to be more careful. The resemblances among our models mean that it is possible to inadvertently use one sort of structure to discuss properties that are more directly modeled by another. And indeed, the recent history of music theory displays some fascinating (and very fruitful) imprecision about this issue. It is striking that Douthett and Steinbach, who first described several of the lattices found in the center of the voice-leading orbifolds—including Figure 6—explicitly presented their work as generalizing the familiar *Tonnetz*.[20] Their lattices, rather than depicting *parsimonious* voice leading among major and minor triads, displayed *single-semitone* voice leadings among major, minor, and augmented triads; and as a result of this small difference, *every distance* can be interpreted as representing voice-leading size. However, this difference only became apparent after it was understood how to embed their discrete structures in the continuous geometrical figures described at the beginning of this paper. Thus one could say that the continuous voice-leading spaces evolved out of the *Tonnetz*, by way of Douthett and Steinbach's discrete lattices, even though the structures now appear to be fundamentally different. Related points could be made about Quinn's "quality space," whose connection to the voice-leading spaces took several years—and the work of several authors—to clarify.

There is, of course, nothing wrong with this: knowledge progresses slowly and fitfully. But our investigation suggests that we may want to think carefully about which model is appropriate for which music-theoretical purpose. I have tried to show that the issues here are complicated and subtle: the mere fact that tonal pieces modulate by fifth does not, for example, require us to use a tuning lattice in which fifths are

[20] See Douthett and Steinbach 1998. The same is true of Tymoczko 2004, which uses the term "generalized *Tonnetz*" to describe another set of lattices appearing in the voice-leading spaces.

smaller than semitones. (Indeed, the "circle of fifths" C-G-D-... can be interpreted either as a one-dimensional tuning lattice incorporating octave equivalence, or as a diagram of the voice-leading relations among diatonic scales, as in Figure 7.) Likewise, there may be close connections between voice-leading spaces and the Fourier transform, even though the latter associates "Z-related" chords while the former does not. The present paper can be considered a down-payment toward a more extended inquiry, one that attempts to determine the relative strengths and weaknesses of our three different-yet-similar conceptions of musical distance.

References

Callender, C.: Continuous Transformations. Music Theory Online 10(3) (2004)

Callender, C.: Continuous Harmonic Spaces. Journal of Music Theory 51(2) (in press) (2007)

Callender, C., Quinn, I., Tymoczko, D.: Generalized Voice-Leading Spaces. Science 320, 346–348 (2008)

Cohn, R.: Properties and Generability of Transpositionally Invariant Sets. Journal of Music Theory 35, 1–32 (1991)

Cohn, R.: Maximally Smooth Cycles, Hexatonic Systems, and the Analysis of Late-Romantic Triadic Progressions. Music Analysis 15(1), 9–40 (1996)

Cohn, R.: Neo-Riemannian Operations, Parsimonious Trichords, and their 'Tonnetz' Representations. Journal of Music Theory 41(1), 1–66 (1997)

Cohn, R.: Introduction to Neo-Riemannian Theory: A Survey and a Historical Perspective. Journal of Music Theory 42(2), 167–180 (1998)

Cohn, R.: As Wonderful as Star Clusters: Instruments for Gazing at Tonality in Schubert. Nineteenth-Century Music 22(3), 213–232 (1999)

Douthett, J., Steinbach, P.: Parsimonious Graphs: a Study in Parsimony, Contextual Transformations, and Modes of Limited Transposition. Journal of Music Theory 42(2), 241–263 (1998)

Hall, R., Tymoczko, D.: Poverty and polyphony: a connection between music and economics. In: Sarhanghi, R. (ed.) Bridges: Mathematical Connections in Art, Music, and Science, Donostia, Spain (2007)

Hoffman, J.: On Pitch-class set cartography (unpublished) (2007)

Lewin, D.: Re: Intervallic Relations between Two Collections of Notes. Journal of Music Theory 3, 298–301 (1959)

Lewin, D.: Special Cases of the Interval Function between Pitch-Class Sets X and Y. Journal of Music Theory 45, 1–29 (2001)

Quinn, I.: General Equal Tempered Harmony (Introduction and Part I). Perspectives of New Music 44(2), 114–158 (2006)

Quinn, I.: General Equal-Tempered Harmony (Parts II and III). Perspectives of New Music 45(1), 4–63 (2007)

Robinson, T.: The End of Similarity? Semitonal Offset as Similarity Measure. In: The annual meeting of the Music Theory Society of New York State, Saratoga Springs, NY (2006)

Straus, J.: Uniformity, Balance, and Smoothness in Atonal Voice Leading. Music Theory Spectrum 25(2), 305–352 (2003)

Straus, J.: Voice leading in set-class space. Journal of Music Theory 49(1), 45–108 (2007)

Tymoczko, D.: Scale Networks in Debussy. Journal of Music Theory 48(2), 215–292 (2004)

Tymoczko, D.: The Geometry of Musical Chords. Science 313, 72–74 (2006)

Tymoczko, D.: Scale Theory, Serial Theory, and Voice Leading. Music Analysis 27(1), 1–49 (2008a)

Tymoczko, D.: Voice leading and the Fourier Transform. Journal of Music Theory 52(2) (in press) (2008b)

Pairwise Well-Formed Scales and a Bestiary of Animals on the Hexagonal Lattice

Jon Wild

Department of Music Research, Schulich School of Music,
McGill University, Montreal, Canada
wild@music.mcgill.ca

Abstract. Some pitch-class collections may be represented as subsets of a two-dimensional lattice or generalised *Tonnetz*. Whereas a well-formed scale of cardinality n is formed as a simple interval chain, and thus defined unambiguously by the size of its generating interval, there are a great number of inequivalent ways of forming connected n-subsets of the two-dimensional lattice defined by a given pair of basis intervals. Only very few of these connected subsets or *lattice animals* ever turn out to correspond to collections that possess the pairwise well-formed property. Pwwf scales are found to correspond to members of a small family of lattice animals that is independent of the generators at the basis of the lattice. Finally a method is shown for constructing a pair of generators that will yield any given heptatonic pwwf scale; the method is easily extended to other cardinalities.

Well-Formed Scales

Carey and Clampitt [1] introduced the concept of a *well-formed scale* to the music-theoretical community; these collections had earlier been investigated by Erv Wilson [2] under the name "Moments of Symmetry". For the purposes of the present paper an important characteristic of a well-formed scale can be expressed as follows: it is a scale with exactly two step sizes, whose "tokenised" cyclic interval list (e.g. *aaabaaabaab*) has the property that each token is maximally evenly distributed among the other tokens. Well-formed scales are *generated*; that is, they are formed by the iteration of a single generating interval, with the resulting pitches collapsed into an octave. The properties of such collections have been extensively researched in recent music-theoretical literature; see for example [3].

Pairwise Well-Formed Scales

Clampitt [4] introduced a generalisation of the well-formed scale: the *pairwise well-formed scale*. In his elegant study Clampitt demonstrates a number of interesting structural and transformational features of pwwf scales and locates examples from world musics as well as from 20th-century Western music. A simple characterisation of these scales that serves our current purposes is as follows: a pwwf scale has exactly three sizes of step (I will call such scales *three-stepped*), and its token list has the

E. Chew, A. Childs, and C.-H. Chuan (Eds.): MCM 2009, CCIS 38, pp. 273–285, 2009.

property that each token is maximally evenly distributed among the others. Only odd cardinalities of pwwf scale are possible, and the multiplicity of each step size must be coprime to the scale cardinality. Clampitt gives the example of the diatonic scale in Zarlino's syntonic tuning [5], whose scale steps enjoy the following frequency ratios relative to C: 1:1, 9:8, 5:4, 4:3, 3:2, 5:3, 15:8, 2:1. Setting the step sizes a=9:8, b=10:9, c=16:15 we obtain the token list *abcabac*, which is easily verified to have the relevant property: the a's are as evenly distributed as three items could be among seven; the b's are as evenly distributed as two items could be; likewise for the c's.

The Syntonic Diatonic as a Generated Collection

Zarlino's syntonic diatonic may be characterised as generated by two intervals, for example 3:2 and 5:4. This is easily seen in a *Tonnetz* representation as in Figure 1. The horizontal arrows show intervals of a perfect fifth, and the vertical arrows a major third.[1]

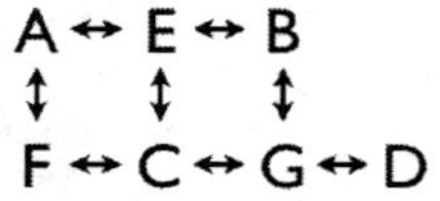

A ↔ E ↔ B
↕ ↕ ↕
F ↔ C ↔ G ↔ D

Fig. 1.

But whereas a singly-generated scale of n notes is unambiguously defined by the size of its generator, a doubly-generated scale like this one needs aspects of its *Tonnetz* geometry to be specified before it is uniquely determined. To illustrate, two other scales of cardinality 7 generated by the same pair of intervals are shown in Figure 2; these others are not pairwise well-formed and while the first at least is three-stepped, the second has more than three step sizes.

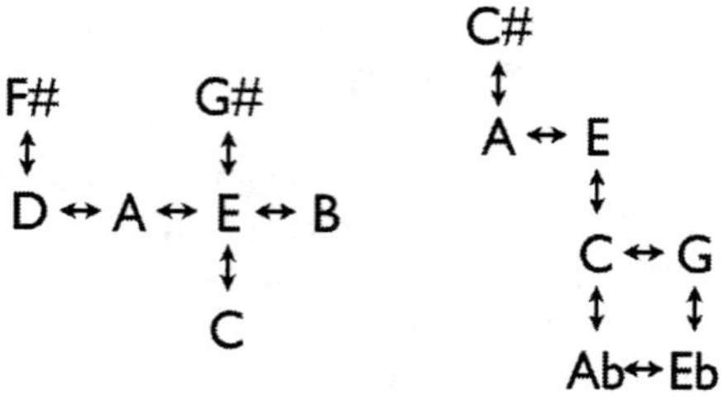

F# G#
↕ ↕
D ↔ A ↔ E ↔ B
↕
C

C#
↕
A ↔ E
↕
C ↔ G
↕ ↕
Ab ↔ Eb

Fig. 2.

Not all doubly-generated scales, then, are three-stepped, much less pairwise well-formed. Later on we shall be interested in the converse question: are all pairwise well-formed scales doubly-generated?

---

[1] This pair is not the only choice of generators; we could just as easily have chosen the pair (3:2, 16:15), in which case the corresponding diagram would have the upper row offset by one position towards the left.

An Alternative Lattice Representation

The syntonic diatonic is often represented on a triangular lattice in order to emphasise the pure triads that characterise this scale. On such a lattice the connections by minor third are evident, as well as those by perfect fifth and major third; this third axis however does not represent an independent lattice dimension, because the minor third is the difference between the two generating intervals. It is well-known that the triangular lattice is dual to the hexagonal lattice: the nodes of the former become the cells of the latter; the edges of the former become connected hexagonal faces of the latter. In this manner the syntonic diatonic is represented as in Figure 3.

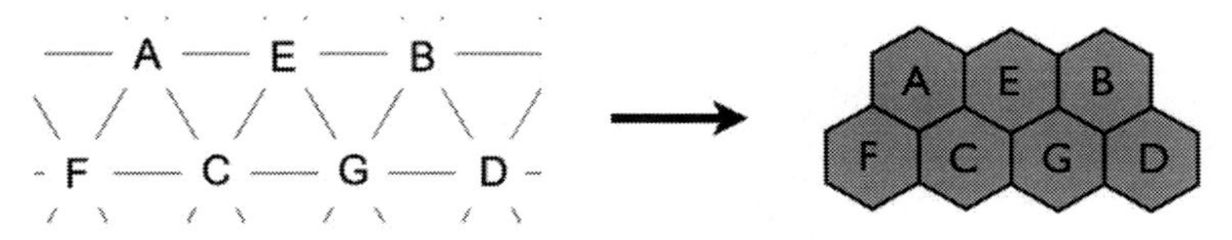

Fig. 3.

Animals on the Hexagonal Lattice—The Heptatonic Case

The first question to consider is: how many different connected configurations of n notes can we construct on this lattice? We shall start by considering n=7. We shall need to count as distinct those configurations that are related by symmetry operations on the hexagonal lattice (except for translation)—this is because rotating or reflecting amounts to permuting or inverting the generating intervals, which results in distinct scales. These configurations are analogous to *polyominoes* on a square grid, which have played an important role in recreational mathematics. Somewhat less studied are the extensions of polyominoes to other lattices such as the hexagonal one here; there are various names for these arrangements, like "polyhexes" but here I prefer the more whimsical name mathematicians sometimes use: *lattice animals*.[2] The number of n-animals on a hexagonal lattice increases surprisingly quickly. As with polyominoes and other polyforms, no formula is known for enumerating them directly; the animals must be explicitly generated by an algorithm whose run time and memory requirements increase exponentially with n. The computation becomes prohibitively expensive when n reaches somewhere in the 30s. For n varying from 3 to 9 we obtain the following sequence for the number of n-animals on the hexagonal lattice: 11, 44, 186, 814, 3652, 16689, 77359. These numbers thus represent the quantity of distinct n-note scales that can be formed by concatenating generators of two given sizes in all possible arrangements.[3] A small number of the 3652 hepta-animals are depicted in Figure 4 to convey the variety of forms that are possible. Note that all connections are

[2] When we count rotations and reflections as distinct, we are enumerating *fixed* as opposed to *free* animals.

[3] These numbers include the animals formed by iterating only one of the two generators, along each of the three axes of the lattice. If we wished to exclude these "one-dimensional" animals, we would subtract 3 from these numbers.

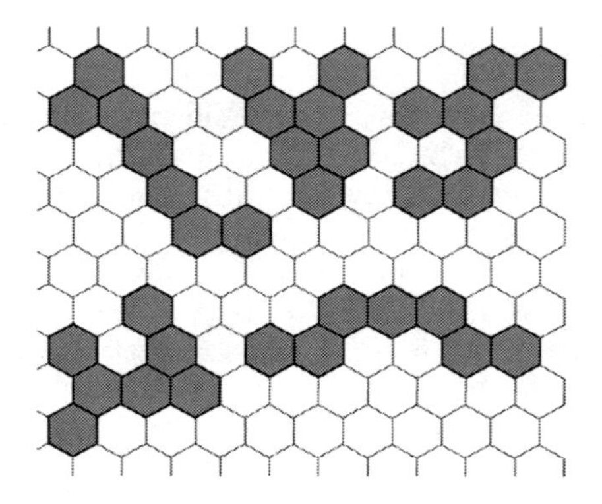

Fig. 4.

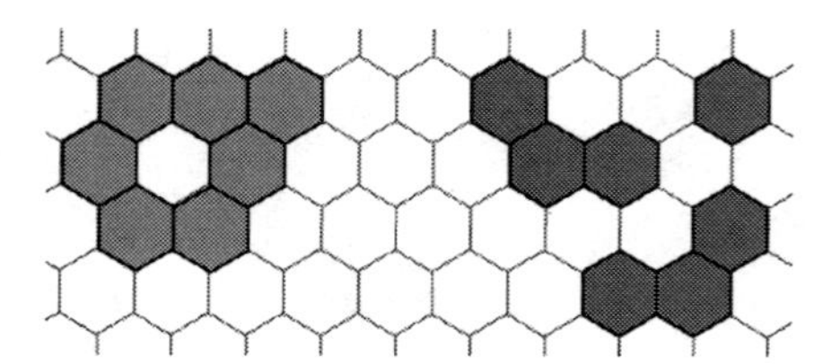

Fig. 5.

face-connections—in other words it is not enough for each note of a scale to be representable as some sum of the generators; the notes must form a connected subset (not necessarily *simply* connected, as the first animal in Figure 5 shows). A disconnected quasi-animal such as the second of Figure 5 is not enumerated here; of course if we relaxed the connectivity constraints to allow Figure 5 we would have an infinite number of scales to consider. Still, the large size of our bestiary of connected lattice animals would initially appear to be a confounding factor in the study of the relationship between animals and pwwf scales.

Three-Stepped Animals on the *Tonnetz* of Fifths and Thirds

The author wrote a computer program that generates all animals of a given size and evaluates the scale corresponding to each. Of the 3649 scales of 7 notes that use both generators (and remember we're dealing only with the *Tonnetz* generated by the pair (3:2, 5:4) for now), just 63 of them are restricted to three step sizes.

Pwwf Animals on the *Tonnetz* of Fifths and Thirds

Of the 63 three-stepped scales generated above, a third of them, 21, have the pairwise well-formed property. The corresponding 21 animals are shown in Figure 6; they are all row-convex for all three orientations of "row" (i.e. they have no gaps in the chains of generators, looked at from whatever angle) and they are "nice" in other ways developed below. As well as the syntonic diatonic mentioned above, we find an

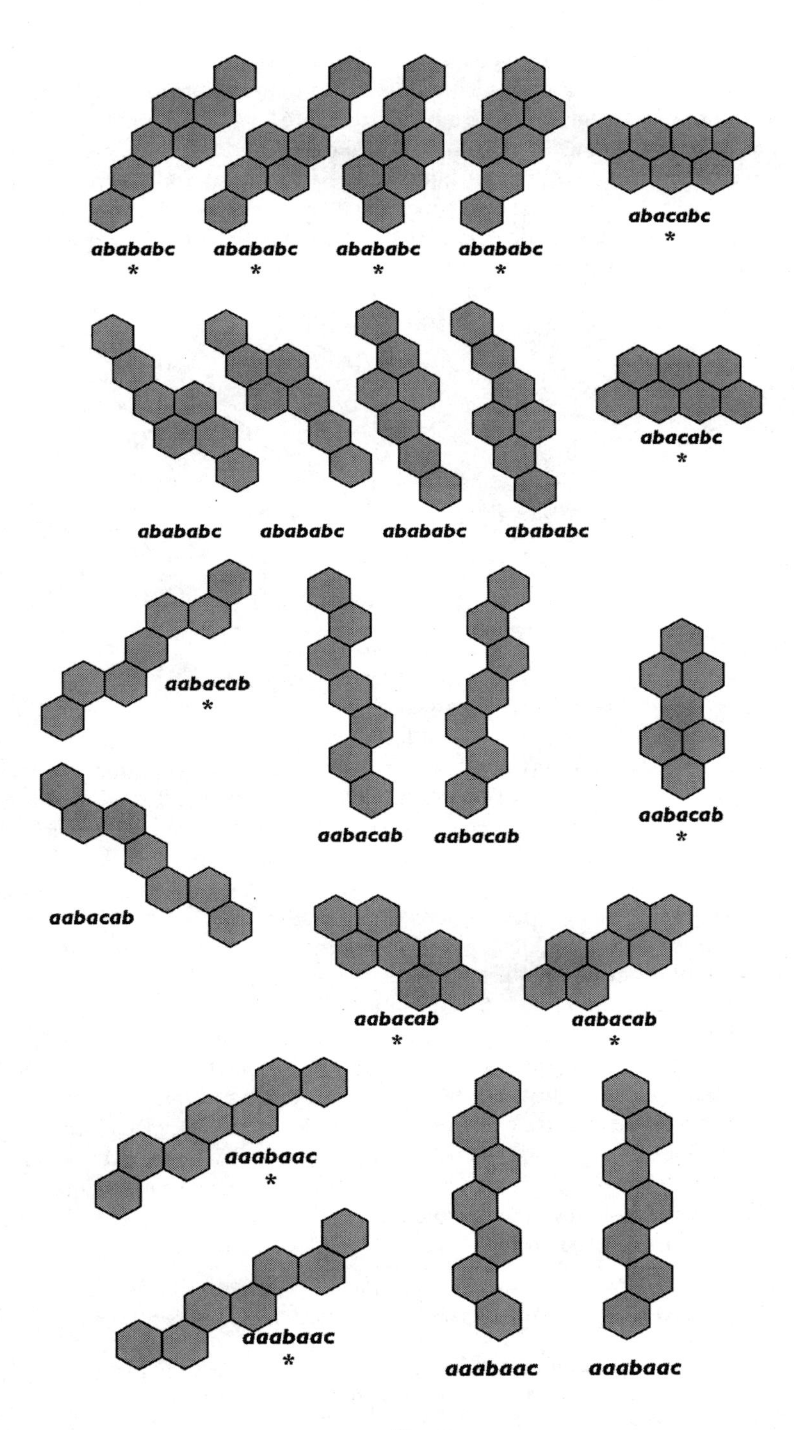
ababab c
*
ababab c
*
ababab c
*
ababab c
*
abacabc
*
ababab c
ababab c
ababab c
ababab c
abacabc
*
aabacab
*
aabacab
aabacab
aabacab
*
aabacab
aabacab
*
aabacab
*
aaabaac
*
aaabaac
*
aaabaac
aaabaac

Fig. 6.

alternative diatonic tuning (this one has also previously been noted by Clampitt as pwwf) where the note D is a syntonic comma lower than in Zarlino's scale; this variant appears in the 16th-century treatise by Fogliano [6] which slightly predates Zarlino's. A third pwwf tuning of a diatonic set, where G, E and B appear a syntonic comma lower than in the Fogliano tuning, is one I have not come across before. It contains four justly-tuned triads, one fewer than the other two diatonic tunings. Its mirror image (reflected in a vertical axis) is not diatonic; it is a tuning for the Hungarian gypsy scale which Clampitt [4] has previously identified as pwwf. Two other pwwf scales on this lattice contain four justly tuned triads; all the scales mentioned in this paragraph are shown labelled with note names in Figure 7.

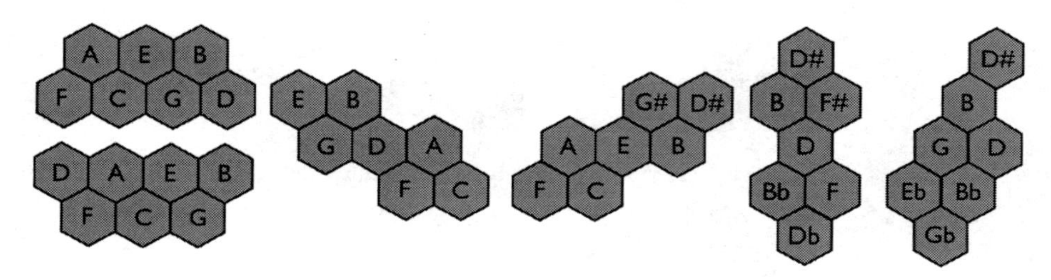

Fig. 7.

In Figure 6 a list of step size tokens was associated with each scale; Clampitt [7] has recently identified this kind of list with the *words* of mathematical word theory. Each word has been put in a sort of prime form: among all the mappings of tokens to step sizes, and all the rotations of the word, the one with lowest lexical position (i.e. soonest in alphabetical order) has been selected as representative of the whole word class. Reducing to these primeform words shows that the 21 pwwf scales on this lattice belong to just four classes: *aaabaac*, *aabacab*, *abababc*, and *abacabc*. These in fact exhaust the pwwf words possible for scales of cardinality 7.

In none of the 21 pwwf animals are all three of the axis intervals—the perfect fifth, major third and minor third—required to construct the scale.[4] This means each of the animals is generated by only two intervals—but they may be *any* two. If we are more strict about the generators, and only enumerate animals that require exclusively the perfect fifth and major third, we will eliminate all those that have "unsupported left-leaning segments", and find there are 12 animals remaining in our bestiary; they are the ones marked with an asterisk in Figure 6, which I call the "strict list" for the generator pair (3:2, 5:4). The unmarked animals also turn out to be generated by just two intervals—but those two intervals are the fifth and *minor* third; or the major and minor thirds. The corresponding scales would appear in the strict lists for the lattices formed by those pairs of generators.[5]

---

[4] To illustrate, an example of a (non-pwwf) collection that *does* require all three intervals is shown in Figure 8.

[5] An alternative formulation for the results of this paper would restrict itself to the strict lists. In that formulation a square rather than hexagonal grid would be appropriate, where scales are represented by polyominoes. This eliminates the possibility of animals that require connections along the third axis of the hexagonal grid.

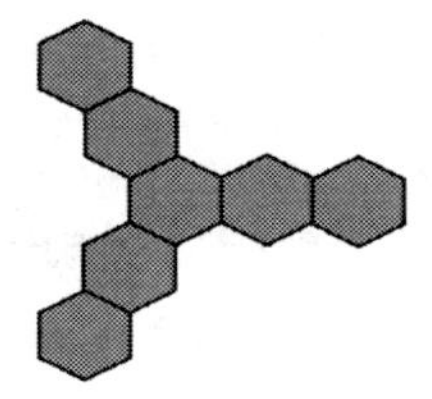

Fig. 8.

Pwwf Animals on Other Lattices

Once we begin to experiment with lattices generated by different intervals, a remarkable fact emerges empirically. Scales exhibiting pairwise well-formedness correspond

Table 1.

Pair of generators	Three-stepped heptatonic scales on this lattice	Pairwise well-formed scales on this lattice	PWWF scales, strict list **
3:2, 5:4	63	21	12
3:2, 7:4	105	44	13
5:4, 7:4	91	42	14
3:2, 11:8	83	22	16
5:4, 11:8	93	23	16
7:4, 11:8	62	19 *	14
3:2, 13:8	95	26	12
5:4, 13:8	72	26	17
7:4, 13:8	115	22	12
11:8, 13:8	70	46	14

*: The {7:4, 11:8} lattice includes one PWWF scale that is *singly*—not doubly—generated. That is, the seven-note chain formed by stacking the interval 11:7 has the pwwf property. The resulting scale has the form *aabacab*. Clampitt has previously remarked that some of these symmetrical pwwf scales may be generated by a single interval. Since a scale formed by iterating a single generator can have at most three step sizes, and since these scales can be pwwf (as in this example) or not (as in every other simply generated scale on any of the lattices in Table 1), such one-dimensional animals that “live” on a single row do not have a place in the healthy/unhealthy opposition scheme for two-dimensional lattice animals. It is trivial to find a lattice where this scale appears also as a two-dimensional animal: one generator is 11:7, and the other is 14641:2401, or 11:7 stacked four times.

**: In the “strict list” column only those scales are enumerated that can be built using the given generators literally, rather than using their combination or difference. This is a stricter requirement than that the collections form connected portions of the resulting lattices—although any collection that forms a connected portion of a lattice is strictly generated in this sense on *some* lattice.

again and again to a small group of recognisable animals from among the more than three thousand candidates, no matter the size of the generators of the lattice. First, a few statistics on the number of three-stepped and pwwf heptatonic scales for various combinations of generators are presented in Table 1.[6] As shown, different generators result in different counts.

As in the case of the lattice of Fifths and Thirds, pwwf animals on other lattices also only ever require two generators out of the three axis intervals. In fact we can make a stronger statement: pwwf animals only ever require two **directed** generators and it is possible to start from one cell and cumulatively generate all others using a pair of directed arrows, as shown in the first animal of Figure 9. Here the animal requires exclusively arrows to the right and arrows at an angle of 60 degrees (measured counter-clockwise from horizontal). The second animal shown also does not require any arrows that point along the third axis—but it **does** need arrows pointing in both directions on the horizontal axis in order to reach everywhere from an initial cell, so it cannot be constructed using only two directed generators. The third animal shown requires only two directed generators, but cannot be generated from a single starting cell.[7]

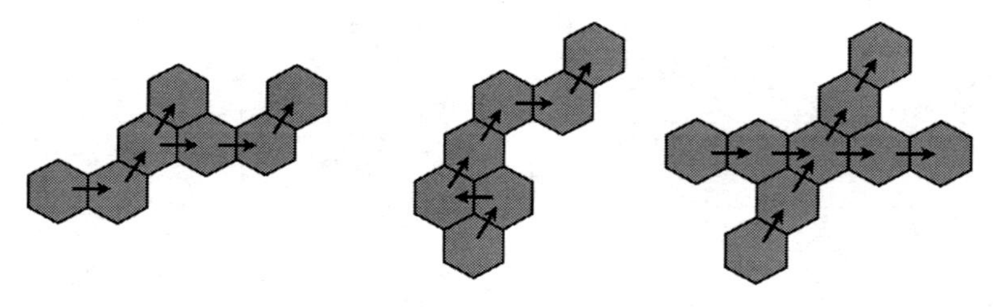

Fig. 9.

Healthy Animals

I designate as *healthy* those two-dimensional animals that, on some lattice, yield pairwise well-formed collections. Thus for example all the animals shown in Figure 6, that yield pwwf scales on the lattice of fifths and thirds, are healthy. Being healthy is not a guarantee that the corresponding scale is pwwf on *every* lattice. But the empirical evidence suggests that being healthy **and three-stepped** *is* sufficient. In other words if one of the healthy animals turns out to have three step sizes in its scale on a certain lattice, then those three step sizes will always be ordered as one of the pwwf words. And conversely, any three-stepped scale corresponding to an unhealthy animal is guaranteed *not* to be pwwf. At the time of submission these results have been

[6] The generators entabulated here happen to be intervals from the harmonic series; any generators may of course be used, with the caveat that degenerate scales may result when it is possible to form identical sums of the two generators in more than one way.

[7] The ways in which the second and third animals shown in Figure 9 fail to meet the condition are in fact interchangeable: the middle animal could be redrawn to only require two directed generators if we had two "starting" cells; and the third animal could be redrawn with a single starting cell if we allowed one of the generating intervals to be used both upwards and downwards.

experimentally verified for a large number of generator-pairs, but the statement remains a conjecture for now.

Healthy heptatonic animals fall into what I identify as three families, shown in Figure 10. **Type I** animals (upper left of Figure 10) have two parallel chains of generators, of lengths 3 and 4, along any of the three axes. **Type II** animals (right-hand side) have three parallel chains of generators, of lengths 2, 3 and 2—in fact there are three different ways of seeing a Type II animal as three parallel chains of generators of lengths 2, 3 and 2, as shown in Figure 11. The **Type III** animal (lower left) has four parallel chains of generators, of lengths 2, 2, 2, and 1. Figure 12 shows there are two ways of seeing a Type III animal as four parallel chains of generators of lengths 2, 2, 2, and 1.

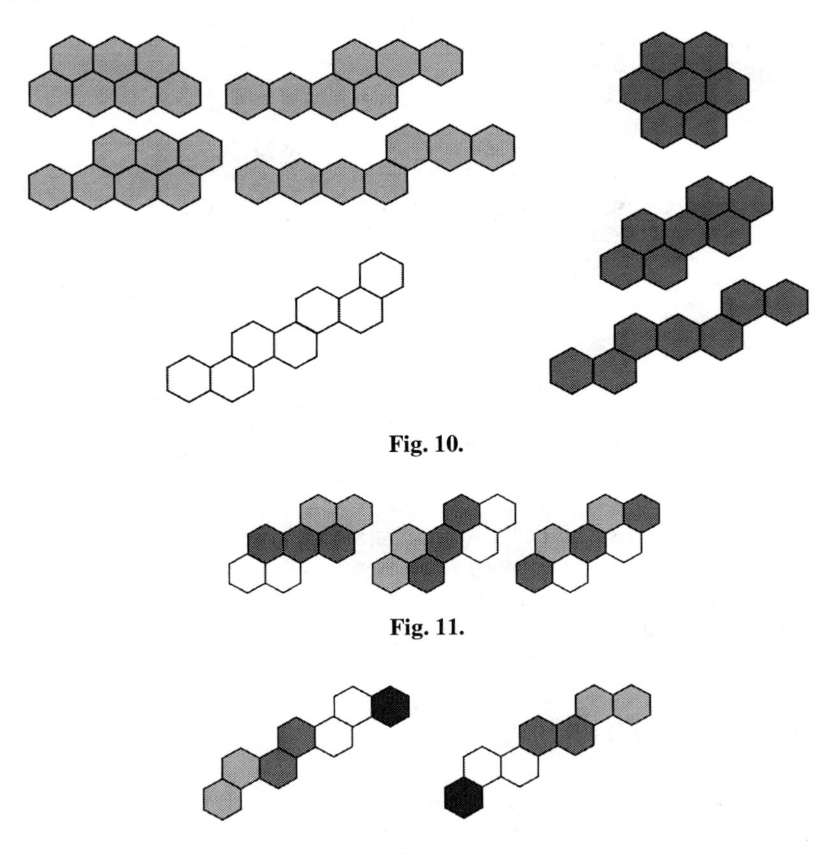

Fig. 10.

Fig. 11.

Fig. 12.

Several symmetries obtain: if a given Type I or Type III animal is pwwf on a given lattice, then so is its rotation by 180 degrees. Type II animals are rotationally symmetrical at 180 degrees; each Type II animal, if pwwf on a given lattice, will be accompanied by its mirror reflection if it is distinct.

While all the examples shown in Figure 10 were positioned so their generator chains run along the horizontal axis, any of these animals can be transformed into any of its rotations or reflections by permuting and/or inverting the generators used—the different orientations, if they are distinct, will represent the same scale on different lattices. Counting all the distinct rotations and reflections of the eight free animals shown in Figure 10 we find a total of 58 healthy animals as an upper limit for the number of heptatonic pwwf scales on any given lattice.

Type I animals support the pwwf words *aaabaac*, *abababc* and *abacabc*. Type II animals support exclusively the word *aabacab* (and I emphasise that this is true independent of the lattice's basis intervals). Type III animals, like Type I animals, support the words *aaabaac*, *abababc* and *abacabc*. In fact we see in Figure 13 how Type III animals may be decomposed into parallel chains of the same generator, of lengths 4 and 3; by substitution of generators, then, they will correspond to Type I animals on a different lattice.

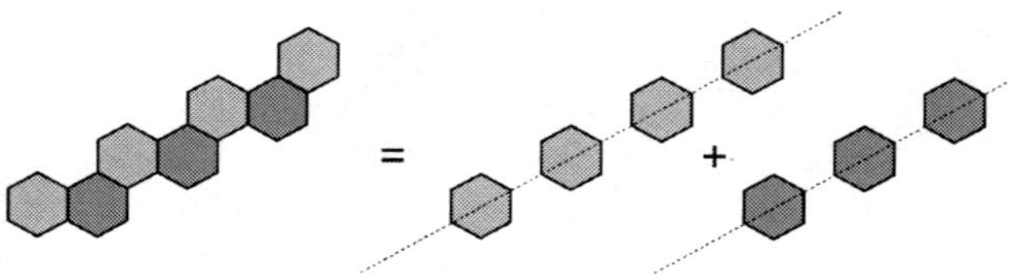

Fig. 13.

Further, it is easy to see how the four varieties of Type I animals are equivalent to one another under a substitution of generator, as are the three varieties of Type II.

We can conclude that if we are allowed to specify the lattice, we can construct every heptatonic pwwf scale discovered so far using only two animals: Type Ia for words *aaabaac*, *abababc* and *abacabc*, and Type IIa for the word *aabacab*.

It must not be imagined that the "unhealthy" heptatonic animals are uniformly misshapen. As the gallery of all non-pwwf three-stepped animals on the lattice of fifths and thirds in Figure 14 shows, many animals that support three-stepped scales that are *not* pairwise well-formed are nonetheless symmetrical, or present double chains of generators of other lengths than 4+3.

Arbitrary Heptatonic Pwwf Scales

Earlier we wondered whether *all* pwwf scales were doubly generated, just as all well-formed scales are generated by iterating a single interval. We could rephrase the question as follows: given the step-sizes *a, b* and *c* and a particular pwwf word, can we select a pair of generators such that the scale sought appears on that lattice? It turns out that the answer is yes: we can use a scale whose generators we already know, and reverse-engineer the required generators from the scalar mapping. For example, let us say we wish to construct a scale of the form *abacabc*, with step sizes a=150.7 cents, b=65.0 cents, and c therefore equal to 381.5 cents. We know the syntonic diatonic exhibits the same word; here that scale is shown on the lattice of fifths and thirds, where

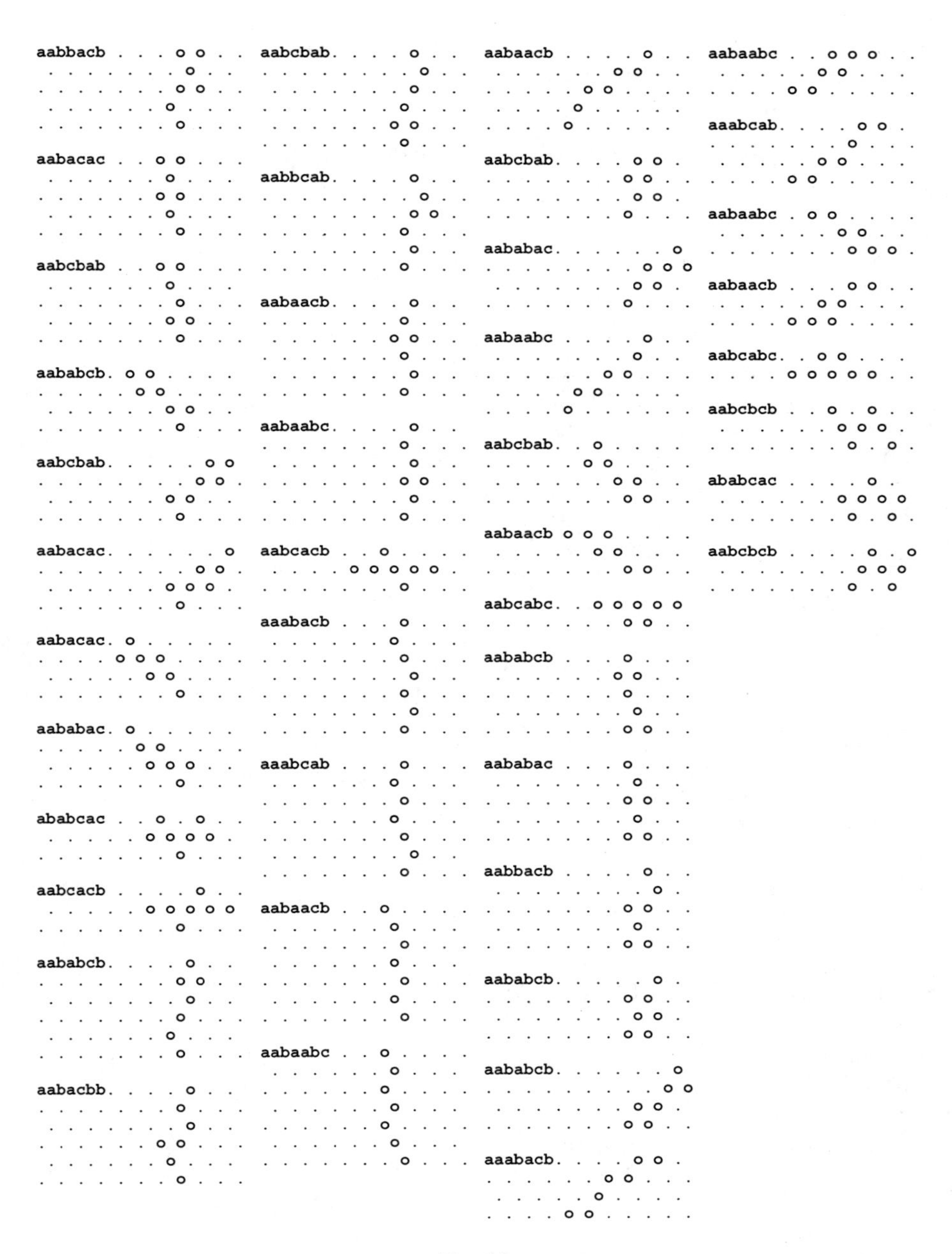
aabbacb
aabacac
aabcbab
aababcb
aabcbab
aabacac
aabacac
aababac
ababcac
aabcacb
aababcb
aabacbb
aabcbab
aabbcab
aabaacb
aabaabc
aabcacb
aaabacb
aaabcab
aabaacb
aabaabc
aabaacb
aabcbab
aababac
aabaabc
aabcbab
aabaacb
aabcabc
aababcb
aababac
aabbacb
aababcb
aababcb
aaabacb
aabaabc
aaabcab
aabaabc
aabaacb
aabcabc
aabcbcb
ababcac
aabcbcb

Fig. 14.

the numbers appearing on the nodes of the lattice indicate the scalar position of the corresponding pitch, when the scale is rotated to match the primeform word *abacabc* (i.e. in Lydian mode):

```
abacabc; a=203.9, b=182.4, c=111.7

. . . . . . . . . . . . . .

 . . . . 3 7 4 . . . . . .

. . . . 1 5 2 6 . . . . . .
```

The step sizes of the scale we seek to generate are completely unrelated to those of the syntonic diatonic, but we can use the pattern traced out by the scalar ordering to calculate new generators. The first scale step, 1-2, is comprised of two horizontal generators—this is apparent in the segment 1-(5)-2 on the lower row. Since the interval between degrees 1 and 2 (i.e., token *a*) is bissected by another scale degree on the lattice, we define GEN1 = $(a+1200)/2$. The other generator, ascending to the right on the lattice, connects scale degrees 1 and 3 (or 5 and 7, or 2 and 4). The word *abacabc* tells us that each of these generic thirds is the sum of step sizes *a* and *b*. So we define GEN2 = $(a+b)$. Substituting the desired step size values $a=150.7$ and $b=65.0$ we obtain GEN1 = 694.05 and GEN2 = 213.8.

Likewise, by examining the mapping between scalar order and lattice arrangement for representative scales of each of the other heptatonic pwwf words, we can obtain suitable expressions for their generators as a function of their step sizes:[8]

abacabc:	GEN1 = $(a+1200)/2$;	GEN2 = $(a+b)$;	animal is Type Ia.
abababc :	GEN1 = $a+b$;	GEN2 = b ;	animal is Type Ia.
aaabaac:	GEN1 = a;	GEN2 = $3a+b$;	animal is Type Ia.
aabacab:	GEN1 = $2a+b+c$;	GEN2 = $-a$;	animal is Type IIa.

Since the four words account for all heptatonic pwwf scales, it follows that *every* heptatonic pwwf scale is doubly generated.[9]

Pwwf Scales of Other Cardinalities

We find similar results for other cardinalities of scale: a given *n*-animal is either healthy or unhealthy, and the families of healthy animals are analogous to the heptatonic ones. The non-singular pwwf scales can be constructed as near-equal pairs of generator chains (for example, of lengths 5 and 4 when the scale cardinality is 9), corresponding to Types I and III. But we find no analogues to Type II animals for non-heptatonic scales: as Clampitt [4] has shown, the heptatonic pattern *aabacab* is unique among all pwwf words in having three different multiplicities of step sizes.

There are considerably fewer pwwf collections of cardinality 9 than for cardinality 7 on a given lattice, especially when considered relative to the hugely expanded num-

[8] In each case these are not the only expressions that would work.

[9] Some pwwf scales have an additional interpretation, as generated by a single interval. See the footnote to Table 1.

ber of candidate animals. The lattice of fifths and thirds, for example, only possesses four pwwf scales of cardinality 9 among over 77,000 candidate animals. A search for scales of cardinality 5 yields 20 pwwf scales among 186 animals on the same lattice. For n=5 and n=9 I have been able to use the above method for arbitrary pwwf scales to find formulae for the generator pairs as a function of the desired step sizes. I conjecture this will be possible for higher odd cardinalities, too.

References

[1] Carey, N., Clampitt, D.: Aspects of Well-Formed Scales. Music Theory Spectrum 11(2), 187–206 (Autumn 1989)

[2] Wilson, E.: On the Development of Intonational Systems by Extended Linear Mapping. Xenharmonikon 3 (1975)

[3] Clough, J., Engebretson, N., Covachi, J.: Scales, sets and interval cycles: a taxonomy. Music Theory Spectrum (1999)

[4] Clampitt, D.: Pairwise well-formed scales: Structural and transformational properties, Ph.D. dissertation, SUNY Buffalo (1997)

[5] Zarlino, G.: Le istitutioni harmoniche (Venice, 1558)

[6] Fogliano, L.: Musica theorica (1529)

[7] Clampitt, D.: Mathematical and Musical Properties of Pairwise Well-Formed Scales. Paper read at MCM 2007 (2007)

[8] Carey, N.: Coherence and sameness in pairwise well-formed scales. Journal of Mathematics and Music (2007)

[9] Apagodou, M. : Counting Hexagonal Lattice Animals. Preprint paper online at arXiv:math/0202295 (last accessed March 1) (2009)

Generalized *Tonnetz* and Well-Formed GTS: A Scale Theory Inspired by the Neo-Riemannians*

Marek Žabka

Department of Musicology, Comenius University, Gondova 2, 81801 Bratislava, Slovakia

Abstract. The paper connects two notions originating from different branches of the recent mathematical music theory: the neo-Riemannian *Tonnetz* and the property of well-formedness from the theory of the generated scales. These notions are mathematized and their properties are rigorously investigated. As the first result, the concepts of the generalized *Tonnetze* and of the multidimensional (i.e. based on multiple generators) generated tone-systems (GTS) are formally defined. Secondly, we prove a theorem stating that a normal two-dimensional GTS is well-formed if and only if it is closed. This is the main mathematical result of the paper and it can be considered a generalization of Carey-Clampitt's work on one-dimensional generated scales to GTS's with two generators. Finally, we illustrate power of the proposed theoretical framework. It covers various theoretical concepts found in different musical contexts. Besides the neo-Riemannian *Tonnetze* and Carey-Clampitt's generated scales, our examples include Mazzola's 'harmonic band,' the pitch helix known from the psychology of hearing, the ancient Chinese system of *lü-lü*, the Arabic 24-*nīm* system, and the ancient Indian 22-*śruti* system. In particular, we give a possible explanation of the number 22 in the Indian system.

1 Inspirations

The neo-Riemannian theory and its use of the concept of the *Tonnetz* (see e.g. Cohn 1998, Gollin 1998) provided a source of inspiration for this paper. In this context, a very special role plays David Lewin's (1998) illuminating analysis of a passage from Bach's $F\sharp$ minor fugue.[1] The analysis is based on an original idea that a *Tonnetz* might also be generated by other intervals than the fifth and the third.[2] This way, Lewin modifies the basic neo-Riemannian concept and applies it in a different, yet very meaningful analytical situation.

Another stream of inspiration comes from Carey-Clampitt's diatonic theory. Carey and Clampitt (1989) defined a very powerful property of the well-formedness and showed that (one-dimensional) generated scales commonly

* This paper was supported by the Fulbright Foundation through a fellowship awarded to the author.

[1] For an interesting visualization of Lewin's structural ideas see (Reed and Bain 2007).

[2] Clough (2002) used other generalized *Tonnetze* in an analysis of Kurtág's music.

E. Chew, A. Childs, and C.-H. Chuan (Eds.): MCM 2009, CCIS 38, pp. 286–298, 2009.

encountered in music usually have this property. As the main result, they proved that there is a direct relation between the acoustical 'closeness' of the end points of the generated scale and its structural well-formedness.[3]

2 Generalized *Tonnetz*

The proposed theory relies on David Lewin's concept of the *Generalized Interval Systems* (GIS) and the concept of labeled directed graphs. For a definition of GIS see e.g. (Lewin 2007).

Definition 1. *We say that an ordered sextuple* $(N, A, L_N, l_N, L_A, l_A)$ *is a* labeled directed graph *if the following conditions hold.*

1. N *is a non-empty set of* nodes.
2. A *is a set of ordered pairs of nodes (i.e. a subset of the direct product* $N \times N$*) and its elements are called* arrows.
3. L_N *and* L_A *are non-empty sets of* node labels *and* arrow labels, *respectively.*
4. $l_N : N \to L_N$ *is a mapping assigning node labels to nodes.*
5. $l_A : A \to L_A$ *is a mapping assigning arrow labels to arrows.*

The quadruple (N, A, L_A, l_A) *is an* arrow-labeled directed graph.

Definition 2. *Let* $G = (N, I, \mathsf{int}^*)$ *be a commutative GIS and assume that the group* I *is generated by a finite subset* X. *A* generalized Tonnetz *(g-*Tonnetz*)* $T(I; X)$ *is the arrow-labeled directed graph* (N, A, X, int) *where* A *denotes the complete inverse image of* X *under* int^*, $A = \mathsf{int}^{*-1}[X]$, *and* int *denotes the restriction of* int^* *to* A, $\mathsf{int} = \mathsf{int}^*|A$. *Further, we say that the dimension of the g-*Tonnetz $T(I; X)$ *is* n *if* X *has exactly* n *distinct elements.*

The complete inverse image of X under $\mathsf{int}^* : N \times N \to I$ is the set of all ordered pairs of nodes (p, q) such that their image $\mathsf{int}^*(p, q)$ is in X. (See also Fig. 2.) The asterisk will be omitted in the notation of int^* if there is no risk of confusion.

Example 1. It is important to recognize the difference between the g-*Tonnetz* and the underlying group of intervals (or GIS). A single group may correspond to different g-*Tonnetze* if different sets of generators are selected. Consider the diatonic system, usually modeled as $\mathbf{Z}_7$. If we think of it as generated by the fifth $f = 4$ we get the g-*Tonnetz* $T(Z_7; 4)$, which gives the usual depiction of the diatonic on a circle with seven points.

On the other hand, if we think of the diatonic system as being generated simultaneously by the fifth $= 4$ and the third $= 2$, the resulting g-*Tonnetz* $T(\mathbf{Z}_7; 4, 2)$ is quite different. It is two dimensional and can be depicted on the Möbius strip. It is shown in Fig. 1. Mazzola (2002) describes this g-*Tonnetz* and calls it a 'harmonic band.' It models the space of simple diatonic harmonies.

[3] For a related, independently formulated concept of 'the moments of symmetry,' see also (Wilson 1975).

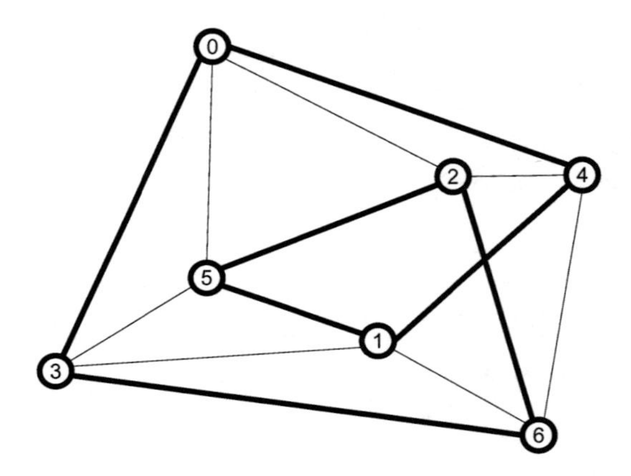

Fig. 1. The diatonic system as a two dimensional g-*Tonnetz* $T(\mathbf{Z}_7; 4, 2)$

As Kolman (2004) showed, two GIS's are isomorphic if and only if their underlying interval groups are isomorphic. Therefore, a g-*Tonnetz* is, up to isomorphism, determined by the group of intervals and the selected set of generators. Thus, a complete study of the commutative groups and their generating subsets is sufficient for the complete understanding of the g-*Tonnetze*.

For our study of the g-*Tonnetze*, we will rely on representations of Abelian groups as quotient groups of free Abelian groups. Any Abelian group I may be represented as a quotient group of the free Abelian group $\mathbf{Z}(X)$ where $\mathbf{Z}(X) = \{\sum_{i=1}^{n} k_i \xi_i \mid k_i \in \mathbf{Z}, \xi_i \in X\}$.

Lemma 1. *Let an Abelian group I be generated by the finite set X of n elements. Then there exists a subset K of $\mathbf{Z}(X)$ with m elements where $m \leq n$ and $\mathbf{Z}(X)/K$ is a representation of I. If I is finite then $m = n$.*

Definition 3. *Let $X = \{\xi_1, \ldots, \xi_n\}$, $K = \{\kappa_1, \ldots, \kappa_m\}$ and $m \leq n$. Assume the g-*Tonnetz *$T = T(\mathbf{Z}(X)/K; X)$. We say that the elements κ_i, $1 \leq i \leq m$ are* commas *and K is a* set of commas *of T.*

According to Lemma 1, the assumptions of the previous definition cover, up to isography, any g-*Tonnetz* and we may limit our investigation of the g-*Tonnetze* to those of this type. In other words, any g-*Tonnetz* is fully determined by the set of generators and a set of commas. For the rest of the paper, let (N, I, int) denote a commutative GIS with the group of intervals $I = \mathbf{Z}(X)/K$, and (N, A, X, int) denote the g-*Tonnetz* $T(I; X)$. Figure 2 shows the mappings related to a g-*Tonnetz*. The arrows with curved tails denote the natural injections of the subsets A and X in their supersets $N \times N$ and I, respectively.

Example 2. We return to our previous example of two g-*Tonnetze* based on $\mathbf{Z}_7$. The first one, applicable to the Pythagorean heptatonic scale, is determined by one generating interval f and one comma $7f$. The other example, applicable to

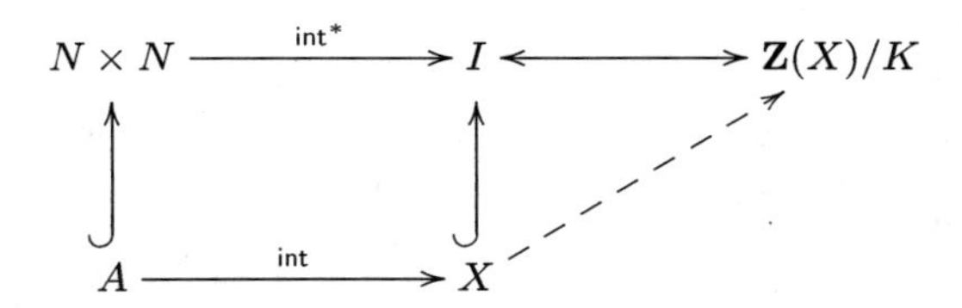

Fig. 2. Diagram of mappings related to a g-*Tonnetz*

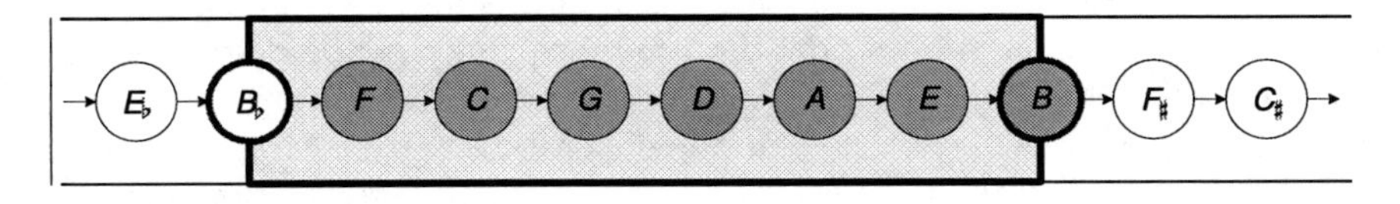

Fig. 3. The diatonic system with one generator

the diatonic heptatonic scale in pure tuning, is determined by two generating intervals f and th and two commas, e.g. $3f + th$ and $f - 2th$. These situations are shown in Figures 3 and 4.

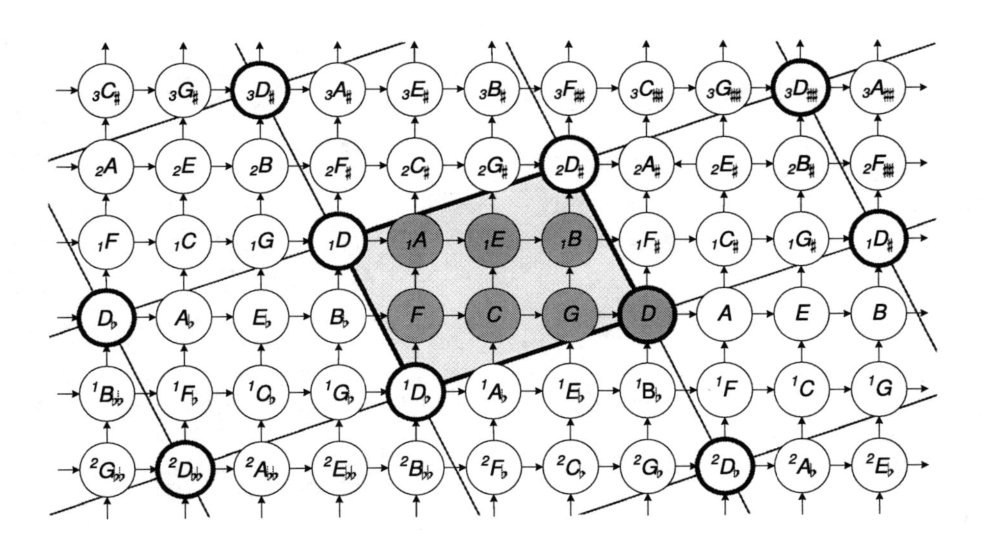

Fig. 4. The diatonic system with two generators

3 Unpitched Generated Tone System

Definition 4. *An* unpitched generated tone system *(UGTS) is a labeled directed graph* $(N, A, X, \mathsf{int}, \mathbf{Z}(X), \mathsf{gen})$ *where the following conditions are satisfied.*

1. (N, A, X, int) *is a finite* g-Tonnetz.
2. $[\mathsf{gen}(t)] + \mathsf{int}^*(t, u) = [\mathsf{gen}(u)]$ *for all pairs* $(t, u) \in A$.

The mapping gen *is called a* generating function. *We say that a UGTS is* n-*dimensional if the underlying* g-Tonnetz *is* n-*dimensional.*

The UGTS differs from a g-*Tonnetz* in that it specifies how the tones are generated from the generators. While the g-*Tonnetz* is a perfectly symmetrical structure, the generating function brings the 'imperfectness' into the picture. E.g., in the case of the Pythagorean heptatonic, the underlying structure is $\mathbf{Z}_7$, in which there is no difference between the perfect fifths and the diminished fifth. The generating function defines how the tones are generated and so yields the information about the imperfectness.

For a given g-*Tonnetz*, the tones may be generated in many ways. However, only some of them are of real interest. A basic requirement is that they are 'compactly' positioned within $\mathbf{Z}(X)$. The compactness means that we take one compact block of elements of the corresponding free group marked by the selected commas. Such blocks are shown also in Figures 3 and 4 where the chosen elements are colored. The following definition formalizes this idea.

Definition 5. *Assume an* n-*dimensional UGTS* $S = (N, A, X, \mathsf{int}, \mathbf{Z}(X), \mathsf{gen})$. *Fix a set of commas* $K = \{\kappa_1, \ldots, \kappa_n\}$ *and a node* $o \in N$. *Let denote:*

$$R(o, K) = \{\mathsf{gen}(o) + z \in \mathbf{Z}(X) \mid z = \sum_{\kappa \in K} r(\kappa)\kappa,\ r(\kappa) \in [0, 1)\}$$

We say that S *is* compact *with respect to* K *and* o *if* $\mathsf{gen}[N] = R(o, K)$.

Example 3. Figure 5 shows a UGTS which is not compact.

Lemma 2. *Let* $S = (N, A, X, \mathsf{int}, \mathbf{Z}(X), \mathsf{gen})$ *be a compact UGTS with respect to* $o \in N$ *and a set of commas* K, *and* $e : K \to \{-1, 1\}$ *be a mapping. Assume the following conditions.*

1. $K_e = \{e(\kappa)\kappa \mid \kappa \in K\}$.
2. *A mapping* $\mathsf{gen}_e : N \to \mathbf{Z}(X)$ *is defined in the following way. For any* $t \in N$, $\mathsf{gen}(t) - \mathsf{gen}(o) = \sum_{\kappa \in K} r_t(\kappa)\kappa$, *consider the set* $K(t) = \{\kappa \in K \mid r_t(\kappa) = 0, e(\kappa) = -1\}$. *The mapping* gen_e *assigns the value* $\mathsf{gen}_e(t) = \mathsf{gen}(t) + \sum_{\kappa \in K(t)} \kappa$.

Then $S_e = (N, A, X, \mathsf{int}, \mathbf{Z}(X), \mathsf{gen}_e)$ *is a compact UGTS with respect to* o *and* K_e.

Definition 6. *Assume the notation from Lemma 2. UGTS's* S *and* S_e *are called* neighboring. *The elements* $\mathsf{gen}_e(o)$ *of* $\mathbf{Z}(X)$ *are called* corners *of* S.

Definition 7. *Let* S *be a UGTS, compact with respect to a node* o *and a set of commas* K. *Further, assume a node* $t \in N$, *i.e.:*

$$\mathsf{gen}(t) = \mathsf{gen}(o) + \sum_{r(\kappa) \in [0,1),\ \kappa \in K} r(\kappa)\kappa.$$

We say that, with respect to o *and* K*:*

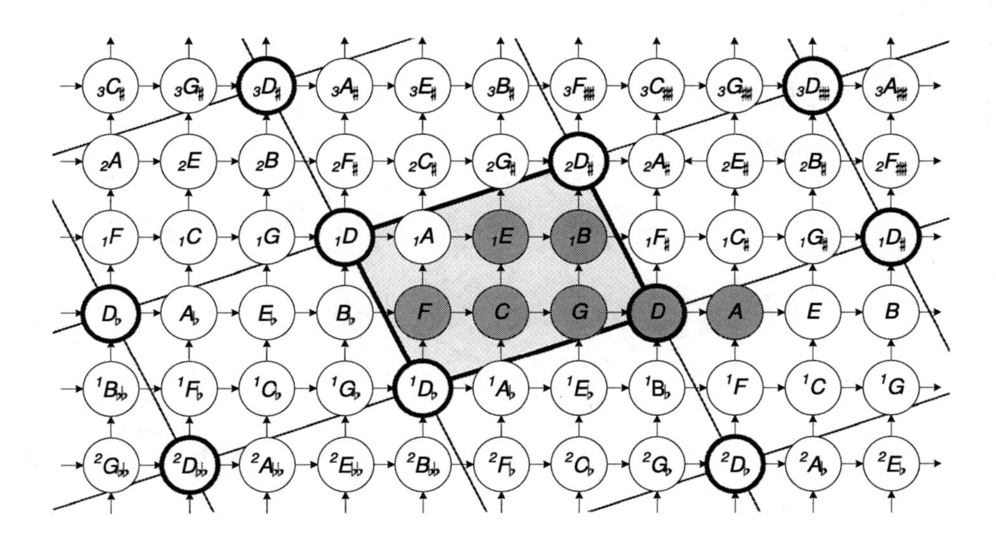

Fig. 5. A non-compact UGTS

1. *o is the* origin.
2. *t is an* edge node *(or a λ-*edge node*) if $r(\kappa) = 0$ for some $\lambda \in K$ and for all $\kappa \in K$, $\kappa \neq \lambda$.*
3. *t is an* inner node *if $r(\kappa) \neq 0$ for all $\kappa \in K$.*

Example 4. Figures 6 and 7 show two of possible interpretations of the 12-tone chromatic scale as a UGTS. In both cases, there are some edge nodes besides the origin: D in the first and C and ${}_1E$ in the second approach.

Definition 8. *A UGTS is called* normal *if it is compact and for any neighboring S_e there exist $\xi \in X$ and an inner node $t \in N$ such that $\mathsf{gen}_e(t) + \xi = \mathsf{gen}_e(o)$ or $\mathsf{gen}_e(t) - \xi = \mathsf{gen}_e(o)$ in $\mathbf{Z}(X)$.*

Example 5. In a normal UGTS, the corners are accessible from an inner tone through a pure generator. It can be shown that in a normal UGTS, any two tones can be connected by a chain of pure generators. Figure 8 shows a UGTS which is not normal.

4 Generated Tone System

Definition 9. *A* generated tone system *(GTS) is a pair (S, pitch) where S is a UGTS and $\mathsf{pitch} : X \to \mathbf{R}_{12}$ is a mapping. The mapping* pitch *is called* pitch function. *A GTS is called compact, neighboring or normal if the underlying UGTS is compact, neighboring or normal, respectively.*

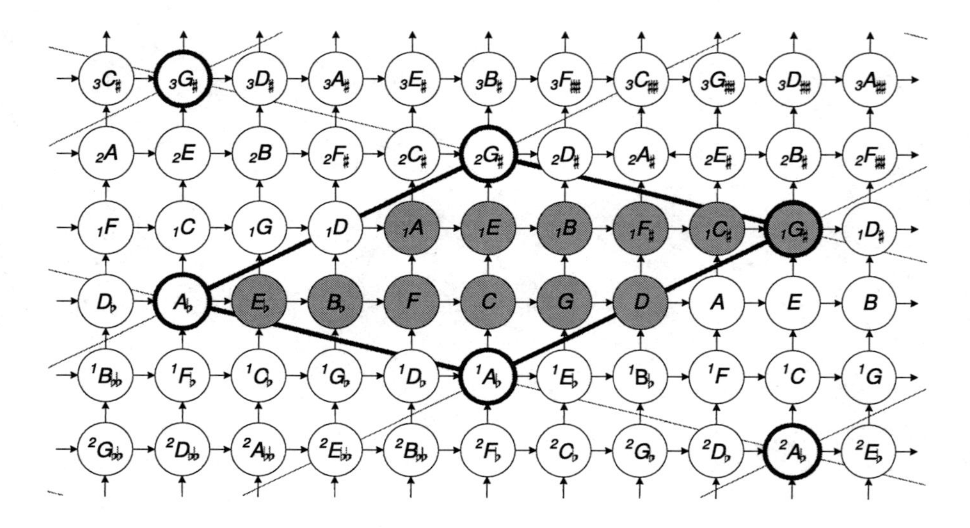

Fig. 6. Chromatic scale as a two-dimensional system

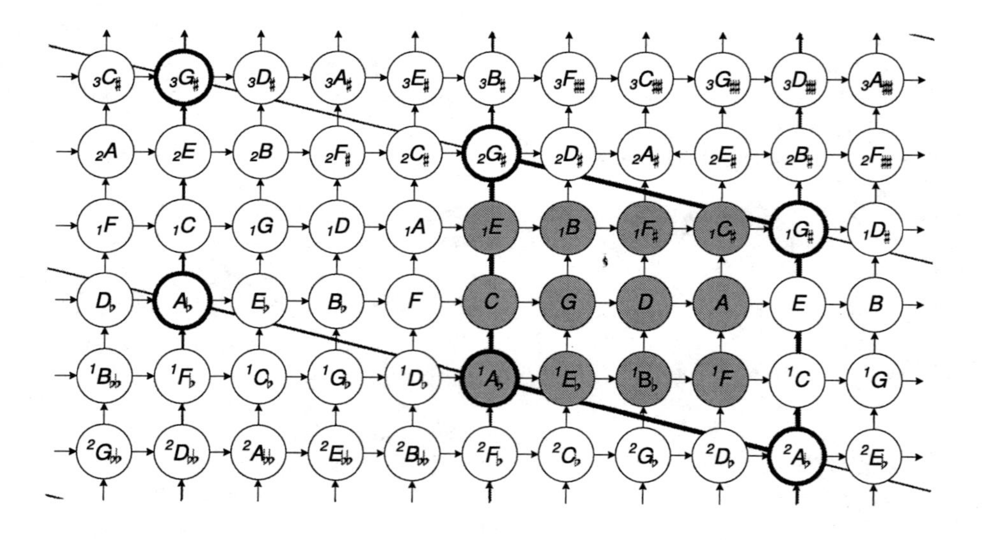

Fig. 7. Chromatic scale as a two-dimensional system – another approach

We notate $\mathbf{R}_{12} = [0, 12)$ the left-closed, right-open interval of real numbers between 0 and 12. From the definition of the free group, it follows that there is a unique group homomorphism $\mathsf{pitch}^* : \mathbf{Z}(X) \to \mathbf{R}_{12}$ such that $\mathsf{pitch}(\xi) = \mathsf{pitch}^*(\xi)$ for all $\xi \in X$. In notating this homomorphism, we will omit the asterisk and also call it a pitch function if there is no risk of confusion. For the rest of the paper we assume a GTS $S = (N, A, X, \mathsf{int}, \mathbf{Z}(X), \mathsf{gen}, \mathsf{pitch})$.

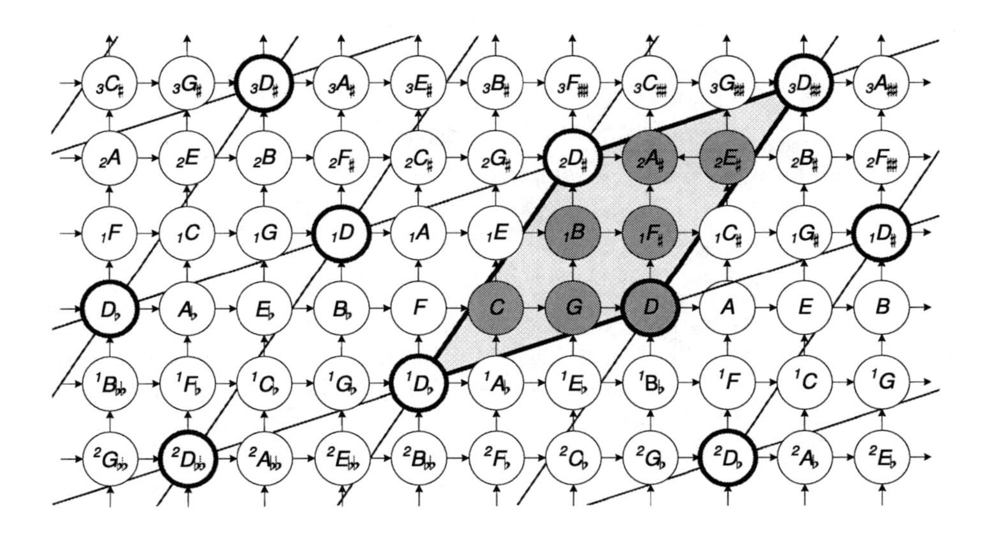

Fig. 8. A non-normal UGTS

Definition 10. *We define a ternary relation* 'between' *on* $\mathbf{Z}(X)$. *Let* $\alpha_1, \alpha_2, \alpha_3$ *be elements of* $\mathbf{Z}(X)$. *We say that* α_2 *is between* α_1 *and* α_3 *and write* $\lhd(\alpha_1, \alpha_2, \alpha_3)$ *if there are real numbers* $0 \leq p, q < 12$ *such that the following conditions hold.*

1. $\mathsf{pitch}(\alpha_1) \oplus p = \mathsf{pitch}(\alpha_2)$.
2. $\mathsf{pitch}(\alpha_2) \oplus q = \mathsf{pitch}(\alpha_3)$.
3. $p + q \leq 12$.

Further, if $\alpha_i = \mathsf{gen}(t_i)$ *for* $t_i \in N$ *and* $i = 1, 2, 3$ *then we also say that* t_2 *is between* t_1 *and* t_3 *and write* $\lhd(t_1, t_2, t_3)$.

Notice that the regular addition of real numbers in the last condition from the previous definition cannot be replaced by addition modulo 12. Addition modulo 12 is distinguished from regular addition by using the symbol '$\oplus$'.

Definition 11. *Consider two nodes* $t, u \in N$. *We say that the* span *of the ordered pair* (t, u) *is* $(k - 1)$ *if there are exactly* k *distinct nodes between* t *and* u. *We denote the span of* (t, u) *by* $\mathsf{span}(t, u)$. *Further, the ordered pair* (t, u) *is called a* step *if* $\mathsf{span}(t, u) = 1$.

Definition 12. *Consider two elements* $\alpha, \beta \in \mathbf{Z}(X)$. *The* size *of the ordered pair* (α, β) *is the real number* $r \in [0, 12)$ *for which* $\mathsf{pitch}(\alpha) \oplus r = \mathsf{pitch}(\beta)$. *We denote the size of* (α, β) *by* $\mathsf{size}(\alpha, \beta)$.

Notice that in general the span is not invariant for neighboring GTS's. However, we require this in the definition of the well-formedness.

Definition 13. *Consider a GTS S. We say that:*

1. *S is* semi-well-formed *if for all $t_1, u_1, t_2, u_2 \in N$:*

$$\mathsf{int}(t_1, u_1) = \mathsf{int}(t_2, u_2) \Rightarrow \mathsf{span}(t_1, u_1) = \mathsf{span}(t_2, u_2)$$

2. *S is* well-formed *(WF) if for any $e : K \to \{-1, 1\}$ and all $t_1, u_1, t_2, u_2 \in N$:*

$$\mathsf{int}(t_1, u_1) = \mathsf{int}(t_2, u_2) \Rightarrow \mathsf{span}(t_1, u_1) = \mathsf{span}_e(t_2, u_2),$$

where span_e denotes the the span function in the neighboring GTS S_e.

Example 6. We give an example of a GTS which is semi-WF but not WF. Assume the GTS depicted in Fig. 6. The GTS which includes the node F is semi-WF. The span of any fifth is 3 and the span of any third is 1. On the other hand, the neighboring GTS containing the corner ${}^1B$ is not semi-WF. The span of the fifth (C, G) is 2 and the span of the fifth (G, D) is 4. Therefore the system is not WF.

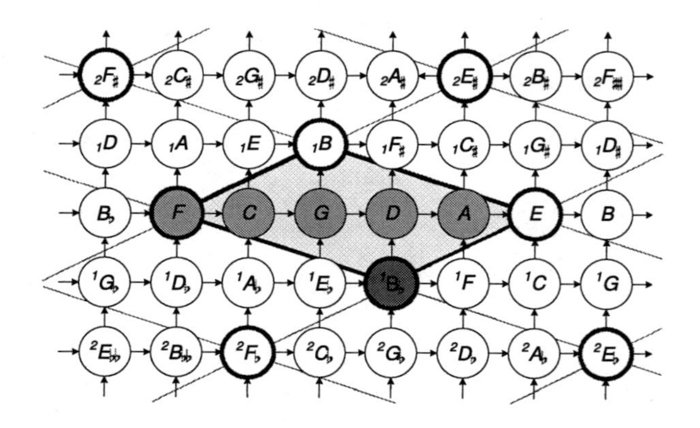

Fig. 9. A semi-WF and not WF GTS

Definition 14. *Consider a GTS S, compact with respect to a node o and a set of commas K. We say that:*

1. *S is* open *if for some $\kappa \in K$ there are two nodes $m, n \in N$ such that $\lhd(\mathsf{gen}(o), \mathsf{gen}(m), \mathsf{gen}(o) + \kappa)$ and $\lhd(\mathsf{gen}(o) + \kappa, \mathsf{gen}(n), \mathsf{gen}(o))$.*
2. *S is* closed *if it is not open.*

Example 7. Loosely speaking, in an open GTS there is a node between two (neighboring) corners whose distance is a comma. Therefore the comma is not sufficiently small. The system from the previous example shown in Fig. 9 is open (i.e. not closed) because the comma $(\mathsf{gen}({}^1B\ ) - \mathsf{gen}(F))$ is too large: G is between F and ${}^1B$ and D is between ${}^1B$ and F. The systems from Figures 3, 4, 6, and 7 are closed. Also the non-normal system from Fig. 8 is closed.

5 The Main Theorem

We are ready to state the main result of the paper. It asserts that a normal two-dimensional GTS is well-formed if and only if it is closed.

Theorem 1. *Let S be a two-dimensional GTS. Assume that S is normal with respect to a node $o \in N$ and a set of commas $K \subseteq \mathbf{Z}(X)$. Then S is closed if and only if S is well-formed.*[4]

As mentioned before, this theorem can be considered a generalization of the famous results of (Carey and Clampitt 1989). Carey and Clampitt formulated a 'closure condition' and a 'symmetry condition' for the category of one-dimensional generated scales. Their main conclusion was that these two conditions are equivalent. They coined the term 'well-formed scales' referring to the scales meeting the two equivalent conditions.

In the present approach, the category of two-dimensional generated tone systems is considered. The closure condition is generalized through the property of being 'closed' as defined in Definition 14. The situation with the symmetry condition is more complex. Carey and Clampitt's symmetry condition can be expressed in several different versions which are equivalent for the one-dimensional case. In our generalization, we consider the following version: The intervals of same generation orders have same scale step orders. (In particular, every generating interval is of the same span.) The generation order of intervals is generalized in the multi-dimensional case through the int function – see Section 2. The scale step order is generalized through the span function, which depends on the pitch function – see Definition 11. This way Carey and Clampitt's symmetry condition is generalized here by the property of 'well-formedness' as defined in Definition 13.

6 A Case Study: The System of *Śrutis*

The concepts of g-*Tonnetz* and GTS apply to surprisingly many phenomena encountered in various musical contexts. They are suitable to model situations where two (or more) basic elements are freely combined to built complex, symmetrical structures. The diatonic scale and the chromatic scale in pure tuning are basic examples of this procedure.

In this section, we want to focus on the systems where the generating elements are the perfect fifth and a small interval of the size approximately a half of semitone. These generators are important for various music cultures, notably for Arabic and Indian music. In the Indian music theory the small interval is usually called *śruti* and we will use this name. So we consider a g-*Tonnetz* with two generators f and s. The basic problem is to specify the commas.

One comma is easy to think of. When we move from a given point by f in opposite directions we arrive to points a whole tone apart (considering the octave

[4] The proof of this theorem can be found in (Žabka 2009).

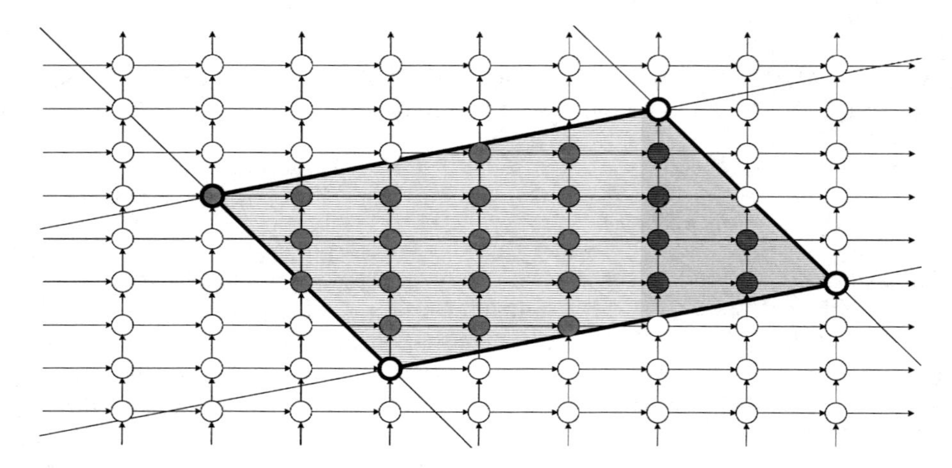

Fig. 10. The g-*Tonnetz* of 24 *nīms*

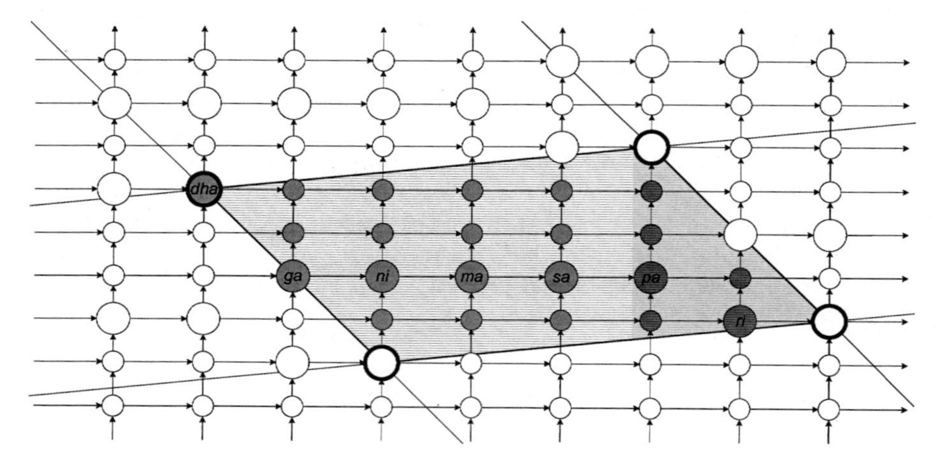

Fig. 11. The g-*Tonnetz* of 22 *śrutis* with high-lighted *sa-grāma*

equivalence, of course). Now if we bend the lower one upwards by two *śrutis* and the upper one by the same amount downwards we obtain almost the same tone. This is the basis of the first comma: $-f + 2s \approx f - 2s$, which gives the comma $\kappa = 4s - 2f$.

The other comma is related to the one underlying the Pythagorean pentatonic. A tone tuned as the fifth perfect fifth is lower than the starting tone just by a small interval. By bending the fifth fifth upwards results in a comma. However, there is an issue: Should it be bent by two or by one *śruti*? In the first case, the other comma is $\lambda_1 = 5f + 2s$. In the second case, it is $\lambda_2 = 5f + 1s$. It is fascinating that both options seem to have been (unconsciously) applied by major music cultures – the Arabic and the Indian. Figures 7 and 8 show the g-*Tonnetze* $T_1(\mathbf{Z}(f, s)/\{\kappa, \lambda_1\}; f, s)$ and $T_2(\mathbf{Z}(f, s)/\{\kappa, \lambda_2\}; f, s)$.

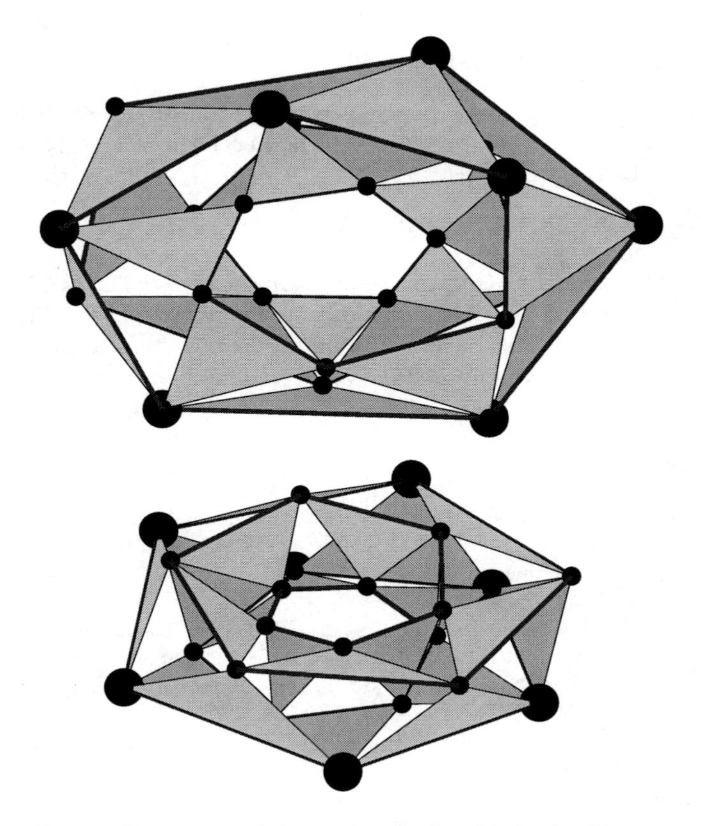

Fig. 12. Two views of the ancient Indian 22-*śruti* g-*Tonnetz*

The first solution leads to a 24-tone GTS. The Arabic music theory knows a system of 24 small intervals called *nīms*. It is usually explained as a result of splitting each tone of the 12-tone chromatic system into two quarter tones. Our approach provides an alternative explanation for the structure of this system. In this explanation, the *nīms* do no have to be (acoustically) uniform.

More striking is the fact that the GTS implied by the second set of commas comprises 22 elements. It seems to model suitably the Indian system of 22 *śrutis*. There is no generally accepted explanation for the number of 22 in this system.[5] Our explanation of this number is very simple and surprisingly accurate. It only relies on four basic assumptions:

1. The perfect fifth and the *śruti* are basal.
2. A fifth down and two *śrutis* up equals approximately a fifth up and two *śrutis* down.

[5] As an example from the recent mathematical music theory, Clough et al. (1993) investigated this system. However, they did not address the question of the total number of *śrutis*.

3. Five fifths up equals approximately one *śruti* bellow.
4. The resulting system is symmetrical.

Notice that we did not have to specify the exact value of a *śruti*. It is sufficient that it is approximately a half of semitone. Then both resulting GTS's are closed and well-formed. As a final illustration, Fig. 9 shows a model of the 22-*śruti* g-*Tonnetz*.

Acknowledgements. I would like to express my thanks to David Clampitt and Richard Cohn for inspiring discussions during the initial phase of my work at the present theory. I received also many useful comments from my anonymous reviewers which helped me improve the final version of the paper.

References

Lewin, D.: Generalized Musical Intervals and Transformations. Oxford University Press, Oxford (2007) (Originally: Yale University Press, 1987)

Lewin, D.: Notes on the Opening of the $F\sharp$ Minor Fugue from WTCI. Journal of Music Theory 42(2), 235–239 (1998)

Reed, J., Bain, M.: A Tetrahelix Animates Bach: Revisualization of David Lewin's Analysis of the Opening of the $F\sharp$ Minor Fugue. Music Theory Online 13(4) (2007)

Cohn, R.: Introduction to Neo-Riemannian Theory: A Survey and a Historical Perspective. Journal of Music Theory 42(2), 167–180 (1998)

Gollin, E.: Some Aspects of Three-Dimensional Tonnetze. Journal of Music Theory 42(2), 195–206 (1998)

Carey, N., Clampitt, D.: Aspects of Well-Formed Scales. Music Theory Spectrum 11(2), 187–206 (1989)

Wilson, E.: Personal corresondence with John Chalmers (1975), `www.anaphoria.com/mos.PDF`

Mazzola, G.: The Topos of Music: Geometric Logic of Concepts, Theory, and Performance. Birkhäuser, Basel (2002)

Kolman, O.: Transfer Principles for Generalized Interval Systems. Perspectives of New Music 42(1), 150–191 (2004)

Clough, J., Douthett, J., Ramanathan, N., Rowell, L.: Early Indian Heptatonic Scales and Recent Diatonic Theory. Music Theory Spectrum 15(1), 36–58 (1993)

Clough, J.: Diatonic Trichords in Two Pieces from Kurtag's 'Kafka-Fragmente': A Neo-Riemannian Approach. Studia Musicologica Academiae Scientiarum Hungaricae 43(3/4), 333–344 (2002)

Žabka, M.: Well-Formed Two-Dimensional Generalized Tone Systems (unpublished paper)

Author Index